Lecture Notes in Artificial Intelligence 11154

Subseries of Lecture Notes in Computer Science

More information about this series at http://www.springer.com/series/1244

Mohadeseh Ganji · Lida Rashidi
Benjamin C. M. Fung · Can Wang (Eds.)

Trends and Applications in Knowledge Discovery and Data Mining

PAKDD 2018 Workshops, BDASC,
BDM, ML4Cyber, PAISI, DaMEMO
Melbourne, VIC, Australia, June 3, 2018
Revised Selected Papers

Editors
Mohadeseh Ganji
University of Melbourne
Melbourne, VIC, Australia

Benjamin C. M. Fung ⓘ
McGill University
Montreal, QC, Canada

Lida Rashidi
University of Melbourne
Melbourne, VIC, Australia

Can Wang
Griffith University
Gold Coast, QLD, Australia

ISSN 0302-9743   ISSN 1611-3349   (electronic)
Lecture Notes in Artificial Intelligence
ISBN 978-3-030-04502-9   ISBN 978-3-030-04503-6   (eBook)
https://doi.org/10.1007/978-3-030-04503-6

Library of Congress Control Number: 2018961439

LNCS Sublibrary: SL7 – Artificial Intelligence

This Springer imprint is published by the registered company Springer Nature Switzerland AG
The registered company address is: Gewerbestrasse 11, 6330 Cham, Switzerland

Preface

On June 3, 2018, five workshops were hosted in conjunction with the 22nd Pacific-Asia Conference on Knowledge Discovery and Data Mining (PAKDD 2018) in Melbourne, Australia. The five workshops were the Pacific Asia Workshop on Intelligence and Security Informatics (PAISI), Workshop on Biologically Inspired Techniques for Knowledge Discovery and Data Mining (BDM), the Workshop on Big Data Analytics for Social Computing (BDASC), Data Mining for Energy Modeling and Optimization (DaMEMO), and the Australasian Workshop on Machine Learning for Cyber-security (ML4Cyber). This volume contains selected papers from five workshops. These workshops provided an informal and vibrant opportunity for researchers and industry practitioners to share their research positions, original research results, and practical development experiences on specific challenges and emerging issues. The workshop topics are focused and cohesive so that participants can benefit from interaction with each other.

In the PAKDD 2018 workshops, each submitted paper was rigorously reviewed by at least two Program Committee members. The workshop organizers received many high-quality publications, but only 32 papers could be accepted for presentation at the workshops and publications in this volume. On behalf of the PAKDD 2018 Organizing Committee, we would like to sincerely thank the workshop organizers for their efforts, resources, and time to successfully deliver the workshops. We would also like to thank the authors for submitting their high-quality work to the PAKDD workshops, the presenters for preparing the presentations and sharing their valuable findings, and the participants for their effort to attend the workshops in Melbourne. We are confident that the presenters and participants alike benefited from the in-person interactions and fruitful discussions at the workshops.

We would also like to especially thank the PC chair, Prof. Dinh Phung, for patiently answering our long e-mails, and the publications chairs, Dr. Mohadeseh Ganji and Dr. Lida Rashidi, for arranging this volume with the publisher.

June 2018

Benjamin C. M. Fung
Can Wang

ML4CYBER 2018 Workshop Program Chairs' Message

Machine learning solutions are now seen as key to providing effective cyber security. A wide range of techniques are used by both researchers and commercial security solutions.

This research includes the use of commonly employed approaches including supervised learning and unsupervised learning.

However, cyber security problems raise challenges to machine learning. The challenges include the problems associated with correctly labelling data, the adversarial nature of security problems, and for supervised tasks the asymmetry between legitimate and malicious activity. Furthermore, malicious attackers have been witnessed to use the machine learning approach to advance their knowledge and capacities. The aim of the workshop is to highlight current research in the cyber security space.

ML4CYBER 2018 attracted 19 submissions, and nine of them were accepted after a single-blind review by at least two reviewers. The overall acceptance rate for the workshop was 47%.

The selected papers looked at a diverse range of security problems, ranging from the whitelisting problem, intrusion detection, spam to specific security issues looking at malware that performed webinjects and securing the network on a smart car. The accepted papers used a range of techniques ranging from the classic approaches such as decision trees and support vector machines to deep learning.

We are thankful to the researchers who made this workshop possible by submitting their work. We congratulate the authors of the submissions who won prizes. We are also thankful to our Program Committee, who provided reviews in a professional and timely way, and the sponsors, the University of Waikato, Trendmicro, and Data61/CSIRO, who provided the prizes. Thanks to the workshop chairs, who provided the environment to put this workshop together.

June 2018

Zun Zhang
Lei Pan
Jonathan Oliver

BDM 2018 Workshop PC Chairs' Message

For the past few years, biologically inspired data mining techniques have been intensively used in different data mining applications such as data clustering, classification, association rule mining, sequential pattern mining, outlier detection, feature selection and bioinformatics. The techniques include neural networks, evolutionary computation, fuzzy systems, genetic algorithms, ant colony optimization, particle swarm optimization, artificial immune system, culture algorithms, social evolution, and artificial bee colony optimization. A huge increase in the number of papers published in the area has been observed in the last decade. Most of these techniques either hybridize optimization with existing data mining techniques to speed up the data mining process or use these techniques as independent data mining methods to improve the quality of patterns mined from the data.

The aim of the workshop is to highlight the current research related to biologically inspired techniques in different data mining domains and their implementation in real-life data mining problems. The workshop provides a platform to researchers from computational intelligence and evolutionary computation and other biologically inspired techniques to get feedback on their work from other data mining perspectives such as statistical data mining, AI and machine learning based data mining.

Following the call for papers, BDM 2018 attracted 16 submissions from five countries, and seven of them were accepted after a blind review by at least three reviewers. The acceptance rate for the workshop was 40%.

The selected papers highlight work in meta association rules discovery using swarm optimization, causal exploration in genome data, citation field learning using RNN, Affine transformation capsule net, frequent itemsets mining in transactional databases, rare events classification based on genetic algorithms, and PSO-based weighted Nadaraya–Watson estimator.

We are thankful to all the authors who made this workshop possible by submitting their work and responding positively to the changes suggested by our reviewers for their work. We are also thankful to our Program Committee, who dedicated their time and provided us with their valuable suggestions and timely reviews. We wish to express our gratitude to the workshop chairs, who were always available to answer our queries and provided us with everything we needed to put this workshop together.

June 2018

Shafiq Alam
Gillian Dobbie

Organization

Workshop of Big Data Analytics for Social Computing (BDASC 2018)

General Chairs

Mark Western	FASSA, Director of Institute for Social Science Research, The University of Queensland, Australia
Junbin Gao	Discipline of Business Analytics, The University of Sydney Business School, The University of Sydney, Australia

Program Chairs

Lin Wu	Institute for Social Science Research, School of Information Technology and Electrical Engineering, The University of Queensland, Australia
Yang Wang	Dalian University of Technology, China
Michele Haynes	Learning Sciences Institute Australia, Australian Catholic University, Brisbane Queensland, Australia

Program Committee

Meng Fang	Tecent AI, China
Zengfeng Huang	Fudan University, China
Liang Zheng	University of Technology Sydney, Australia
Xiaojun Chang	Carnegie Mellon University, USA
Qichang Hu	The University of Adelaide, Australia
Zongyuan Ge	IBM Research, Australia
Tong Chen	The University of Queensland, Australia
Hongxu Chen	The University of Queensland, Australia
Baichuan Zhang	Facebook, USA
Xiang Zhao	National University of Defence Technology, China
Hongcai Ma	Chinese Academic of Science, China
Yifan Chen	University of Amsterdam, The Netherlands
Xiaobo Shen	Nanyang Technological University, Singapore
Xiaoyang Wang	Zhejiang Gongshang University, China
Ying Zhang	Dalian University of Technology, China
Xiaofeng Gong	Dalian University of Technology, China
Wenda Zhao	Dalian University of Technology, China

Chengyuan Zhang   Central Southern University, China
Kim Betts   The University of Queensland, Australia

Sponsorship: ARC Centre of Excellence for Children and Families Over the Life Course

Pacific Asia Workshop on Intelligence and Security Informatics (PAISI)

Workshop Co-chairs

Michael Chau   The University of Hong Kong, SAR China
Hsinchun Chen   The University of Arizona, USA
G. Alan Wang   Virginia Tech, USA

Webmaster

Philip T. Y. Lee   The University of Hong Kong, SAR China

Program Committee

Victor Benjamin   Arizona State University, USA
Robert Weiping Chang   Central Police University, Taiwan
Vladimir Estivill-Castro   Griffith University, Australia
Uwe Gläesser   Simon Fraser University, Canada
Daniel Hughes   Massey University, Australia
Eul Gyu Im   Hanyang University, South Korea
Da-Yu Kao   Central Police University, Taiwan
Siddharth Kaza   Towson University, USA
Wai Lam   The Chinese University of Hong Kong, SAR China
Mark Last   Ben-Gurion University of the Negev, Israel
Ickjai Lee   James Cook University, Australia
Xin Li   City University of Hong Kong, SAR China
You-Lu Liao   Central Police University, Taiwan
Hsin-Min Lu   National Taiwan University, Taiwan
Byron Marshall   Oregon State University, USA
Dorbin Ng   The Chinese University of Hong Kong, SAR China
Shaojie Qiao   Southwest Jiaotong University, China

Shrisha Rao	International Institute of Information Technology - Bangalore, India
Srinath Srinivasa	International Institute of Information Technology - Bangalore, India
Aixin Sun	Nanyang Technological University, Singapore
Paul Thompson	Dartmouth College, USA
Jau-Hwang Wang	Central Police University, Taiwan
Jennifer J. Xu	Bentley University, USA
Baichuan Zhang	Facebook, USA
Yilu Zhou	Fordham University, USA

Workshop on Data Mining for Energy Modelling and Optimization (DaMEMO)

Workshop Chairs

Irena Koprinska	University of Sydney, Australia
Alicia Troncoso	University Pablo de Olavide, Spain

Publicity Chair

Jeremiah Deng	University of Otago, New Zealand

Program Committee

Vassilios Agelidis	Technical University of Denmark, Denmark
Jeremiah Deng	University of Otago, New Zealand
Irena Koprinska	University of Sydney, Australia
Francisco Martínez-Álvarez	University Pablo de Olavide, Spain
Michael Mayo	University of Waikato, New Zealand
Lilian de Menezes	City University London, UK
Lizhi Peng	Jinan University, China
Bernhard Pfahringer	University of Auckland, New Zealand
Martin Purvis	University of Otago, New Zealand
Mashud Rana	University of Sydney, Australia
Alicia Troncoso	University Pablo de Olavide, Spain
Jun Zhang	South China University of Technology, China

Australasian Workshop on Machine Learning for Cyber-Security (ML4Cyber)

General and Program Chairs

Jonathan Oliver   TrendMicro
Jun Zhang   Swinburne University of Technology

Program Committee

Jonathan Oliver   TrendMicro
Jun Zhang   Swinburne University of Technology, Australia
Ian Welch   Victoria University of Wellington, New Zealand
Wanlei Zhou   University of Technology, Sydney, Australia
Lei Pan   Deakin University, Australia
Islam Rafiqul   Charles Sturt University, Australia
Ryan Ko   Waikato University, New Zealand
Iqbal Gondal   Federation University
Yang Xiang   Swinburne University of Technology, Australia
Vijay Varadharajan   University of Newcastle
Paul Black   Federation University
Lili Diao   TrendMicro
Chris Leckie   University of Melbourne, Australia
Matt Byrne   Fireeye
Goce Ristanoski   Data61
Baichuan Zhang   Facebook

Sponsors

University of Waikato, TrendMicro

Contents

**Australasian Workshop on Machine Learning
for Cyber-Security (ML4Cyber)**

**Biologically-Inspired Techniques for Knowledge Discovery
and Data Mining (BDM)**

Pacific Asia Workshop on Intelligence and Security Informatics (PAISI)

Data Mining for Energy Modeling and Optimization (DaMEMO)

Big Data Analytics for Social Computing (BDASC)

A Decision Tree Approach to Predicting Recidivism in Domestic Violence

Senuri Wijenayake[1], Timothy Graham[2], and Peter Christen[2(✉)]

[1] Faculty of Information Technology, University of Moratuwa, Moratuwa, Sri Lanka
senuri.wijenayake@gmail.com
[2] Research School of Computer Science, The Australian National University,
Canberra, Australia
{timothy.graham,peter.christen}@anu.edu.au

Abstract. Domestic violence (DV) is a global social and public health issue that is highly gendered. Being able to accurately predict DV recidivism, i.e., re-offending of a previously convicted offender, can speed up and improve risk assessment procedures for police and front-line agencies, better protect victims of DV, and potentially prevent future re-occurrences of DV. Previous work in DV recidivism has employed different classification techniques, including decision tree (DT) induction and logistic regression, where the main focus was on achieving high prediction accuracy. As a result, even the diagrams of trained DTs were often too difficult to interpret due to their size and complexity, making decision-making challenging. Given there is often a trade-off between model accuracy and interpretability, in this work our aim is to employ DT induction to obtain both interpretable trees as well as high prediction accuracy. Specifically, we implement and evaluate different approaches to deal with class imbalance as well as feature selection. Compared to previous work in DV recidivism prediction that employed logistic regression, our approach can achieve comparable area under the ROC curve results by using only 3 of 11 available features and generating understandable decision trees that contain only 4 leaf nodes.

Keywords: Crime prediction · Re-offending · Class imbalance
Feature selection

1 Introduction

Domestic violence (DV), defined as violence, intimidation or abuse between individuals in a current or former intimate relationship, is a major social and public health issue. A World Health Organisation (WHO) literature review across 35 countries revealed that between 10% and 52% of women reported at least one instance of physical abuse by an intimate partner, and between 10% and 30% reported having experienced sexual violence by an intimate partner [15].

In Australia, evidence shows that one in six women (17%) and one in twenty men (6%) have experienced at least one incidence of DV since the age of 15 [2,7],

© Springer Nature Switzerland AG 2018
M. Ganji et al. (Eds.): PAKDD 2018, LNAI 11154, pp. 3–15, 2018.
https://doi.org/10.1007/978-3-030-04503-6_1

while a recent report found that one woman a week and one man a month have
been killed by their current or former partner between 2012–13 and 2013–14, and
that the costs of DV are at least $22 billion per year [3]. Whilst DV can affect
both partners in a relationship, these statistics emphasise the highly gendered
nature of this problem. More broadly, gender inequality has been identified as
an explaining factor of violence against women [27].

Worryingly, there has been a recent trend of increasing DV in Australia,
particularly in Western Australia (WA) where family related offences (assault
and threatening behaviour) have risen by 32% between 2014–15 and 2015–16 [28].
Furthermore, police in New South Wales (NSW) responded to over 58,000 call-
outs for DV related incidents in 2014 [6], and DV related assault accounted for
about 43% of crimes against persons in NSW in 2014–15 [20].

Given this context, besides harm to individuals, DV results in an enormous
cost to public health and the state. Indeed, it is one of the top ten risk factors
contributing to disease burden among adult women, correlated with a range of
physical and mental health problems [3,4].

Despite the importance of this issue, there has been relatively little research
on the risk of family violence and DV offending in Australia [5,10]. Recent calls
have been made to develop and evaluate risk assessment tools and decision support systems (DSS) to manage and understand the risk of DV within populations
and to improve targeted interventions that aim to prevent DV and family violence before it occurs. Whilst risk assessment tools have attracted criticism in
areas such as child protection [11], recent studies suggest that DV-related risk
assessment tools can be highly effective in helping police and front-line agencies
to make rapid decisions about detention, bail and victim assistance [17,18].

Contributions: As discussed in the next section, there are a number of challenges and opportunities for using data science techniques to improve the accuracy and interpretability of predictive models in DV risk assessment. In this
paper, we make a contribution to current research by advancing the use of a DT
approach in the context of DV recidivism to provide predictions about the risk
of re-offending that can be easily interpreted and used for decision-making by
front-line agencies and practitioners. We develop and experimentally evaluate
a technique to reduce the size and complexity of trained DTs whilst aiming to
maintain a high degree of predictive accuracy. Our approach is not limited to
specialised data collection or single use cases, but generalises to predicting any
DV-related recidivism using administrative data.

2   Related Work

A key factor when evaluating risk assessment tools and DSS is determining
whether they are able to accurately predict future DV offending amongst a
cohort of individuals under examination. A standard practice is to measure the
accuracy of risk assessment tools using Receiver Operating Characteristic (ROC)
curve analysis [9], which we discuss in more detail later in this paper. Whilst some
tools have been shown to provide reasonably high levels of predictive accuracy

with ROC scores in the high 0.6 to low 0.7 range [24], there are a number of limitations to current approaches.

In summarising the main limitations with current approaches to predicting DV offences using risk assessment tools, Fitzgerald and Graham [10] argue that such tools "often rely on detailed offender and victim information that must form part of specialised data collection, either through in-take or self-report instruments, clinical assessment, or police or practitioner observation" (p. 2). In this way, the cost in terms of time and money for developing these tools is prohibitively high and moreover they do not generalise easily across multiple agencies, social and geographical contexts.

Although there are presumptions in the literature that the accuracy and generalisability of such tools may be increased by combining both official and clinical data sets, studies suggest that the benefits may be negligible, particularly given the high associated costs [25]. Fitzgerald and Graham [10] post that readily available administrative data may be preferable as opposed to data sets that are more difficult and costly to generate. They evaluated the potential of existing administrative data drawn from the NSW Bureau of Crime Statistics and Research (BOCSAR) Re-offending Database (ROD) to accurately predict violent DV-related recidivism [21]. Recidivism is a criminological term that refers to the rate at which individuals who, after release from prison, are subsequently re-arrested, re-convicted, or returned to prison (with or without a new sentence) during a specific time range following their release [26]. In this way, a recidivist can be regarded as someone who is a 'repeat' or 'chronic' offender.

Fitzgerald and Graham [10] used logistic regression to examine the future risk of violent DV offending among a cohort of individuals convicted of any DV offence (regardless of whether it is violent or not) over a specific time period. Using ten-fold cross validation they found the average AUC-ROC (described in Sect. 4) of the models to be 0.69, indicating a reasonable level of predictive accuracy on par with other risk assessment tools described previously. A question that arises from the study is whether more sophisticated statistical models might be able to: (1) improve the accuracy for predicting DV recidivism using administrative data; (2) help to determine and highlight risk factors associated with DV recidivism using administrative data; and (3) provide easily interpretable results that can be readily deployed within risk assessment frameworks.

A particular approach that has received recent attention is decision tree (DT) induction [23], as we describe in more detail in the following section. In the context of predicting violent crime recidivism and in particular DV-related recidivism, Neuilly et al. [19] undertook a comparative study of DT induction and logistic regression and found two main advantages of DTs. First, it provides outputs that more accurately mimic clinical decisions, including graphics (i.e., tree drawings) that can be adapted as questionnaires in decision-making processes. Secondly, the authors found that DTs had slightly lower error rates of classification compared to logistic regression [19], suggesting that DT induction might provide higher predictive accuracy compared to logistic regression. Notably, the related random forest algorithm has recently been used in DV risk prediction and management, with reasonably good predictive performance [13] (however, not considering interpretability which is difficult with random forests).

While existing work on predicting DV recidivism using logistic regression and DT induction is able to obtain results of reasonably high accuracy, the important aspect of *interpretability*, i.e., being able to easily understand and explain the prediction results, has so far not been fully addressed (this is not just the case in predicting DV recidivism, but also in other areas where data science techniques are used to predict negative social outcomes (e.g., disadvantage) [29]). In our study, described next, we employ DT induction which will provide both accurate as well as interpretable results, as we show in our evaluation in Sect. 4.

3   Decision Tree Based Recidivism Prediction

In this study we aim to develop an approach to DV recidivism prediction that is both accurate and interpretable. We believe interpretability is as important as high predictive accuracy in a domain such as crime prediction, because otherwise any prediction results would not be informative and actionable for users who are not experts in prediction algorithms (such as criminologists, law makers, and police forces). We now describe the three major aspects of our work, DT induction, class balancing, and feature selection, in more detail.

Decision Tree Induction: Decision tree (DT) induction [16] is a supervised classification and prediction technique with a long history going back over three decades [23]. As with any supervised classification method, a training data set, $\mathbf{D}_R$, is required that contains ground-truth data, where each record $r = (\mathbf{x}, y) \in \mathbf{D}_R$ consists of a set of input features, $x_i \in \mathbf{x}$ (with $1 \leq i \leq m$ and $m = |\mathbf{x}|$ the number of input features), and a class label y. Without loss of generality we assume $y = \{0, 1\}$ (i.e. a binary, two-class, classification problem). The aim of DT induction is, based on the records in $\mathbf{D}_R$, to build a model in the form of a tree that is able to accurately represent the characteristics of the records in $\mathbf{D}_R$. An example DT trained on our DV data set is shown in Fig. 2.

A DT is a data structure which starts with a root node that contains all records in $\mathbf{D}_R$. Using a heuristic measure such as information gain or the Gini index [16], the basic idea of DT induction algorithms is to identify the best input feature in $\mathbf{D}_R$ that splits $\mathbf{D}_R$ into two (or more, depending upon the actual algorithm used) partitions of highest purity, where one partition contains those records in $\mathbf{D}_R$ where most (ideally all) of their class label is $y = 0$ while the other partition contains those records in $\mathbf{D}_R$ where most (ideally all) of their class label is $y = 1$. This process of splitting is continued recursively until either all records in a partition are in one class only (i.e., the partition is pure), or a partition reaches a minimum partition size (in order to prevent over-fitting [16]).

At the end of this process, each internal node of a DT corresponds to a test on a certain input feature, each branch refers to the outcomes of such a test, and each leaf node is assigned a class label from y based on the majority of records that are assigned to this leaf node. For example, in Fig. 2, the upper-most branch classifies records to be in class $y = 0$ based on tests on only two input features (PP and PC, as described in Table 1).

A trained DT can then be applied on a testing data set, $\mathbf{D}_S$, where the class labels y of records in $\mathbf{D}_S$ are unknown or withheld for testing. Based on the feature values x_i of a test record $r \in \mathbf{D}_S$, a certain path in the tree is followed until a leaf node is reached. The class label of the leaf node is then used to classify the test record r into either class $y = 0$ or $y = 1$. For detailed algorithms the interested reader is referred to [16, 23]. As we describe in more detail in Sect. 4, we will explore some parameters for DT induction in order to identify small trees that are interpretable but achieve high predictive accuracy.

Class Balancing: Many prediction problems in areas such as criminology suffer from a class imbalance problem, where there is a much smaller number of training records with class label $y = 1$ (e.g., re-offenders) versus a much larger number of training records with $y = 0$ (e.g., individuals who do not re-offend). In our DV data set, as described in detail in Sect. 4, we have a class imbalance of around 1:11, i.e., there are 11 times less re-offenders than those who did not re-offend. Such a high class imbalance can pose a challenge for many classification algorithms, including DTs [14], because high prediction accuracy can be achieved by simply classifying all test records as being in the majority class. From a DV risk prediction perspective, this is highly problematic because it means that the classifier would predict every offender as not re-offending [13]. The accuracy would be high, but such a risk prediction tool would not be useful in practice.

Two approaches can be employed to overcome this class imbalance challenge: under-sampling of the majority class and over-sampling of the minority class [8]:

- *Under-sampling of majority class:* Assuming there are $n_1 = |\{r = (\mathbf{x}, y) \in \mathbf{D}_R : y = 1\}|$ training records in class $y = 1$ and $n_0 = |\{r = (\mathbf{x}, y) \in \mathbf{D}_R : y = 0\}|$ training records in class $y = 0$, with $n_1 + n_0 = |\mathbf{D}_R|$. If $n_1 < n_0$, we can generate a balanced training data set by using all training records in $\mathbf{D}_R$ where $y = 1$, but we sample n_1 training records from $\mathbf{D}_R$ where $y = 0$. As a result we obtain a training data set of size $2 \times n_1$ that contains the same number of records in each of the two classes $y = 0$ and $y = 1$.
- *Over-sampling of minority class:* One potential challenge with under-sampling is that the resulting size of the training set can become small if the number of minority class training records (n_1) is small. Under-sampling can also lead to a significant loss of detailed characteristics of the majority class as only a small fraction of its training records is used for training. As a result such an under-sampled training data set might not contain enough information to achieve high prediction accuracy. An alternative is to over-sample the training records from the minority class [8,14]. The basic idea is to replicate (duplicate) records from the minority class until the size of the minority class (n_1) equals the size of the majority class (n_0), i.e., $n_1 = n_0$.

We describe in Sect. 4 how we applied these two class balancing methods to our DV data set in order to achieve accurate prediction results.

Feature Selection: Another challenge to interpretable prediction results is the often increasing number of features in data sets used in many domains. While having more detailed information about DV offenders, for example, will likely be useful to improve predictive accuracy, it potentially can also lead to more complex models such as larger DTs that are more difficult to employ in practice.

Identifying which available input features are most useful for a given prediction or classification problem is therefore an important aspect to obtain interpretable prediction outcomes. As Hand [12] has shown, the first few most important features are also those that are often able to achieve almost as high prediction accuracy as the full set of available features. Any additional, less predictive feature, included in a model can only increase prediction accuracy incrementally. There is thus a trade-off between model complexity, interpretability, and predictive accuracy. Furthermore, using less features will likely also result in less time-consuming training times.

Besides interpretability, a second advantage of DTs over other classification techniques is that the recursive generation of a DT using the input features available in a training data set is actually based on a ranking of the importance of the available features according to a heuristic measure such as information gain or the Gini index [16]. The feature with the best value according to the used measure is the one that is best suited to split the training data sets into smaller sub-sets of highest purity, as described above.

Therefore, to identify a ranking of all available input features we can train a first DT using all available features, and then remove the least important feature (which has the smallest information gain or the highest Gini index value [16]) before training the next DT, and repeat this process until only one (the most important) features is left. Assuming a data set contains m input features, we can generate a sequence of $m - 1$ DTs that use from m to only 1 feature. For each of these trees we calculate its predictive accuracy and assess its complexity as the size of the generated tree. Depending upon the requirements of an application with regard to model complexity (tree size, which affects the tree's interpretability), and predictive accuracy, a suitable tree can then be selected.

We illustrate in Algorithm 1 our overall approach which incorporates both class balancing and iterative feature selection. The output of the algorithm is a list of tuples, each containing a trained DT, the set of used input features, the size of the tree, and the DT's predictive quality (calculated as the AUC-ROC and the F-measure [9] as described below). As we discuss in the following section, from the eleven features available in our data set, not all will be important to predict DV recidivism. We apply the approach described in Algorithm 1 and investigate both the sizes of the resulting DTs as well as their predictive accuracy.

4   Experimental Evaluation

We now describe in more detail the data set we used to evaluate our DT based prediction approach for recidivism in DV, explain the experimental setup, and then present and discuss the obtained results.

Algorithm 1. *Decision tree learning with class balancing and feature selection*

Input:
- $\mathbf{D}_R$: Training data set
- $\mathbf{D}_S$: Testing data set
- M: Set of all input features in $\mathbf{D}_R$ and $\mathbf{D}_S$
- cb: Class-balancing (sampling) method (*under* or *over*)

Output:
- C: List of classification result tuples

1:      $\mathbf{D}_R^0 = \{r = (\mathbf{x}, y) \in \mathbf{D}_R : y = 0\}$ // All training records in class $y = 0$
2:      $\mathbf{D}_R^1 = \{r = (\mathbf{x}, y) \in \mathbf{D}_R : y = 1\}$ // All training records in class $y = 1$
3:      $n_0 = |\mathbf{D}_R^0|, n_1 = |\mathbf{D}_R^1|$ // Number of training records in the two classes
4:      **if** $cb = under$ **then:**
5:          $\mathbf{D}_R^s = \mathbf{D}_R^1 \cup sample(\mathbf{D}_R^0, n_1)$ // Sample n_1 training records from the majority class
6:      **else:**
7:          $\mathbf{D}_R^s = \mathbf{D}_R^0 \cup replicate(\mathbf{D}_R^1, n_0)$ // Replicate training records from minority class
8:      $\mathbf{C} = []$ // Initialise classification results list
9:      $\mathbf{M}_u = \mathbf{M}$ // Initialise the set of features to use as all features
10:     **while** $|\mathbf{M}_u| \geq 1$ **do:**
11:         $\mathbf{dt}_u, lif_u = TrainDecisTree(\mathbf{D}_R^s, \mathbf{M}_u)$ // Train tree and get the least important feature
12:         $s_u = GetTreeSize(\mathbf{dt}_u)$
13:         $auc_u, fmeas_u = GetPredictionAccuracy(\mathbf{dt}_u, \mathbf{D}_S)$
14:         $\mathbf{C}.append([\mathbf{dt}_u, \mathbf{M}_u, s_u, auc_u, fmeas_u])$ // Append results to results list
15:         $\mathbf{M}_u = \mathbf{M}_u \setminus lif_u$ // Remove least important feature from current features
16:     **return C**

Data Set: The data set of administrative data extracted from the NSW Bureau of Crime Statistics and Research (BOCSAR) Re-offending Database (ROD) [21] consists of $n = 14,776$ records, each containing the eleven independent variables (features) shown in Table 1 as well as the dependent class variable. The considered features are grouped to represent the offender, index offence, and criminal history related characteristics of the offenders.

This study aims to predict whether an offender would re-commit a DV related offence within a duration of 24 months since the first court appearance finalisation date (class $y = 1$) or not (class $y = 0$). DV related offences in class $y = 1$ include any physical, verbal, emotional, and/or psychological violence or intimidation between domestic partners.

The Australian and New Zealand Standard Offence Classification (ANZSOC) [1] has recognised murder, attempted murder and manslaughter (ANZSOC 111-131), serious assault resulting in injury, serious assault not resulting in injury and common assault (ANZSOC 211-213), aggravated sexual assault and non-aggravated sexual assault (ANZSOC 311-312), abduction and kidnapping and deprivation of liberty/false imprisonment (ANZSOC 511-521), stalking (ANZSOC 291), and harassment and private nuisance and threatening behaviour (ANZSOC 531-532) as different forms of violent DV related offences.

Experimental Setup: As the aim of the study was to provide a more interpretable prediction to officers involved, a DT classifier along with a graphical representation of the final decision tree was implemented using Python version 3.4, where the *scikit-learn* (http://scikit-learn.org) machine learning library [22] was used for the DT induction (with the Gini index as feature selection measure [16]), and tree visualisations were generated using *scikit-learn* and the *pydotplus* (https://pypi.python.org/pypi/pydotplus) package.

Table 1. Independent variables (features) in the ROD data set used in the experiments as described in Sect. 4. Variable name abbreviations (in bold) are used in the text.

Variable	Description
Offender demographic characteristics	
Gender (**G**)	Whether the offender was recorded in ROD as male or female
Age (**A**)	The age category of the offender at the index court finalisation was derived from the date of birth of the offender and the date of finalisation for the index court appearance
Indigenous status (**IS**)	Recorded in ROD as 'Indigenous' if the offender had ever identified as being of Aboriginal or Torres Strait Islander descent, otherwise 'non-Indigenous'
Disadvantage areas index (quartiles) (**DA**)	Measures disadvantage of an offenders residential postcode at the index offence. Based on the Socio-Economic Index for Areas (SEIFA) score (Australian Bureau of Statistics)
Index conviction characteristics	
Concurrent offences (**CO**)	Number of concurrent proven offences, including the principal offence, at the offenders index court appearance
AVO breaches (**AB**)	Number of proven breach of Appended Violence Order (AVO) offences at the index court appearance
Criminal history characteristics	
Prior juvenile or adult convictions (**PC**)	Number of Youth Justice Conferences or finalised court appearances with any proven offence(s) as a juvenile or adult prior to the index court appearance
Prior serious violent offence conviction past 5 years (**P5**)	Number of Youth Justice Conferences or finalised court appearances in the 5 years prior to the reference court appearance with any proven homicide or serious assault
Prior DV-related property damage offence conviction past 2 years (**P2**)	Number of Youth Justice Conferences or finalised court appearances in the 2 years prior to the reference court appearance with any proven DV property damage offence
Prior bonds past 5 years (**PO**)	Number of finalised court appearances within 5 years of the reference court appearance at which given a bond
Prior prison or custodial order (**PP**)	Number of previous finalised court appearances at which given a full-time prison sentence/custodial order

In a preliminary analysis we identified that only 8% ($n_1 = 1,182$) of the 14,776 offenders recommitted a DV offence within the first 24 months of the finalisation of their index offence. The data set was thus regarded as imbalanced and we applied the two class balancing approaches discussed in Sect. 3:

– *Under-sampling of majority class:* The re-offender and non-re-offender records were separated into two groups, with 1,182 re-offenders and 13,594 non-re-offenders respectively. Next, we randomly sampled 1,182 non-re-offender records, resulting in a balanced data set containing 2,364 records.
– *Over-sampling minority class:* In this approach we duplicated re-offender records such that their number ended up to be the same as the number of non-re-offender records. The resulting balanced data set containing 27,188 records was then shuffled so that the records were randomly distributed.

Each of the two balanced data sets were randomly split into a training and testing set with 70% of all records used to train a DT and the remaining 30% for testing. We applied the iterative feature elimination approach described in Algorithm 1, resulting in a sequence of DTs trained using from 11 and 1 features.

To further explore the ability of DTs of different sizes to obtain high predictive accuracy, we also varied the *scikit-learn* DT parameter *max_leaf_nodes*, which explicitly stops a DT from growing once it has reached a certain size. As shown in Fig. 1, we set the value for this parameter from 2 (i.e., a single decision on one input feature) to 9,999 (which basically means no limitation in tree size). While a DT of limited size might result in reduced predictive accuracy, our aim was to investigate this accuracy versus interpretability trade-off which is an important aspect of employing data science techniques in practical applications such as DV recidivism prediction.

We evaluated the predictive accuracy of the trained DTs using the commonly used measures of Area Under the Receiver Operator Characteristic Curve (AUC-ROC), which is calculated as the area under the curve generated when plotting the true positive rate (TPR) versus the false positive rate (FPR) at a varying threshold of the probability that a test record is classified as being in class $y = 1$ [9]. Because the TPR and FPR are always between 0 and 1, the resulting AUC-ROC will also be between 0 and 1. An AUC-ROC of 0.5 corresponds to a random classifier while and AUC-ROC of 1.0 corresponds to perfect classification.

As a second measure of predictive accuracy we also calculated the F-measure [9], the harmonic mean of precision and recall, which is commonly used in classification applications. The F-measure considers and averages both types of errors, namely false positives (true non-re-offenders wrongly classified as re-offenders) and false negatives (true re-offenders wrongly classified as non-re-offenders).

Table 2. Baseline AUC-ROC results as presented in Fitzgerald and Graham [10] using logistic regression on the same data set used in our study.

Experimental approach	ROC AUC	95% Confidence interval
Internal validation (on full data set)	0.701	0.694–0.717
Ten-fold cross validation	0.694	0.643–0.742

Results and Discussion: In Table 2 we show the baseline results obtained by a state-of-the-art logistic regression based approach using the same data set as the one we used. As can be seen, an AUC-ROC of 0.694 was obtained, however this approach does not allow easy interpretation of results due to the logistic regression method being used.

The detailed results of our DT based approach are shown in Fig. 1, where tree sizes, AUC-ROC and F-measure results can be seen for different number of input features used. As can be seen, with large trees (i.e., no tree growing limits) and using all input features, exceptionally high prediction results (with AUC-ROC and F-measure of up to 0.9) can be achieved. However, the corresponding DTs, which contain over 4,000 nodes, will not be interpretable. Additionally, such large trees would likely overfit the given testing data set.

As can be seen, almost independent of the number of input features (at least until only around two features were used), a DT can be trained with an AUC-ROC of around 0.65, which is less than 5% below the logistic regression based state-of-the-art baseline approach shown in Table 2.

As is also clearly visible, the under-sampling approach (resulting in a much smaller data set than over-sampling) led to worse prediction accuracy results when using all input features, but also to much smaller trees. When using only the few most important features the prediction accuracy of both class balancing methods are very similar. As can be seen in Table 3, the overall ranking of features according to their importance is quite similar, with criminal history features being prominent in the most important set of features.

We show one small example DT learned using the over-sampling method based on only three features in Fig. 2. Such a small tree will clearly be quite easy to interpret by DV experts in criminology or by police forces.

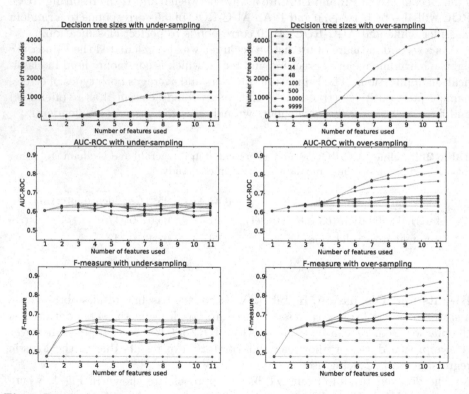

Fig. 1. Results for different number of features used to train a DT: Left with under- and right with over-sampling. The top row shows the sizes of the generated DTs, the middle row shows AUC-ROC and the bottom row shows F-measure results. The parameter varied from 2 to 9, 999 was the maximum number of leaf nodes as described in Sect. 4.

Table 3. Feature importance rankings for over- and under-sampling as discussed in Sect. 3 averaged over all parameter settings used in the experiments. The first ranked feature is the most important one. Feature codes are from Table 1

Sampling approach	1	2	3	4	5	6	7	8	9	10	11
Over-sampling	PP	PC	A	DA	PO	CO	IS	P2	G	AV	P5
Under-sampling	PC	PO	PP	CO	A	AV	DA	P2	IS	G	P5

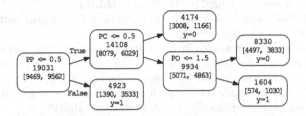

Fig. 2. A small example decision tree (rotated to the left) learned using only three input features (PP, PC, PO, see Table 1 for descriptions) and achieving an AUC-ROC of 0.64. This is almost the same as achieved by much larger trees as shown in Fig. 1, and less than 5% below a previous logistic regression based state-of-the-art approach [10].

5   Conclusion and Future Work

Domestic Violence (DV) is displaying a rising trend worldwide with a significant negative impact on the mental and physical health of individuals and society at large. Having decision support systems that can assist police and other front-line officers in the assessment of possible re-offenders is therefore vital.

With regard to predictive tools that can be used by non-technical users, interpretability of results is as important as high accuracy. Our study has shown that even small decision trees (DTs), that are easily interpretable, trained on balanced training data sets and using only a few input features can achieve predictive accuracy almost as good as previous state-of-the-art approaches.

As future work, we aim to investigate the problem of producing interpretable DT models when DV data sets are linked with external administrative data. This could provide access to additional features to gain improved insights to the decision making process, resulting in higher accuracy. While here we have used eleven features only, future studies could deploy hundreds or even thousands of features derived from administrative data sources. The experiments conducted in this study provide a basis to develop methods for maximising both the accuracy and interpretability of DV risk assessment tools using Big Data collections.

References

1. Australian and New Zealand Society of Criminology. http://www.anzsoc.org
2. Australian Bureau of Statistics: Personal safety survey 2016 (2017)

3. Australian Institute of Health and Welfare: Family, domestic and sexual violence in Australia (2018)
4. Australian Institute of Health and Welfare; Australia's National Research Organisation for Women's Safety Limited (ANROWS): Examination of the health outcomes of intimate partner violence against women: state of knowledge paper (2016)
5. Boxall, H., Rosevear, L., Payne, J.: Identifying first time family violence perpetrators: the usefulness and utility of categorisations based on police offence records. Trends Issues Crime Crim. Justice (487) (2015). https://search.informit.com.au/documentSummary;dn=370297168229940;res=IELHSS
6. Bulmer, C.: Australian police deal with a domestic violence matter every two minutes. ABC News (2015)
7. Cox, P.: Violence against women in Australia: additional analysis of the Australian Bureau of Statistics' personal safety survey. Horizons Research Report, Australia's National Research Organisation for Women's Safety (ANROWS) (2012)
8. Drummond, C., Holte, R.C.: C4.5, class imbalance, and cost sensitivity: why undersampling beats over-sampling. In: ICML Workshop (2003)
9. Fawcett, T.: ROC graphs: notes and practical considerations for researchers. Technical Report HPL-2003-4, HP Laboratories, Palo Alto (2004)
10. Fitzgerald, R., Graham, T.: Assessing the risk of domestic violence recidivism. Crime and Justice Bulletin, NSW Bureau of Crime Statistics and Research (2016)
11. Gillingham, P.: Risk assessment in child protection: problem rather than solution? Aust. Soc. Work 59(1), 86–98 (2006)
12. Hand, D.: Classifier technology and the illusion of progress. Stat. Sci. 21(1), 1–14 (2006)
13. Hsieh, T., Wang, Y.H., Hsieh, Y.S., Ke, J.T., Liu, C.K., Chen, S.C.: Measuring the unmeasurable: a study of domestic violence risk prediction and management. J. Technol. Hum. Serv. 36, 56–68 (2018)
14. Japkowicz, N., Stephen, S.: The class imbalance problem: a systematic study. Intell. Data Anal. 6(5), 429–449 (2002)
15. Krug, E., Mercy, J., Dahlberg, L., Zwi, A.: The world report on violence and health (2002)
16. Lior, R., Maimon, O.: Data Mining with Decision Trees: Theory and Applications, vol. 81, 2nd edn. World Scientific, Singapore (2014)
17. Mason, R., Julian, R.: Analysis of the Tasmania police risk assessment screening tool (RAST). Final report, Tasmanian Institute of Law Enforcement Studies, University of Tasmania (2009)
18. Messing, J.T., Campbell, J., Sullivan Wilson, J., Brown, S., Patchell, B.: The lethality screen: the predictive validity of an intimate partner violence risk assessment for use by first responders. J. Interpers. Violence 32(2), 205–226 (2017)
19. Neuilly, M.A., Zgoba, K.M., Tita, G.E., Lee, S.S.: Predicting recidivism in homicide offenders using classification tree analysis. Homicide Stud. 15(2), 154–176 (2011)
20. New South Wales Police Force: NSW Police Force Annual Report, 2014–15 (2015)
21. NSW Bureau of Crime Statistics and Research: Re-offending statistics for NSW (2018)
22. Pedregosa, F., Varoquaux, G., Gramfort, A., Michel, V., et al.: Scikit-learn: machine learning in Python. J. Mach. Learn. Res. 12(Oct), 2825–2830 (2011)
23. Quinlan, J.R.: Induction of decision trees. Mach. Learn. 1(1), 81–106 (1986)
24. Rice, M.E., Harris, G.T., Hilton, N.: The violence risk appraisal guide and sex offender risk appraisal guide for violence risk assessment. In: Handbook of Violence Risk Assessment, pp. 99–119. Routledge, Abington (2010)

25. Ringland, C., et al.: Measuring recidivism: police versus court data. BOC-SAR NSW Crime Justice Bull. (175), 12 (2013). https://search.informit.com.au/documentSummary;dn=344533455169983;res=IELHSS
26. Ronda, M.: Recividism. In: Encyclopedia of Social Problems, pp. 756–757. Sage Publications, Inc. (2008)
27. Wall, L.: Gender equality and violence against women: what's the connection? Australian Institute of Family Studies, ACSSA Research Summary, no. 7 (2014)
28. Western Australia Police Force: Crime in Western Australia (2018)
29. Wu, L., Haynes, M., Smith, A., Chen, T., Li, X.: Generating life course trajectory sequences with recurrent neural networks and application to early detection of social disadvantage. In: Cong, G., Peng, W.-C., Zhang, W.E., Li, C., Sun, A. (eds.) ADMA 2017. LNCS (LNAI), vol. 10604, pp. 225–242. Springer, Cham (2017). https://doi.org/10.1007/978-3-319-69179-4_16

Unsupervised Domain Adaptation Dictionary Learning for Visual Recognition

Zhun Zhong[✉], Zongmin Li, Runlin Li, and Xiaoxia Sun

College of Computer and Communication Engineering,
China University of Petroleum, Qingdao, China
zhunzhong007@gmail.com

Abstract. Over the last years, dictionary learning method has been extensively applied to deal with various computer vision recognition applications, and produced state-of-the-art results. However, when the data instances of a target domain have a different distribution than that of a source domain, the dictionary learning method may fail to perform well. In this paper, we address the cross-domain visual recognition problem and propose a simple but effective unsupervised domain adaptation approach, where labeled data are only from source domain. In order to bring the original data in source and target domain into the same distribution, the proposed method forcing nearest coupled data between source and target domain to have identical sparse representations while jointly learning dictionaries for each domain, where the learned dictionaries can reconstruct original data in source and target domain respectively. So that sparse representations of original data can be used to perform visual recognition tasks. We demonstrate the effectiveness of our approach on standard datasets. Our method performs on par or better than competitive state-of-the-art methods.

1 Introduction

In the past decade, machine learning has been widely used for various computer vision applications, such as multimedia retrieval [1–3], image classification [4–9], object detection [10–13], person re-identification [14–18], etc. Traditional machine learning methods often learn a model from the training data, and then apply it to the testing data. The fundamental assumption here is that the training data and testing data have the same distribution. However, in real-world applications, it cannot always guarantee that training data share the same distribution with testing data. Therefore, it may produce very poor results when the testing data and training data have the different distributions since the training model is no longer optimal on testing data. For example, applies image classification classifier trained on amazon dataset to phone photos in real life. Face recognition model trained on frontal and well-illumination images to recognize non-frontal poses and less-illumination images. This often viewed as visual

© Springer Nature Switzerland AG 2018
M. Ganji et al. (Eds.): PAKDD 2018, LNAI 11154, pp. 16–26, 2018.
https://doi.org/10.1007/978-3-030-04503-6_2

domain adaptation problem which has been increasing interest in understanding and overcoming.

Domain adaptation aims at learning an adaptive classifier by utilizing the information between source domain with a plenty of labeled data and target domain which is collected from a different distribution. Generally, we can divide domain adaptation into two settings depending on the availability of labels in the target domain data: semi-supervised domain adaptation, and unsupervised domain adaptation. In scenario of semi-supervised domain adaptation, labeled data is available in both source domain (with a plenty of labeled data) and target domain (with a few labeled data), while in scenario of unsupervised domain adaptation labeled data are only available from source domain. In this paper, we mainly focus on unsupervised domain adaptation which is a more challenging task, and more in line with the real-world applications.

Many recent works [19–21] focus on subspace based method to tackle visual domain adaptation problems. In [21], Li et al. determined a feature subspace via canonical correlation analysis (CCA) [22] for recognizing faces with different poses. In [19], Gopalan et al. using geodesic flows to generate intermediate subspaces along the geodesic path between source domain subspace and target domain subspace on the Grassmann manifold. In [20], Gong et al. proposed Geodesic Flow Kernel (GFK), which computes a symmetric kernel between source and target points based on geodesic flow along a latent manifold.

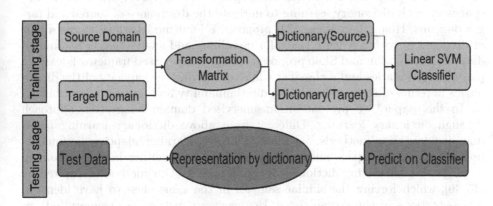

Fig. 1. The overall schema of the proposed framework.

In last few years, the study of dictionary learning based sparse representation has received extensive attention. It has been successfully used for a variety of computer vision applications. For example, classification [23], recognition [24] and denoising [25]. Using an over-complete dictionary, signal or image can be approximated by the combination of only a few number of atoms, that are chosen from the learned dictionary. One of the early dictionary learning algorithms was proposed by Olshausen and Field [26], where a maximum likelihood (ML) learning method was used to sparsely encode images upon a redundant dictionary.

Based on the same ML objective function as in [26], Engan et al. [27] developed a more efficient algorithm, called the method of optimal directions (MOD), in which a closed-form solution for the dictionary update has been proposed. More recently, in [28], Aharon, Elad and Bruckstein proposed the K-SVD algorithm by generalizing k-means clustering and efficiently learns an over-complete dictionary from a set of training signals. This method has been implemented in a variety of image processing problems.

The most existing dictionary based methods assuming that training data and testing data come from the same distribution. However, the learned dictionary may not be optimal if the testing data has different distribution from the data used for training. Learning dictionaries under different domain is a challenging task, and gradually become a hot research over the last few years. In [29], Jia et al. considered a special case where corresponding samples from each domain were available, and learn a dictionary for each domain. Qiu et al. [30] presented a general joint optimization function that transforms a dictionary learned from one domain to the other, and applied such a framework to applications such as pose alignment, pose illumination estimation, and face recognition. Zheng et al. [31] proposed a method achieved promising results on the cross-view action recognition problem with pairwise dictionaries constructed using correspondences between the target view and the source view. In [32], Shekhar et al. learn a latent dictionary which can succinctly represent both the domains in a common projected low-dimensional space. Ni et al. [33] learn a set of subspaces through dictionary learning to mitigate the divergence of source and target domains. Huang and Wang [34] proposed a joint model which learns a pair of dictionaries with a feature space for describing and associating cross-domain data. In [35,36], Zhu and Shao proposed a weakly-supervised framework learns a pairwise dictionaries and a classifier while considering the capacity of the dictionaries in terms of reconstructability, discriminability and domain adaptability.

In this paper, we present an unsupervised domain adaptation approach through dictionary learning. Different from above dictionary learning based domain adaptation methods, our method directly learning adaptive dictionaries in low-level feature space and with no need for labels either in source domain or target domain during dictionary learning process. Our method is inspired by [35,36], which forcing the similar samples in the same class to have identical representations in the sparse space. However, our method is unsupervised, we assume that the nearest coupled low-level features in the original space should maintain their relationship in the sparse space (i.e. these coupled features have the same sparse representation). According to this main idea, we learn a transformation matrix, which selected the nearest data in source domain to each target data. Then the dictionaries for each domain are jointly learned by these selected source data and target data. The data from each domain can be encoded by their dictionaries and then represented by sparse features. Thus, SVM classifier can be trained using these sparse features, and predicting test data on the learned classifier. The learning framework is performed by a classic and efficient dictionary learning method, K-SVD [28]. We demonstrate the effectiveness of our approach on standard cross-domain datasets, and we get state-of-the-art results. An overall schema of the proposed framework is shown in Fig. 1.

1.1 Organization of the Paper

The structure of the rest of the paper is as follows: In Sect. 2, we present our unsupervised domain adaptation dictionary learning algorithm and introduce the classification scheme for the learned dictionary. Experimental results on object recognition are presented in Sect. 3. Finally, the conclusion of this work is given in Sect. 4.

2   Proposed Method

2.1   Problem Notation

Let $I_s = \{I_{s,i}\}_{i=1}^{N_s}$, and $I_t = \{I_{t,j}\}_{j=1}^{N_t}$ be the data instances from the source and target domain respectively, where N_s and N_t denote the number of samples. Each sample from I_s and I_t has a set of d-dimensional local features, thus each sample can represented by $I_{s,i} = \{I_{s,i}^1, I_{s,i}^2, ..., I_{s,i}^{M_i}\}$ and $I_{t,j} = \{I_{t,j}^1, I_{t,j}^2, ..., I_{t,j}^{M_j}\}$ in source and target domain respectively, where M_i and M_j denote the number of local features. Then, the set of local features of source and target domain can be denoted as $Y_s \in \mathbb{R}^{d*L_s}$, and $Y_t \in \mathbb{R}^{d*L_t}$ respectively, where L_s and L_t denote the number of local features in the source and target domain.

2.2   Dictionary Learning

Here, we give a brief review of classical dictionary learning approach. Given a set of d-dimensional input signals, $Y \in \mathbb{R}^{d*L}$, where L is denoted as the number of input signals. Then, learning a K-atoms dictionary of the signals Y, $D \in \mathbb{R}^{d*K}$, can be obtained by solving the following optimization problem:

$$\{D, X\} = argmin_{D,X} \|Y - DX\|_F^2$$
$$s.t. \ \forall_i, \|x_i\|_0 \le T_0 \tag{1}$$

where $D = [d_1, d_2, ..., d_K] \in \mathbb{R}^{d*K}$ denotes the dictionary, $X = [x_1, x_2, ..., x_L] \in \mathbb{R}^{K*L}$ denotes the sparse coefficients of Y decomposed with D, and T_0 is the sparsity level that constraint the number of nonzero entries in x_i.

The performance of sparse representation strictly lie on dictionary learning method. The K-SVD algorithm [28] is a highly effective dictionary learning method that focuses on minimizing the reconstruction error. In this paper, we will solve our formulation of unsupervised domain adaptation dictionary learning based on the K-SVD algorithm.

2.3   Unsupervised Domain Adaptation Dictionary Learning

Now, consider a more general scenario, where we have data from two domains, source domain $Y_s \in \mathbb{R}^{d*L_s}$, and target domain $Y_t \in \mathbb{R}^{d*L_t}$. We wish to jointly learning corresponding dictionaries for each domain. Formally, we desire to minimize the following cost function:

$$
\begin{aligned}
\{D_s, & D_t, X_s, X_t\} \\
&= argmin_{D_s, D_t, X_s, X_t} \|Y_s - D_s X_s\|_F^2 \\
&+ \|Y_t - D_t X_t\|_F^2 \quad s.t. \quad \forall_i, [\|x_i^s\|_0, \|x_i^t\|_0] \le T_0
\end{aligned}
\tag{2}
$$

In addition, in order to maintain the relationship in original feature space, we assume that the nearest coupled low-level features in the original space should also be the nearest couple in the sparse space. Now the new cost function is given by:

$$
\begin{aligned}
\{D_s, & D_t, X_s, X_t\} \\
&= argmin_{D_s, D_t, X_s, X_t} \|Y_s - D_s X_s\|_F^2 \\
&+ \|Y_t - D_t X_t\|_F^2 + C([X_s X_t]) \\
& s.t. \quad \forall_i, [\|x_i^s\|_0, \|x_i^t\|_0] \le T_0
\end{aligned}
\tag{3}
$$

where $D_s = [d_1^s, d_2^s, ..., d_K^s] \in \mathbb{R}^{d*K}$ is the learned source domain dictionary, $X_s = [x_1^s, x_2^s, ..., x_{L_s}^s] \in \mathbb{R}^{K*L_s}$ is the sparse coefficients of source domain, $D_t = [d_1^t, d_2^t, ..., d_K^t] \in \mathbb{R}^{d*K}$ is the learned target domain dictionary, and $X_t = [x_1^t, x_2^t, ..., x_{L_t}^t] \in \mathbb{R}^{K*L_t}$ is the sparse coefficients of target domain. The function $C(\cdot)$ is defined as the distance in the new sparse space of original nearest couples, a small $C(\cdot)$ indicates the data maintain more relationship in new sparse space. This idea is inspired by [35,36], in their method, this function is designed to measure the distances of similar cross-domain instances of the same class. However, our method is exactly unsupervised and directly perform on low-level feature. Thus, the function $C([X_s X_t])$ is defined as:

$$
C([X_s X_t]) = \|X_t - X_s P\|_F^2
\tag{4}
$$

where $P \in \mathbb{R}^{L_s * L_t}$ is the transformation matrix which records the nearest couples between the original data in source and target domain, P can be represented by:

$$
P = \begin{pmatrix}
\Phi(y_1^s, y_1^t) & \cdots & \cdots & \Phi(y_1^s, y_{L_t}^t) \\
\vdots & \ddots & & \vdots \\
\vdots & & \ddots & \vdots \\
\Phi(y_{L_s}^s, y_1^t) & \cdots & \cdots & \Phi(y_{L_s}^s, y_{L_t}^t)
\end{pmatrix}
\tag{5}
$$

where $\Phi(y_i^s, y_j^t)$ is the Gaussian distance between data in original feature space:

$$
\Phi(y_i^s, y_j^t) = \frac{1}{\sqrt{2\pi}} e^{(-\frac{y_i^{s2} - y_j^{t2}}{2})}
\tag{6}
$$

Then, P can be computed by selecting the maximum entry in each column and set to 1 while the other entries are set to 0:

$$P = (i, j) = \begin{cases} 1 & if \quad P(i,j) = max(P(:,j)) \\ 0 & \text{otherwise.} \end{cases} \qquad (7)$$

Thus, Eq. (3) can be written as:

$$\begin{aligned} \{D_s, D_t, X_s, X_t\} \\ = argmin_{D_s, D_t, X_s, X_t} \|Y_s - D_s X_s\|_F^2 \\ + \|Y_t - D_t X_t\|_F^2 + \|X_t - X_s P\|_F^2 \\ s.t. \quad \forall_i, [\|x_i^s\|_0 \|x_i^t\|_0] \leq T_0 \end{aligned} \qquad (8)$$

Assuming P leads to a perfect mapping across the sparse codes X_t and X_s, and each nearest couple has an identical representation after encoding, then $\|X_t - X_s P\|_F^2 = 0$. Thus $X_t = X_s P$, we can rewritten Eq. (8) as:

$$\begin{aligned} \{D_s, D_t, X_s, X_t\} \\ = argmin_{D_s, D_t, X_s, X_t} \|(Y_s - D_s X_s)P\|_F^2 + \|Y_t - D_t X_t\|_F^2 \\ = argmin_{D_s, D_t, X_s, X_t} \|Y_s P - D_s X_s P\|_F^2 + \|Y_t - D_t X_t\|_F^2 \\ = argmin_{D_s, D_t, X_s, X_t} \|Y_s P - D_s X_t\|_F^2 + \|Y_t - D_t X_t\|_F^2 \\ s.t. \quad \forall_i, \|x_i^t\|_0 \leq T_0 \end{aligned} \qquad (9)$$

2.4   Optimization

We can written Eq. (9) as:

$$\begin{aligned} \{\tilde{D}, \tilde{X}\} = argmin_{\tilde{D}, \tilde{X}} \|\tilde{Y} - \tilde{D}\tilde{X}\|_F^2 \\ s.t. \quad \forall_i, \|\tilde{x}_i\|_0 \leq T_0 \end{aligned} \qquad (10)$$

where $\tilde{Y} = \begin{pmatrix} Y_s P \\ Y_t \end{pmatrix}$, $\tilde{D} = \begin{pmatrix} D_s \\ D_t \end{pmatrix}$, and $\tilde{X} = X_t$. Thus, such optimization problem can be solved using the K-SVD algorithm [28].

2.5   Object Recognition

Given the learned D_s and D_t, we obtain sparse representations of the training data in source domain and testing data in target domain respectively. For each image, we obtain a set of sparse representation $X_i = [x_{i,1}, x_{i,2}, ..., x_{i,M_i}] \in \mathbb{R}^{K*M_i}$, where $X_{i,j}$ is the sparse representation of jth feature in image i, K denotes the dictionary size, and M_i is the number of local feature in image i. Then each image represented by a K-vector global representation through max pooling the sparse codes of local features, and then we use linear SVM classifier for cross-domain recognition.

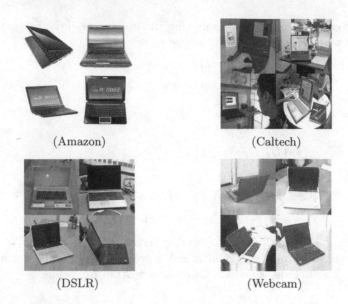

(Amazon)   (Caltech)

(DSLR)   (Webcam)

Fig. 2. Example images from the LAPTOP category on four datasets.

3   Experiments

In this section, we evaluate our domain adaptation approach on 2D object recognition across different datasets.

Experimental Setup: Following the experiment setting in [20], we evaluate our domain adaptation approach on four datasets: Amazon (images downloaded from online merchants), Webcam (low resolution images by a web camera), Dslr (high-resolution images by a SLR camera), and Caltech-256 [37]. We regard each dataset as a domain. Figure 2 shows sample images from these datasets, and clearly highlights the differences between them. We extract 10 classes common to all four datasets: BACKPACK, TOURING-BIKE, CALCULATOR, HEADPHONES, COMPUTER-KEYBOARD, LAPTOP-101, COMPUTER-MONITOR, COMPUTER-MOUSE, COFFEEMUG, AND VIDEO-PROJECTOR. There are 2533 images in total. Each class has 8 to 151 images in a dataset. We use a SURF detector [38] to extract local features over all images. For each pair of source and target domains, we use 20 training samples per class for Amazon/Caltech, and 8 samples per class for DSLR/Webcam when used as source. To draw complete comparison with existing domain adaptation methods, we also carried out experiments on the semi-supervised setting where we additionally sampled 3 labeled images per class from the target domain. We ran 20 different trials corresponding to different selections of labeled data from the source and target domains and testing all unlabeled data in target domain. Our baseline is BOW, where all the images were represented by 800-bin histograms over the codebooks trained from a subset of Amazon images. Our

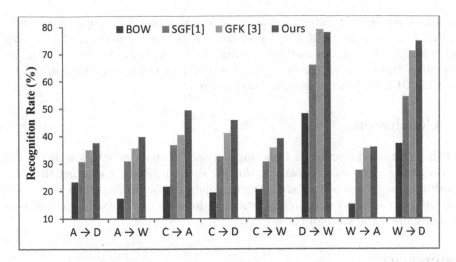

Fig. 3. Cross dataset object recognition accuracies on target domains with unsupervised adaptation over the four datasets (A: Amazon, C: Caltech, D: Dslr, W: Webcam).

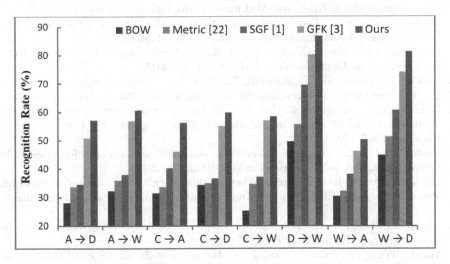

Fig. 4. Cross dataset object recognition accuracies on target domains with semi-supervised adaptation over the four datasets (A: Amazon, C: Caltech, D: Dslr, W: Webcam).

method is also compared with Metric [39], SGF [19] and GFK [20]. Note that, Metric [39] is limited to the semi-supervised setting.

Parameter Settings: For our method, we set dictionary size $K = 512$, and sparse level $T_0 = 5$ for each domain.

Results: The average recognition rate is reported in Figs. 3 and 4 for unsupervised and supervised settings respectively. It is seen that the baseline BOW has the lowest recognition rate, all domain adaptation methods improve accuracy over it. Furthermore, GFK [20] based method clearly outperforms Metric [39] and SGF [19]. Overall, our method consistently demonstrates better performance over all methods except for one pair of source and target combination a little less than GFK [20] in the unsupervised setting.

4 Conclusions

In this paper, we presented a fully unsupervised domain adaptation dictionary learning method to jointly learning domain dictionaries by capturing the relationship between the source and target domain in the original data space. We evaluated our method on publicly available datasets and obtain improved performance upon the state of the art.

References

1. Zheng, L., Wang, S., Liu, Z., Tian, Q.: Fast image retrieval: query pruning and early termination. IEEE Trans. Multimed. **17**(5), 648–659 (2015)
2. Zheng, L., Wang, S., Tian, Q.: Coupled binary embedding for large-scale image retrieval. IEEE Trans. Image Process. **23**(8), 3368–3380 (2014)
3. Kuang, Z., Li, Z., Jiang, X., Liu, Y., Li, H.: Retrieval of non-rigid 3D shapes from multiple aspects. Comput.-Aided Des. **58**, 13–23 (2015)
4. Sánchez, J., Perronnin, F., Mensink, T., Verbeek, J.: Image classification with the fisher vector: theory and practice. Int. J. Comput. Vis. **105**(3), 222–245 (2013)
5. Krizhevsky, A., Sutskever, I., Hinton, G.E.: ImageNet classification with deep convolutional neural networks. In: Advances in Neural Information Processing Systems, pp. 1097–1105 (2012)
6. Simonyan, K., Zisserman, A.: Very deep convolutional networks for large-scale image recognition. arXiv preprint arXiv:1409.1556 (2014)
7. He, K., Zhang, X., Ren, S., Sun, J.: Deep residual learning for image recognition. In: Proceedings of the IEEE Conference on Computer Vision and Pattern Recognition, pp. 770–778 (2016)
8. Zhong, Z., Zheng, L., Kang, G., Li, S., Yang, Y.: Random erasing data augmentation. arXiv preprint arXiv:1708.04896 (2017)
9. Wu, L., Wang, Y., Pan, S.: Exploiting attribute correlations: a novel trace lasso-based weakly supervised dictionary learning method. IEEE Trans. Cybern. **47**(12), 4497–4508 (2017)
10. Girshick, R., Donahue, J., Darrell, T., Malik, J.: Rich feature hierarchies for accurate object detection and semantic segmentation. In: Proceedings of the IEEE Conference on Computer Vision and Pattern Recognition, pp. 580–587 (2014)
11. Girshick, R.: Fast R-CNN. In: Proceedings of the IEEE International Conference on Computer Vision (2015)
12. Ren, S., He, K., Girshick, R., Sun, J.: Faster R-CNN: towards real-time object detection with region proposal networks. In: Advances in Neural Information Processing Systems, pp. 91–99 (2015)
13. Zhong, Z., Lei, M., Cao, D., Fan, J., Li, S.: Class-specific object proposals re-ranking for object detection in automatic driving. Neurocomputing **242**, 187–194 (2017)

14. Zhong, Z., Zheng, L., Cao, D., Li, S.: Re-ranking person re-identification with k-reciprocal encoding. In: 2017 IEEE Conference on Computer Vision and Pattern Recognition (CVPR), pp. 3652–3661. IEEE (2017)
15. Zhong, Z., Zheng, L., Zheng, Z., Li, S., Yang, Y.: Camera style adaptation for person re-identification. In: 2018 IEEE Conference on Computer Vision and Pattern Recognition (CVPR). IEEE (2018)
16. Zheng, L., Yang, Y., Hauptmann, A.G.: Person re-identification: past, present and future. arXiv preprint arXiv:1610.02984 (2016)
17. Wu, L., Wang, Y., Li, X., Gao, J.: What-and-where to match: deep spatially multiplicative integration networks for person re-identification. Pattern Recognit. **76**, 727–738 (2018)
18. Wu, L., Wang, Y., Gao, J., Li, X.: Deep adaptive feature embedding with local sample distributions for person re-identification. Pattern Recognit. **73**, 275–288 (2018)
19. Gopalan, R., Li, R., Chellappa, R.: Domain adaptation for object recognition: an unsupervised approach. In: Proceedings of the IEEE International Conference on Computer Vision, pp. 999–1006. IEEE (2011)
20. Gong, B., Shi, Y., Sha, F., Grauman, K.: Geodesic flow kernel for unsupervised domain adaptation. In: Proceedings of the IEEE Conference on Computer Vision and Pattern Recognition, pp. 2066–2073. IEEE (2012)
21. Li, A., Shan, S., Chen, X., Gao, W.: Maximizing intra-individual correlations for face recognition across pose differences. In: Proceedings of the IEEE Conference on Computer Vision and Pattern Recognition (2009)
22. Hotelling, H.: Relations between two sets of variates. Biometrika **28**, 321–377 (1936)
23. Huang, K., Aviyente, S.: Sparse representation for signal classification. In: Advances in neural information processing systems, pp. 609–616 (2006)
24. Wright, J., Yang, A.Y., Ganesh, A., Sastry, S.S., Ma, Y.: Robust face recognition via sparse representation. IEEE Trans. Pattern Anal. Mach. Intell. **31**(2), 210–227 (2009)
25. Elad, M., Aharon, M.: Image denoising via sparse and redundant representations over learned dictionaries. IEEE Trans. Image Process. **15**(12), 3736–3745 (2006)
26. Olshausen, B.A., Field, D.J.: Sparse coding with an overcomplete basis set: a strategy employed by V1? Vis. Res. **37**(23), 3311–3325 (1997)
27. Engan, K., Aase, S.O., Husoy, J.H..: Method of optimal directions for frame design. In: Acoustics, Speech, and Signal Processing, vol. 5, pp. 2443–2446. IEEE (1999)
28. Aharon, M., Elad, M., Bruckstein, A.: K-SVD: an algorithm for designing overcomplete dictionaries for sparse representation. IEEE Trans. Signal Process. **54**(11), 4311–4322 (2006)
29. Jia, Y., Salzmann, M., Darrell, T.: Factorized latent spaces with structured sparsity. In: Advances in Neural Information Processing Systems, pp. 982–990 (2010)
30. Qiu, Q., Patel, V.M., Turaga, P., Chellappa, R.: Domain adaptive dictionary learning. In: Fitzgibbon, A., Lazebnik, S., Perona, P., Sato, Y., Schmid, C. (eds.) ECCV 2012. LNCS, vol. 7575, pp. 631–645. Springer, Heidelberg (2012). https://doi.org/10.1007/978-3-642-33765-9_45
31. Zheng, J., Jiang, Z., Phillips, P.J., Chellappa, R.: Cross-view action recognition via a transferable dictionary pair. In: BMVC (2012)
32. Shekhar, S., Patel, V.M., Nguyen, H.V., Chellappa, R.: Generalized domain-adaptive dictionaries. In: Proceedings of the IEEE Conference on Computer Vision and Pattern Recognition, pp. 361–368. IEEE (2013)

33. Ni, J., Qiu, Q., Chellappa, R.: Subspace interpolation via dictionary learning for unsupervised domain adaptation. In: Proceedings of the IEEE Conference on Computer Vision and Pattern Recognition, pp. 692–699. IEEE (2013)
34. Huang, D.A., Wang, Y.C.F.: Coupled dictionary and feature space learning with applications to cross-domain image synthesis and recognition. In: Proceedings of the IEEE International Conference on Computer Vision, pp. 2496–2503. IEEE (2013)
35. Zhu, F., Shao, L.: Enhancing action recognition by cross-domain dictionary learning. In: BMVC (2013)
36. Zhu, F., Shao, L.: Weakly-supervised cross-domain dictionary learning for visual recognition. Int. J. Comput. Vis. **109**(1–2), 42–59 (2014)
37. Griffin, G., Holub, A., Perona, P.: Caltech-256 object category dataset (2007)
38. Bay, H., Ess, A., Tuytelaars, T., Van Gool, L.: Speeded-up robust features (SURF). Comput. Vis. Image Underst. **110**(3), 346–359 (2008)
39. Saenko, K., Kulis, B., Fritz, M., Darrell, T.: Adapting visual category models to new domains. In: Daniilidis, K., Maragos, P., Paragios, N. (eds.) ECCV 2010. LNCS, vol. 6314, pp. 213–226. Springer, Heidelberg (2010). https://doi.org/10. 1007/978-3-642-15561-1_16

On the Large-Scale Transferability
of Convolutional Neural Networks

Liang Zheng[1]([⊠]), Yali Zhao[2], Shengjin Wang[2], Jingdong Wang[3], Yi Yang[4],
and Qi Tian[5]

[1] Singapore University of Technology and Design, Singapore, Singapore
liangzheng06@gmail.com
[2] Tsinghua University, Beijing, China
[3] Microsoft Research, Beijing, China
[4] University of Technology Sydney, Sydney, Australia
[5] UTSA, San Antonio, USA

Abstract. Given the overwhelming performance of the Convolutional Neural Network (CNN) in the computer vision and machine learning community, this paper aims at investigating the effective transfer of the CNN descriptors in generic and fine-grained classification at a large scale. Our contribution consists in providing some simple yet effective methods in constructing a competitive baseline recognition system. Comprehensively, we study two facts in CNN transfer. (1) We demonstrate the advantage of using images with a properly large size as input to CNN instead of the conventionally resized one. (2) We benchmark the performance of different CNN layers improved by average/max pooling on the feature maps. Our evaluation and observation confirm that the Conv5 descriptor yields very competitive accuracy under such a pooling strategy. Following these good practices, we are capable of producing improved performance on seven image classification benchmarks.

1 Introduction

This work studies the use of a pre-trained Convolutional Neural Network (CNN) on the tasks of image classification. Specifically, the CNN descriptors extracted from these pre-trained models are employed for visual representation on these target tasks. The target and source domains are typically different in their contents, *e.g.,* generic classification and fine-grained classification. In the community, The CNN models have been record-leading in a number of vision tasks, *e.g.,* image recognition [1], instance retrieval [2–4], person re-identification [5–7], *etc.* Transferring the CNN model from its training source to a target dataset is a common practice due to the expense in collecting a sufficient amount of training data. An example that illustrates the difficulty in data collection consists of instance-level image retrieval. In this task, it is infeasible to collect large amount of training data due to the wide variety in query content. Another example includes the fine-grained classification: experts are needed for class annotation,

© Springer Nature Switzerland AG 2018
M. Ganji et al. (Eds.): PAKDD 2018, LNAI 11154, pp. 27–39, 2018.
https://doi.org/10.1007/978-3-030-04503-6_3

so it is prohibitive to generate large amount of training data. Based on the challenge of data collection and the advantage of CNN models, this paper is devoted to the effective usage of pre-trained CNN models in transfer tasks such as image classification.

Our study is motivated by three aspects. First, in most current works, images are resized to a fixed size, *e.g.,* 227×227 for AlexNet [1], before being used as input to CNN. It ensures that the output CNN descriptor is a $1 \times 1 \times N$ vector, where N is the number of channels in the fully connected layer. Here we use $1 \times 1 \times N$ instead of N because the former reflects the fact that the fully connected layer is also the output of the convolution operation. However, the resizing process may suffer from information loss during image down-sampling, such that some fine details of the images are not included in the feature representation. In previous works, Simonyan *et al.* [8] merge classification results from multi-scale image inputs on the ILSVRC'12 validation set [9]. In this work, we initialize a comprehensive study of this issue on a number of transfer datasets.

Second, previous works on transferring CNN models usually extract features from the fully connected (FC) layers [2,10–12],. The used CNN models learned towards assigning category labels to training images and the FC features are to some extent invariant to illumination, rotation, *etc.* However, the FC features are limited in their description of local patterns, a critical problem when occlusion or truncation exists [12–14]. As with the description to local stimulus, CNN features from bottom or intermediate layers have shown promises [10,15]. The bottom and intermediate layers contain discriminatively trained convolutional kernels which respond to specific visual patterns that evolve from bottom to top layers. Although they capture local activations, the intermediate CNN features are less invariant to image translations due to their small receptive fields. On the effectiveness usage of the intermediate CNN features, this paper finds that with simple pooling steps, the recognition accuracy these intermediate features undergoes impressive improvement. Two contemporary works [16,17] provide similar insights to ours in image retrieval. This work is carried out independently and evaluates the effectiveness of intermediate features more comprehensively on 7 datasets for image classification.

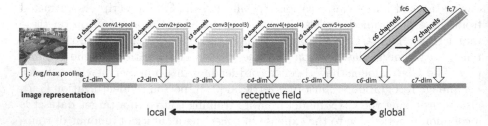

Fig. 1. Feature extraction scheme. An input image of arbitrary size is fed forward in the CNN network. Feature maps in layer k consist of c_k channels. Then, average or max pooling is employed to generate a c_k-dim feature vector. We use for image classification the pooled feature from a single layer.

Taking the above issues into account, this paper suggests some good practices when transferring a pre-trained CNN model to the target datasets. While there exists sporadic reports on these aspects, we extensively evaluate three techniques contributing to a competitive recognition baseline on a serial of benchmarks. The contributions of this paper are listed below.

- Evidence accumulates that using images with larger sizes as input to the CNN model is yields consistent improvement during CNN transfer.
- Average/max pooling of features from intermediate layers is confirmed to be effective in improving invariance to image translations. Specifically, the pooled Conv5 feature, with much lower dimensionality, is shown to yield competitive accuracy with the FC features.

Following the above-mentioned practices, we report results on 7 image classification benchmarks. The classification tasks include generic classification, fine-grained classification, and scene classification. We present improved classification accuracy. We present the pipeline of the feature extraction in Fig. 1. In a nutshell, using scale 1.0 as CNN input, max/avg pooling is performed on the feature maps within each convolutional layer to yield a single vector. Linear SVM is used for all the classification datasets. Throughout this paper, unless specified, we use term "$ConvX$" to refer to layer "$PoolX$ ($ConvX$ + max pooling)", to avoid confusion with our pooling step.

2 Related Works

We will provide a brief literature review from several closely related aspects, *i.e.,* comparison of CNN features from different layers, combining features across multiple scales, and image retrieval/classification using CNN.

Comparison of CNN Features from Different Layers. In most cases, features from the fully connected (FC) layers are preferred, with sporadic reports on intermediate features. For the former, good examples include "Regions with Convolutional Neural Network Features" (R-CNN) [11], CNN baselines for recognition [12], Neural Codes [14], *etc.* The prevalent usage of FC features is mainly attributed to its strong generalization and semantics-descriptive ability. Regarding intermediate features, on the other hand, results of He *et al.* [10] on the Caltech-101 dataset [18] suggest that the Conv5 feature is superior if Spatial Pyramid Pooling (SPP) is used, and is inferior to FC6 if no pooling step is taken. Xu *et al.* [19] find that the VLAD [20] encoded Conv5 features produce higher accuracy on the MEDTest14 dataset in event detection. In [21], Ng *et al.* observe that better performance in image retrieval appears with intermediate layers of GoogLeNet [22] and VGGNet [8] when VLAD encoding is used. Xie *et al.* [23] builds on this work and propose the Inter-layer Activeness Propagation on the intermediate layers. This paper extensively evaluates on 10 datasets the competitiveness of Conv5 with FC features using simple pooling techniques. In two contemporary works, Mousavian *et al.* [16] and Tolias *et al.* [17] draw similar insights in image retrieval. Our work is carried out independently and provides a comprehensive evaluation of intermediate features on both image retrieval and classification.

Image Classification Using CNN. In image classification has been greatly advanced by the introduction of CNN [1]. CNN features are mainly used as global [1] or part [24] descriptors, and are shown to outperform classic hand-crafted ones in both transfer and non-transfer classification tasks. While base-line results with FC features have been reported on small- or medium-sized datasets [12], a detailed evaluation of different layers as well as their combination is still lacking. This paper aims at filling this gap by conducting comprehensive empirical studies.

3   Method

In this section we first describe the pooling step (see Fig. 1) in Sect. 3.1. The proposed image resizing protocol will be presented in Sect. 3.2.

3.1   Pooling

Given an image d, we denote the feature maps of layer k as $f^k (k = 1, \ldots, K)$. Assume that f^k takes the size of $w_d^k \times h_d^k \times c^k$, where w_d^k and h_d^k are the width and height of each channel, respectively, and c^k denotes the number of channels (or convolutional kernels) of layer L_k. Note that, for input images with different sizes, the size of the convolutional maps can be different. Then, we exert average or max pooling steps on the maps, $i.e.$,

$$p_{avg}^k(i) = \frac{1}{w \times h} \sum f^k(\cdot, \cdot, i), i = 1, 2, \ldots c^k, \tag{1}$$

$$p_{max}^k(i) = \max f^k(\cdot, \cdot, i), i = 1, 2, \ldots c^k, \tag{2}$$

where p_{avg}^k and p_{max}^k are both $1 \times 1 \times c^k$ dimensional, representing the output of average and max pooling, respectively. In VGGNet, for example, the result of average/max pooling on Conv5 is of dimension $1 \times 1 \times 512$.

Discussion. By pooling intermediate convolutional maps, there are two main advantages. First, using the intermediate features, the local structures (instead of the global cues by the commonly used FC features) are paid more attention to. This is because the convolutional filters are sensitive to specific visual pat-terns, ranging from low-level bars to mid-level parts. Second, by pooling, the resulting vectors have higher invariance to translation, occlusion, and trunca-tion, which greatly improves the effectiveness of the intermediate features. An additional advantage of pooling is the computational efficiency brought about by low-dimensional feature vectors (512-dim for VGGNet Conv5 feature).

Figure 2 illustrates some examples where conv5 feature (followed with average pooling) captures common local structures in relevant image pairs which are

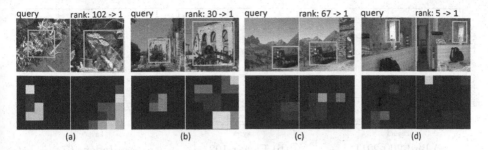

Fig. 2. Illustration of the advantage of Conv5 feature. Four image pairs are shown. To the left of each pair is the query image, and to the right the relevant database image. The second row depicts the feature maps from a certain channel of the Conv5 feature of the corresponding images. For each relevant image, its ranks obtained by both the FC6 and Conv5 features (after average pooling) are shown. Clearly, the pooled Conv5 feature captures mid-level cues and improves recognition accuracy.

largely lost in the FC6 representation. For example, Fig. 2(c) and (d) each show two images containing a similar pattern at distinct positions. The convolutional maps (second row in Fig. 2) that respond to such patterns are very different due to the intensive image variations. In this example, average pooling alleviates the influence of translation variance and improves the rank of the relevant image. In another example, when truncation (Fig. 2(a)) exists, the top-layer feature (FC6) is less effective because the high-level information is to some extent blocked. In this case, low-level features may be of great value by detecting common local patterns between the two images. Then, average pooling improves truncation invariance by capturing the local similarities.

Fig. 3. Sample images of the experimental datasets. For each dataset, one or two groups of images are listed that belong to the same class (image classification). The class labels are also shown above the columns of classification datasets.

3.2   The Role of Image Size

In conventional cases, images are resized so that all images have the same size. In AlexNet, for example, during CNN training, all images are resized to 227×227

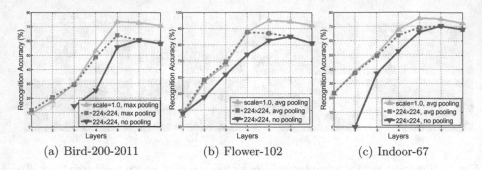

(a) Bird-200-2011   (b) Flower-102   (c) Indoor-67

Fig. 4. Image classification accuracy on three benchmarks. We plot the accuracy against features extracted from different CNN layers (VGGNet). For each dataset, we compare two image sizes, *i.e.*, 224 × 224 and scale 1.0 (see Sect. 3.2); also, we compare the cases when no pooling is used and when max or avg pooling is used.

before being fed into the network. During testing, images are typically resized to 227 × 227 for feature extraction and classification. In transfer tasks, however, the pre-trained CNN model does not provide direct classification results as implied in the FC layers, and it may not fit well the distribution of an unseen dataset. The network mainly discovers low-level and mid-level (more important) cues for the target dataset. Under such cases, the simple image resizing does not have a solid back up. Resizing large images into 227 × 227 (or 224 × 224 in VGGNet) may suffer from substantial information loss such as the missed details and object distortion. This problem is not trivial, as in object recognition, the object-of-interest may take up only a small region in the target image, and in a larger image, details can be more clearly observed. It is the finer details that makes a difference towards more accurate visual recognition, especially in datasets containing building, fine-grained classes, *etc.* Also, object distortion may compromise image matching between the query and database images. In recognition, keeping aspect ratio of an image will also help preserve the shape of objects/scene, thus being beneficial for accurate classification.

Table 1. Average side lengths of the 7 datasets.

Datasets	Bird	Flower	Indoor	SUN	Cal-101	Cal-256	VOC'07
Long side	490.8	664.0	521.3	1,002.8	319.7	398.6	496.4
Short side	364.3	500.0	399.1	733.2	227.0	296.9	358.2

Consequently, in this paper, relatively large images (compared with 224×224) are taken as input. Specifically, given a dataset, we calculate the average image height and width of the training set. Then, the larger value between the average height and width is taken as the long side of all training and testing images. For example, if the average image size of the training set is 400 × 300, we will accordingly resize all images into a longer size of 400 pixels, and keep the aspect ratio. In this paper, the image size calculated in this manner is termed **scale 1.0**. The other scales are defined as the ratio between their long side to that of scale 1.0. For example, if a long size of 400 pixels is defined as scale 1.0, then scaling it to 300 pixels yields a scale of 0.75. In Sect. 4, several image scales will be tested to show the advantage of this protocol.

As with image size, the original size of an image is the result of camera resolution; it does not depend on the image pattern or the distance from the observer to the object. It is hard to tell the distance from the observer to the object only from the image size. However, the size of an image does affect the pattern in the image. If we fix the distance from the observer to the object/scene, a larger size of an image (higher resolution) corresponds to finer details of the object/scene. There is not a clear link between image size and image content. Typically, the image size reflects the resolution of the data collection source. For example, early datasets such as Caltech-101 [18] may have small average image sizes, while more recent datasets such as SUN-397 [25] has larger images.

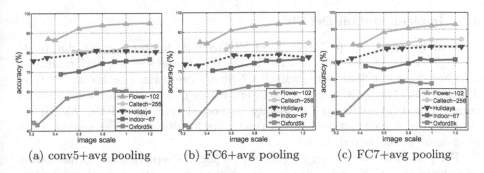

| (a) conv5+avg pooling | (b) FC6+avg pooling | (c) FC7+avg pooling |

Fig. 5. Impact of image size on five datasets. For scale = 1.0, all images are resized with equal long size determined in the training set, and the aspect ratio is preserved. The smallest scale corresponds to 224 × 224. We observe that scale 1.0 yields competitive or superior accuracy to the other scales.

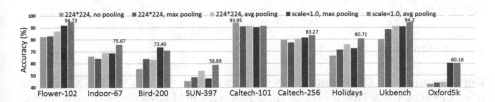

Fig. 6. Comparison of average and max pooling of **Conv5 features** on the 9 datasets. Two image sizes are shown, *i.e.*, 224 × 224 and scale 1.0 (Sect. 3.2). For each image size, max and avg pooling are compared. In most cases, scale 1.0 with average pooling yields the best results.

4   Experiments

4.1   Datasets

In this paper, 7 datasets are tested. For generic classification, we use **Caltech-101** [18], and **Caltech-256** [26] and **PASCAL VOC'07** [27] datasets. 30 and 60 images per category are randomly selected for training on Caltech-101 and -256, respectively. For VOC'07, we use the standard train/test split, and calculate the mean Average Precision (mAP) over 20 classes. Then, for scene classification, we evaluate on **Indoor-67** [28] and **SUN-397** [25] datasets, in which 80 and 50 training images per category are selected, respectively. For fine-grained classification, Oxford **Flower-102** [29] and Caltech-UCSD **Bird-200-2011** [30] datasets are used. 20 and 30 training images per category are randomly chosen, respectively. For all the datasets except VOC'07, we repeat the random partitioning of training/test images for 10 times, and the averaged classification accuracy is reported. Sample images of the 10 datasets are shown in Fig. 3. The average side lengths of the datasets are presented in Table 1.

4.2   Implementation Details

We mainly report the performance of VGGNet [8] for brevity and due to its state-of-the-art accuracy in ILSVRC'14 [9]. We also present the performance of AlexNet [1] in our final system. Both CNNs are pre-trained on the ILSVRC'12 dataset [31]. The VGGNet has 19 layers which can be divided into 8 major convolutional layers (including 3 fully connected layers), and the AlexNet is composed of 8 convolutional layers (including 3 fully connected layers). Both models are available online [32].

For pooled features from each layer, we use the square root normalization introduced in [33]. Namely, we exert a square root operator on each dimension, and then l_2-normalize the vector. In our preliminary experiment, the addition of the square root operator improves the classification accuracy marginally compared with the direct l_2 normalization. For VGGNet and AlexNet, the dimensions of the pooled Conv5 features are 256 and 512, respectively. For VGGNet, the five convolutional layers each have 2, 2, 4, 4, and 4 sub-layers, respectively. We take the last sub-layer as the representative convolutional layer (see Sect. 3.1).

4.3   Evaluation

Performance of Different CNN Layers. As shown in [34], CNN features from different CNN layers vary in their receptive field sizes and semantic levels. To evaluate the impact of different CNN layers on classification accuracy, we test the CNN features (with or without pooling) extracted from the seven layers on six datasets, *i.e.,* Bird-200, Flower-102, Indoor-67, Holidays, Ukbench, and Oxford5k, and results are presented in Fig. 4 and Table 2.

Across the six testing dataset, our observations are consistent: CNN features extracted from the bottom layers are generally inferior to those from the top layers. This finding is expected because the bottom-layer filters respond to low-level visual patterns that generally do not have enough discriminative ability. To some extent, these low-level filters are similar to the visual words in the Bag-of-Words model with the SIFT descriptor [35] which do not have much semantic meanings. On the Bird-200-2011 dataset, for example, the Conv1 feature with max pooling only yields a classification accuracy of 11.62% by the conventional 224×224 resizing. As a consequence, it is desirable that further discriminative cues be incorporated in the bottom features, *e.g.,* spatial constraints [36], or multi-feature fusion [37].

A noticeable observation is that **the "Conv5 + a/m pooling" feature yields very competitive performance to the FC features.** For image classification, Conv5 feature is very competitive in 4 out of 7 datasets, *i.e.,* on fine-grained and scene classification tasks. In fact, fine-grained and scene classification are representative transfer tasks in which the target classes are more distant from the source ImageNet [31]. On Flower-102, for example, Wilcoxon signed-rank test tells a significant difference between Conv5 and FC6, and from Fig. 4 we can tell with high significance level that Conv5 yields higher accuracy than FC6. Moreover, in both classification tasks, local or mid-level elements are very informative cues for discriminating different classes. In this scenario, the Conv5 feature is advantageous to encodes mid-level activations which further acquire translation/occlusion invariance through the pooling step. Experimental results confirm our assumption: using the Conv5 features, classification accuracy improves by +0.62%, +0.66%, +0.35%, and +1.12% on Bird, Flower, Indoor, and SUN, respectively.

For generic recognition as the case in Caltech-101, Caltech-256, and PASCAL VOC'07, our observation is that FC6 outperforms conv5, and that FC6 somehow is superior to FC7 as well as supported by the Wilcoxon test. This can be attributed to two reasons. First, objects in the first two datasets are well-positioned in the center of the image: translation and occlusion are not severe. Second, the VGGNet is trained on generic ImageNet classification dataset: it is less a transfer problem for generic classification task. However, the "transfer" property can still be observed by telling FC7 is inferior to FC6 on Caltech-101, and Caltech-256 datasets; otherwise, it should be expected that FC7 could produce better features.

Impact of Image Sizes. We then evaluate the impact of image size on classification accuracy. According to Sect. 3.2, various image scales are generated. Recall that scale 1.0 is so defined that images keep their aspect ratio and have the same long side as calculated in the training set or the database.

Using different image scales, we report results on 5 datasets, *i.e.,* Flower-102, Caltech-256, and Indoor-67, which are shown in Fig. 5. Note that for all the testing datasets, the smallest size corresponds to 224×224. From Fig. 5 we can see that an increase in image scale brings consistent improvement in classification accuracy. The results validate our assumption that using a larger image (*e.g.,* scale 1.0) and keeping the aspect ration benefit for higher accuracy. Moreover, after reaching scale 1.0, the classification performance tends to stay stable when further increasing the image scale. In fact, when simply resizing an image to a scale larger than 1.0, there is no information gain, which explains why there would not be much improvement. Another limitation of using scale larger than 1.0 is that the memory consumption will be increased.

To further validate our proposal, we have conducted experiments on the ILSVRC'12 validation set [9] In addition to the above evidence, we also collect results on the ILSVRC'12 validation set [9] which is not a transfer dataset. Using VGGNet, we average the output of the "softmax" layer, and sort the scores in descending order. Results are presented in Table 3, which indicate that a larger image size leads to lower classification error and that a scale larger than 1.0 does not help decrease classification error further. In summary, the results are supporting our proposal that using properly large images and keeping aspect ration bring about decent improvement.

Table 2. Results on 10 benchmarks *w.r.t* different CNN layers with average/max (a/m) pooling. Max pooling is used on the Bird dataset.

Datasets	Bird	Flower	Indoor	SUN	Cal-101	Cal-256	VOC'07
conv4+a/m pool.	53.20	88.01	67.81	50.71	80.44	63.86	67.55
conv5+a/m pool.	**73.40**	**94.73**	**75.67**	**58.88**	91.07	83.29	81.78
FC6+a/m pool.	72.78	94.07	75.32	57.76	92.24	84.20	82.31
FC7+a/m pool.	70.64	92.05	71.4	58.31	89.28	83.82	82.57

Table 3. Recognition error (%) on ILSVRC'12 validation set. Scale $= 0.45$ corresponds to image resized to 224×224.

Scale	1.1	1.0	0.95	0.8	0.6	0.45
Top-1 error	25.91	**25.45**	25.53	25.98	30.41	34.12
Top-5 error	7.81	**7.62**	7.79	8.04	10.71	13.17

Impact of Pooling. This paper suggests the use of average/max pooling to aggregate the CNN feature maps. Here, we demonstrate the effectiveness of such pooling steps in Figs. 4 and 6. Two major conclusions are drawn. First, Fig. 4 indicate that on all the CNN layers, avg/max pooling typically leads to higher accuracy than directly using raw CNN features. This is because raw features from the bottom layers are very sensitive to image variances such as translation or occlusion (FC features also have some sensitivity). Pooling addresses this problem by aggregating local activations into a global representation that is similar to the traditional Bag-of-Words model, so that the visual representations are more robust. Pooling also improves computational efficiency by a lower feature dimension.

Second, we evaluate the performance difference between average pooling and max pooling in Fig. 6. The Conv5 CNN feature is used in the comparison on 9 testing datasets. Our results indicate that in most cases (8 out of 9 datasets), average pooling is superior to max pooling, except for the Bird-200 dataset. That is, for Conv5 features, average pooling works well on datasets containing scenes or large objects. In the Bird-200 dataset (see Fig. 3), however, the target objects (birds) only takes up a very small region. As a consequence, the activation maps in the Conv5 layer should be sparse compared with the other datasets in which the objects (or scenes) are larger. So for Bird-200, the effectiveness of average pooling is compromised by zeros in the maps.

5 Conclusion

This paper aims to provide a solid and competitive baseline for image classification using the Convolutional Neural Network. With the off-the-shelf CNN models, we share several good practises with the community of the effective usage of CNN features. First, we find that using larger images (scale 1.0) as input to CNN, rather than 227×227, leads to superior accuracy. Second, with simple pooling steps such as the average/max pooling, the activation maps of the intermediate CNN features can be aggregated to improve the robustness against image variances. We observe that during transfer, the Conv5 features yield competitive accuracy compared with the FC features.

Acknowledgement. This work was supported in part to Dr. Qi Tian by ARO grant W911NF-15-1-0290 and Faculty Research Gift Awards by NEC Laboratories of America and Blippar. This work was supported in part by National Science Foundation of China (NSFC) 61429201.

References

1. Krizhevsky, A., Sutskever, I., Hinton, G.E.: Imagenet classification with deep convolutional neural networks. In: NIPS (2012)
2. Sharif Razavian, A., Sullivan, J., Maki, A., Carlsson, S.: A baseline for visual instance retrieval with deep convolutional networks. In: ICLR (2015)
3. Zheng, L., Wang, S., Liu, Z., Tian, Q.: Packing and padding: coupled multi-index for accurate image retrieval. In: CVPR (2014) 1947–1954
4. Zheng, L., Wang, S., Tian, Q.: Coupled binary embedding for large-scale image retrieval. IEEE Trans. Image Process. **23**(8), 3368–3380 (2014)
5. Wu, L., Shen, C., van den Hengel, A.: Deep linear discriminant analysis on fisher networks: a hybrid architecture for person re-identification. Pattern Recognit. **65**, 238–250 (2017)
6. Wu, L., Wang, Y., Li, X., Gao, J.: What-and-where to match: deep spatially multiplicative integration networks for person re-identification. Pattern Recognit. **76**, 727–738 (2018)
7. Wu, L., Shen, C., van den Hengel, A.: Deep recurrent convolutional networks for video-based person re-identification: an end-to-end approach. arXiv preprint arXiv:1606.01609 (2016)

8. Simonyan, K., Zisserman, A.: Very deep convolutional networks for large-scale image recognition. arXiv preprint arXiv:1409.1556 (2014)
9. Russakovsky, O., et al.: ImageNet large scale visual recognition challenge. IJCV 1–42 (2015)
10. He, K., Zhang, X., Ren, S., Sun, J.: Spatial pyramid pooling in deep convolutional networks for visual recognition. In: Fleet, D., Pajdla, T., Schiele, B., Tuytelaars, T. (eds.) ECCV 2014. LNCS, vol. 8691, pp. 346–361. Springer, Cham (2014). https://doi.org/10.1007/978-3-319-10578-9_23
11. Girshick, R., Donahue, J., Darrell, T., Malik, J.: Rich feature hierarchies for accurate object detection and semantic segmentation. In: CVPR (2014)
12. Razavian, A.S., Azizpour, H., Sullivan, J., Carlsson, S.: CNN features off-the-shelf: an astounding baseline for recognition. In: CVPR Workshops (2014)
13. Zheng, L., Wang, S., Wang, J., Tian, Q.: Accurate image search with multi-scale contextual evidences. Int. J. Comput. Vis. **120**(1), 1–13 (2016)
14. Babenko, A., Slesarev, A., Chigorin, A., Lempitsky, V.: Neural codes for image retrieval. In: Fleet, D., Pajdla, T., Schiele, B., Tuytelaars, T. (eds.) ECCV 2014. LNCS, vol. 8689, pp. 584–599. Springer, Cham (2014). https://doi.org/10.1007/978-3-319-10590-1_38
15. Zheng, L., Yang, Y., Tian, Q.: SIFT meets CNN: a decade survey of instance retrieval. IEEE Trans. Pattern Anal. Mach. Intell. **40**(5), 1224–1244 (2017)
16. Mousavian, A., Kosecka, J.: Deep convolutional features for image based retrieval and scene categorization. arXiv preprint arXiv:1509.06033 (2015)
17. Tolias, G., Sicre, R., Jégou, H.: Particular object retrieval with integral max-pooling of CNN activations. arXiv preprint arXiv:1511.05879 (2015)
18. Fei-Fei, L., Fergus, R., Perona, P.: One-shot learning of object categories. IEEE Trans. Pattern Anal. Mach. Intell. **28**(4), 594–611 (2006)
19. Xu, Z., Yang, Y., Hauptmann, A.G.: A discriminative CNN video representation for event detection. In: CVPR (2015)
20. Jégou, H., Douze, M., Schmid, C.: Product quantization for nearest neighbor search. TPAMI **33**(1), 117–128 (2011)
21. Ng, J., Yang, F., Davis, L.: Exploiting local features from deep networks for image retrieval. In: CVPR Workshops (2015)
22. Szegedy, C., et al.: Going deeper with convolutions. arXiv preprint arXiv:1409.4842 (2014)
23. Xie, L., Zheng, L., Wang, J., Yuille, A., Tian, Q.: Interactive: inter-layer activeness propagation. arXiv preprint arXiv:1605.00052 (2016)
24. Zhang, N., Donahue, J., Girshick, R., Darrell, T.: Part-based R-CNNs for fine-grained category detection. In: Fleet, D., Pajdla, T., Schiele, B., Tuytelaars, T. (eds.) ECCV 2014. LNCS, vol. 8689, pp. 834–849. Springer, Cham (2014). https://doi.org/10.1007/978-3-319-10590-1_54
25. Xiao, J., Hays, J., Ehinger, K., Oliva, A., Torralba, A., et al.: Sun database: large-scale scene recognition from abbey to zoo. In: CVPR
26. Griffin, G., Holub, A., Perona, P.: Caltech-256 object category dataset (2007)
27. Everingham, M., Van Gool, L., Williams, C.K.I., Winn, J., Zisserman, A.: The pascal visual object classes (VOC) challenge. Int. J. Comput. Vis. **88**(2), 303–338 (2010)
28. Quattoni, A., Torralba, A.: Recognizing indoor scenes. In: CVPR (2009)
29. Nilsback, M.E., Zisserman, A.: Automated flower classification over a large number of classes. In: Sixth Indian Conference on Computer Vision, Graphics & Image Processing (2008)

30. Wah, C., Branson, S., Welinder, P., Perona, P., Belongie, S.: The caltech-UCSD birds-200-2011 dataset (2011)
31. Deng, J., Dong, W., Socher, R., Li, L.J., Li, K., Fei-Fei, L.: ImageNet: a large-scale hierarchical image database. In: CVPR (2009)
32. Jia, Y., et al.: Caffe: convolutional architecture for fast feature embedding. In: ACM Multimedia (2014)
33. Relja, A., Zisserman, A.: Three things everyone should know to improve object retrieval. In: CVPR (2012)
34. Zeiler, M.D., Fergus, R.: Visualizing and understanding convolutional networks. In: Fleet, D., Pajdla, T., Schiele, B., Tuytelaars, T. (eds.) ECCV 2014. LNCS, vol. 8689, pp. 818–833. Springer, Cham (2014). https://doi.org/10.1007/978-3-319-10590-1_53
35. Lowe, D.G.: Distinctive image features from scale invariant keypoints. IJCV **60**(2), 91–110 (2004)
36. Zhang, Y., Jia, Z., Chen, T.: Image retrieval with geometry-preserving visual phrases. In: CVPR (2011)
37. Zheng, L., Wang, S., Tian, L., He, F., Liu, Z., Tian, Q.: Query-adaptive late fusion for image search and person re-identification. In: CVPR (2015)

Call Attention to Rumors: Deep Attention Based Recurrent Neural Networks for Early Rumor Detection

Tong Chen[1](✉), Xue Li[1], Hongzhi Yin[1], and Jun Zhang[2]

[1] The University of Queensland, Brisbane, Australia
tong.chen@uq.edu.au, xueli@itee.uq.edu.au, db.hongzhi@gmail.com
[2] Swinburne University of Technology, Melbourne, Australia
junzhang@swin.edu.au

Abstract. The proliferation of social media in communication and information dissemination has made it an ideal platform for spreading rumors. Automatically debunking rumors at their stage of diffusion is known as *early rumor detection*, which refers to dealing with sequential posts regarding disputed factual claims with certain variations and highly textual duplication over time. Thus, identifying trending rumors demands an efficient yet flexible model that is able to capture long-range dependencies among postings and produce distinct representations for the accurate early detection. However, it is a challenging task to apply conventional classification algorithms to rumor detection in earliness since they rely on hand-crafted features which require intensive manual efforts in the case of large amount of posts. This paper presents a deep attention model based on recurrent neural networks (RNNs) to *selectively* learn temporal representations of sequential posts for rumor identification. The proposed model delves soft-attention into the recurrence to simultaneously pool out distinct features with particular focus and produce hidden representations that capture contextual variations of relevant posts over time. Extensive experiments on real datasets collected from social media websites demonstrate that the deep attention based RNN model outperforms state-of-the-art baselines by detecting rumors more quickly and accurately than competitors.

Keywords: Early rumor detection · Recurrent neural networks
Deep attention models

1 Introduction

The explosive use of contemporary social media in communication has witnessed the widespread of rumors which can pose a threat to the cyber security and social stability. For instance, on April 23rd 2013, a fake news claiming two explosions happened in the White House and Barack Obama got injured was posted by a hacked Twitter account named Associated Press. Although the White House and

M. Ganji et al. (Eds.): PAKDD 2018, LNAI 11154, pp. 40–52, 2018.
https://doi.org/10.1007/978-3-030-04503-6_4

Associated Press assured the public minutes later the report was not true, the fast diffusion to millions of users had caused severe social panic, resulting in a loss of \$136.5 billion in the stock market[1]. This incident of a false rumor showcases the vulnerability of social media on rumors, and highlights the practical value of automatically predicting the veracity of information.

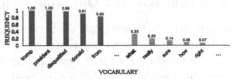

(a) Streaming posts in regards to an event   (b) Statistics on textual phrases

Fig. 1. For social media posts regarding a specific event, *i.e.,* "Trump being disqualified from U.S. election", tokens like "Donald Trump", "Obama" and "disqualified" appear extremely frequently in disputed postings.

Debunking rumors at their formative stage is particularly crucial to minimizing their catastrophic effects. Most existing rumor detection models employ learning algorithms that incorporate a wide variety of features and formulate rumor detection into a binary classification task. They commonly craft features manually from the content, sentiment [1], user profiles [2], and diffusion patterns of the posts [3–5]. Embedding social graphs into a classification model also helps distinguish malicious user comments from normal ones [6,7]. However, feature engineering is extremely time-consuming, biased, and labor-intensive. Moreover, hand-crafted features are data-dependent, making them incapable of resolving contextual variations in different posts.

Recent examinations on rumors reveal that social posts related to an event under discussion are coming in the form of time series wherein users forward or comment on it continuously over time. Meanwhile, as shown in Fig. 1, during the discussion of arbitrary topics, users' posts exhibit high duplication in their textual phrases due to the repeated forwarding, reviews, and/or inquiry behavior [8]. This poses a challenge on efficiently distilling distinct information from duplication and timely capturing textual variations from posts.

The propagation of information on social media has temporal characteristics, whilst most existing rumor detection methodologies ignore such a crucial property or are not able to capture the temporal dimension of data. One exception is [9] where Ma *et al.* uses an RNN to capture the dynamic temporal signals of rumor diffusion and learn textual representations under supervision. However, as the rumor diffusion evolves over time, users tend to comment differently in various stages, such as from expressing surprise to questioning, or from believing to debunking. As a consequence, textual features may change their patterns

---

[1] http://www.dailymail.co.uk/news/article-2313652/AP-Twitter-hackers-break-news-White-House-explosions-injured-Obama.html.

with time and we need to determine which of them are more important to the detection task. On the other hand, the existence of duplication in textual phrases impedes the efficiency of training a deep network. In this sense, two aspects of temporal long-term characteristic and dynamic duplication should be addressed simultaneously in an early rumor detection model.

1.1   Challenges and Our Approach

In summary, there are three challenges in early rumor detection to be addressed: (1) automatically learning representations for rumors instead of using labor-intensive hand-crafted features; (2) the difficulty of maintaining the long-range dependency among variable-length post series to build their internal representations; (3) the issue of high duplication compounded with varied contextual focus. To combat these challenges, we propose a novel deep attention based recurrent neural network (RNN) for early detection on rumors, namely *CallAtRumors* (**Call Attention to Rumors**). The overview of our framework is illustrated in Fig. 2. For one event (i.e., topic) our model converts posts related to one event into feature matrices. Then, the RNN with soft attention mechanism automatically learns latent representations by feed-forwarding each input weighted by attention weights. Finally, an additional hidden layer with *sigmoid* activation function using the learned latent representations to classify whether this event is a rumor or not.

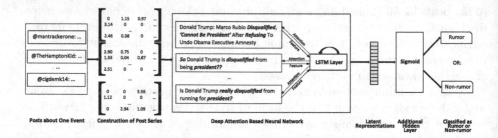

Fig. 2. Schematic overview of our framework.

1.2   Contributions

The main contributions of our work are summarized in three aspects:

- We propose a deep attention neural network that learns to perform rumor detection automatically in earliness. The model is capable of learning continuous hidden representations by capturing long-range dependency an contextual variations of posting series.
- The deterministic soft-attention mechanism is embedded into recurrence to enable distinct feature extraction from high duplication and advanced importance focus that varies over time.

- We quantitatively validate the effectiveness of attention in terms of detection accuracy and earliness by comparing with state-of-the-arts on two real social media datasets: Twitter and Weibo.

2 Related Work

Our work is closely connected with early rumor detection and attention mechanism. We will briefly introduce these two aspects in this section.

2.1 Early Rumor Detection

The problem of rumor detection [10] can be viewed as binary classification tasks. The extraction and selection of discriminative features significantly affects the performance of the classifier. Hu *et al.* first conducted a study to analyze the sentiment differences between spammers and normal users and then presented an optimization formulation that incorporates sentiment information into a novel social spammer detection framework [11]. Also the propagation patterns of rumors were developed by Wu *et al.* through utilizing a message propagation tree where each node represents a text message to classify whether the root of the tree is a rumor or not [3]. In [4], a dynamic time series structure was proposed to capture the temporal features based on the time series context information generated in every rumor's life-cycle. However, these approaches requires daunting manual efforts in feature engineering and they are restricted by the data structure.

Early rumor detection is to detect viral rumors in their formative stages in order to take early action [12]. In [8], some very rare but informative enquiry phrases play an important role in feature engineering when combined with clustering and a classifier on the clusters as they shorten the time for spotting rumors. Manually defined features has shown their importance in the research on real-time rumor debunking by Liu *et al.* [5]. By contrast, Wu *et al.* proposed a sparse learning method to automatically select discriminative features as well as train the classifier for emerging rumors [13]. As those methods neglect the temporal trait of social media data, a time-series based feature structure [4] is introduced to seize context variation over time. Recently, recurrent neural network was first introduced to rumor detection by Ma *et al.* [9], utilizing sequential data to spontaneously capture temporal textual characteristics of rumor diffusion which helps detecting rumor earlier with accuracy. However, without abundant data with differentiable contents in the early stage of a rumor, the performance of these methods drops significantly because they fail to distinguish important patterns.

2.2 Attention Mechanism

As a rising technique in natural language processing problems [14, 15] and computer vision tasks [16–18], attention mechanism has shown considerable discriminative power for neural networks. For instance, Bahdanau *et al.* extended the basic encoder-decoder architecture of neural machine translation with attention mechanism to allow the model to automatically search for parts of a source sentence that are relevant to predicting a target word [19], achieving a comparable performance in the English-to-French translation task. Vinyals *et al.* improved the attention model in [19], so their model computed an attention vector reflecting how much attention should be put over the input words and boosted the performance on large scale translation [20]. In addition, Sharma *et al.* applied a location softmax function [21] to the hidden states of the LSTM (Long Short-Term Memory) layer, thus recognizing more valuable elements in sequential inputs for action recognition. In conclusion, motivated by the successful applications of attention mechanism, we find that attention-based techniques can help better detect rumors with regards to both effectiveness and earliness because they are sensitive to distinctive textual features.

3   CallAtRumors: Early Rumor Detection with Deep Attention Based RNN

In this section, we present the details of our framework with deep attention for classifying social textual events into rumors and non-rumors.

3.1 Problem Statement

Individual posts contain very limited content due to their nature of shortness in context. On the other hand, an event is generally associated with a number of posts making similar claims. These related posts can be easily collected to describe an event more faithfully. Hence, we are interested in detecting rumor on an aggregate (event) level instead of identifying each single posts [9], where sequential posts related to the same topics are batched together to constitute an event, and our model determines whether the event is a rumor or not.

Let $E = \{E_i\}$ denote a set of given events, where each event $E_i = \{(p_{i,j}, t_{i,j})\}_{j=1}^{n_i}$ consists of all relevant posts $p_{i,j}$ at time stamp $t_{i,j}$, and the task is to classify each event as a rumor or not.

3.2 Constructing Variable-Length Post Series

Algorithm 1 describes the construction of variable-length post series. To ensure a similar word density for each time step within one event, we group posts into batches according to a fixed post amount N rather than slice the event time span evenly. Specifically, for every event $E_i = \{(p_{i,j}, t_{i,j})\}_{j=1}^{n_i}$, post series are constructed with variable lengths due to different amount of posts relevant to

```
   Input   : Event-related posts $E_i = \{(p_{i,j}, t_{i,j})\}_{j=1}^{n_i}$, post amount $N$, minimum series length
             $Min$
   Output: Post Series $S_i = \{T_1, ..., T_v\}$
 1 /*Initialization*/;
 2 $v = 1$; $x = 0$; $y = 0$;
 3 while true do
 4    if $n_i \geq N \times Min$ then
 5       while $v \leq \lfloor \frac{n_i}{N} \rfloor$ do
 6          $x = N \times (v - 1) + 1$;
 7          $y = N \times v$;
 8          $T_v \leftarrow (p_{i,x}, ..., p_{i,y})$;
 9          $v++$;
10       end
11       $T_v \leftarrow (p_{i,y+1}, ..., p_{i,n_i})$;
12    else
13       while $v < Min$ do
14          $x = \lfloor \frac{n_i}{Min} \rfloor \times (v - 1) + 1$;
15          $y = \lfloor \frac{n_i}{Min} \rfloor \times v$;
16          $T_v \leftarrow (p_{i,x}, ..., p_{i,y})$;
17          $v++$;
18       end
19       $T_v \leftarrow (p_{i,y+1}, ..., p_{i,n_i})$;
20    end
21 end
22 return $S_i$;
```

Algorithm 1. Constructing Variable-Length Post Series

different events. We set a minimum series length Min to maintain the sequential property for all events.

To model different words in the post series, we calculate the tf-idf for the most frequent K vocabularies within all posts. Finally, every post is encoded by the corresponding tf-idf vector, and a matrix of $K \times N$ for each time step can be constructed as the input of our model. If there are less than N posts within an interval, we will expand it to the same scale by padding with 0s. Hence, each set of post series consists of at least Min feature matrices with a same size of K (number of vocabularies) $\times$ N (vocabulary feature dimension).

3.3 Long Short-Term Memory (LSTM) with Deterministic Soft Attention Mechanism

To capture the long-distance temporal dependencies among continuous time post series, we employ following Long Short-Term Memory (LSTM) unit which plays an important role in language sequence modelling and time series processing [22–26] to learn high-level discriminative representations for rumors:

$$
\begin{aligned}
i_t &= \sigma(U_i h_{t-1} + W_i x_t + V_i c_{t-1} + b_i), \\
f_t &= \sigma(U_f h_{t-1} + W_f x_t + V_f c_{t-1} + b_f), \\
c_t &= f_t c_{t-1} + i_t \tanh(U_c h_{t-1} + W_c x_t + b_c), \\
o_t &= \sigma(U_o h_{t-1} + W_o x_t + V_o c_t + b_o), \\
h_t &= o_t \tanh(c_t),
\end{aligned}
\tag{1}
$$

where $\sigma(\cdot)$ is the logistic sigmoid function, and i_t, f_t, o_t, c_t are the input gate, forget gate, output gate and cell input activation vector, respectively. In each of them, there are corresponding input-to-hidden, hidden-to-output, and hidden-to-hidden matrices: $U_\bullet$, $V_\bullet$, $W_\bullet$ and the bias vector $b_\bullet$.

In Eq. (1), the context vector x_t is a dynamic representation of the relevant part of the social post input at time t. To calculate x_t, we introduce an attention weight $a_t[i], i = 1, \ldots, K$, corresponding to the feature extracted at different element positions in a tf-idf matrix d_t. Specifically, at each time stamp t, our model predicts a_{t+1}, a softmax over K positions, and y_t, a softmax over the binary class of rumors and non-rumors with an additional hidden layer with $sigmoid(\cdot)$ activations (see Fig. 3(c)). The location softmax [21] is thus, applied over the hidden states of the last LSTM layer to calculate a_{t+1}, the attention weight for the next input matrix d_{t+1}:

$$a_{t+1}[i] = P(L_{t+1} = i|h_t) = \frac{e^{W_i^\top h_t}}{\sum_{j=1}^{K} e^{W_j^\top h_t}} \qquad i \in 1, \ldots, K, \qquad (2)$$

where $a_{t+1}[i]$ is the attention weight for the i-th element (word index) at time step $t + 1$, W_i is the weight allocated to the i-th element in the feature space, and L_{t+1} represents the word index and takes 1-of-K values.

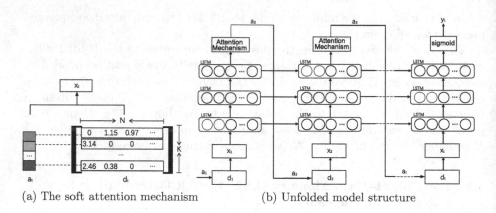

(a) The soft attention mechanism   (b) Unfolded model structure

Fig. 3. (a) The attention module computes the current input x_t as an average of the tf-idf features weighted according to the attention softmax a_t. (b) At each time stamp, the proposed model takes the feature slice x_t as input and propagates x_t through stacked layers of LSTM and predicts next location weight a_{t+1}. The class label y_t is calculated at the last time step t.

The attention vector a_{t+1} consists of K weight scalars for each feature dimension, representing the importance attached to each word in the input matrix d_{t+1}. Our model is optimized to assign higher focus to words that are believed to be distinct in learning rumor/non-rumor representations. After calculating these weights, the **soft deterministic attention** mechanism [19] computes the expected value of the input at the next time step x_{t+1} by taking weighted sums over the word matrix at different positions:

$$x_{t+1} = \mathbb{E}_{P(L_{t+1}|h_t)}[d_{t+1}] = \sum_{i=1}^{K} a_{t+1}[i]d_{t+1}[i], \tag{3}$$

where d_{t+1} is the input matrix at time step $t + 1$ and $d_{t+1}[i]$ is the feature vector of the i-th position in the matrix d_{t+1}. Thus, Eq.(3) formulates a deterministic attention model by computing a soft attention weighted word vector $\sum_i a_{t+1}[i]d_{t+1}[i]$. This corresponds to feeding a soft-a-weighted context into the system, whilst the whole model is smooth and differential under the deterministic attention, and thus learning end-to-end is trivial by using standard back-propagation.

3.4   Loss Function and Model Training

In model training, we employ cross-entropy loss coupled with $l2$ regularization. The loss function is defined as follows:

$$\mathcal{L} = -\sum_{c=1}^{C} y_{t,c} \log \hat{y}_{t,c} + \gamma\phi^2, \tag{4}$$

where y_t is the one hot label represented by 0 and 1, $\hat{y}_t$ is the predicted binary class probabilities at the last time step l, $C = 2$ is the number of output classes (rumors or non-rumors), γ is the weight decay coefficient, and ϕ represents all the model parameters.

The cell state and the hidden state for LSTM are initialized using the input tf-idf matrices for faster convergence:

$$\begin{aligned}
c_0 &= f_c\left(\frac{1}{\tau}\sum_{t=1}^{\tau}\left(\frac{1}{K}\sum_{i=1}^{K}d_t[i]\right)\right), \\
h_0 &= f_h\left(\frac{1}{\tau}\sum_{t=1}^{\tau}\left(\frac{1}{K}\sum_{i=1}^{K}d_t[i]\right)\right),
\end{aligned} \tag{5}$$

where f_c and f_h are two multi-layer perceptrons, and τ is the number of time steps for each event sequence. These values are used to compute the first location softmax a_1 which determines the initial input x_1.

4   Experiments

In this section, we evaluate the performance of our proposed methodology in early rumor detection using real-world data collected from two different social media platforms.

4.1 Datasets

We use two public datasets published by [9]. The datasets are collected from Twitter[2] and Sina Weibo[3] respectively. Both of the datasets are organised at event-level with the ground truth verified via Snopes[4] and Sina Community Management Center[5]. In addition, we follow the criteria from [9] to manually gather 4 non-rumors from Twitter and 38 rumors from Weibo for comprehensive class balancing. Note for Tweet datasets, some posts are no longer available when we crawled those tweets, causing a 10% shrink on the scale of data compared with the original Twitter dataset and this is a main cause for a slight performance fluctuation compared with the results in other papers.

Table 1 gives statistical details of the two datasets. We observe that more than 76% of the users tend to repost the original news with very short comments to reflect their attitudes towards those news. As a consequence, the contents of the posts related to one event are mostly duplicate, which can be rather challenging for early rumor detection tasks.

Table 1. Statistical details of datasets. PPE stands for posts per event.

Dataset	Total users	Total posts	Events	Rumors	Non-rumors	Avg. PPE	Min. PPE	Max. PPE
Twitter	466,577	1,046,886	996	498	498	1,051	8	44,316
Weibo	2,755,491	3,814,329	4,702	2,351	2,351	811	10	59,318

4.2 Settings and Baselines

The model is implemented using Tensorflow[6]. All parameters are set using cross-validation. To generate the input variable-length post series, we set the amount of posts N for each time step as 5 and the minimum post series length Min as 2. We selected $K = 10,000$ top words for the construction tf-idf matrices. We randomly split our datasets with the ratio of 70%, 10% and 20% for training, validation and test respectively. We apply a three-layer LSTM model with descending amount of hidden states (specifically 1,024, 512 and 128). The learning rate is set as 0.001 and the γ is set to be 0.005. Our model is trained through back-propagation [27] algorithm, namely Adam [28]. We iterate the whole training process until the loss value converges.

We evaluate the effectiveness and efficiency of CallAtRumors by comparing with the following state-of-the-art approaches in terms of precision and recall:

---

[2] www.twitter.com.

[3] www.weibo.com.

[4] www.snopes.com.

[5] http://service.account.weibo.com.

[6] https://www.tensorflow.org.

- DT-Rank [8]: This is a decision-tree based ranking model using enquiry phrases which is able to identify trending rumors by recasting the problem as finding entire clusters of posts whose topic is a disputed factual claim.
- SVM-TS [4]: SVM-TS can capture the temporal characteristics of from contents, users and propagation patterns based on the time series of rumors' lifecycle with time series modelling technique applied to incorporate carious social context information.
- LK-RBF [12]: We choose this link-based approach and combine it with the RBF (Radial Basis Function) kernel as a supervised classifier because it achieved the best performance in their experiments.
- ML-GRU [9]: This method utilizes basic recurrent neural networks for early rumor detection. Following the settings in their work, we choose the multi-layer GRU (gated recurrent unit) as it performs the best in the experiment.
- CERT [13]: This is a cross-topic emerging rumor detection model which can jointly cluster data, select features and train classifiers by using the abundant labeled data from prior rumors to facilitate the detection of an emerging rumor.

4.3   Effectiveness and Earliness Analysis

In this experiment, we take different ratios of the posts starting from the first post within all events for model training, ranging from 10% to 80% in order to test how early CallAtRumors can detect rumors successfully when there are limited amount of posts available. Through incrementally adding training data in the chronological order, we are able to estimate the time that our method can detect emerging rumors. The results on earliness are shown in Fig. 4. At the early stage with 10% to 60% training data, CallAtRumors outperforms four comparative methods by a noticeable margin. In particular, compared with the most relevant method of ML-GRU, as the data proportion ranging from 10% to 20%, CallAtRumors outperforms ML-GRU by 5% on precision and 4% on recall on both Twitter and Weibo datasets. The result shows that attention mechanism is more effective in early stage detection by focusing on the most distinct features in advance. With more data applied into test, all methods are approaching their best performance. For Twitter dataset and Weibo Dataset with highly noticable duplicate contents in each event, our method starts with 74.02% and 71.73% in precision while 68.75% and 70.34% in recall, which means an average time lag of 20.47 h after the emerge of one event. This result is promising because the average report time over the rumors given by Snopes and Sina Community Management Center is 54 h and 72 h respectively [9], and we can save much manual effort with the help of our deep attention based early rumor detection technique.

Apart from numerical results, Fig. 4(e) visualises the varied attention effects on a detected rumor. Different color degrees reflect various attention degrees paid to each word in a post. In the rumor "School Principal Eujin Jaela Kim banned the Pledge of Allegiance, Santa and Thanksgiving", most of the vocabularies closely connected with the event itself are given less attention weight than words

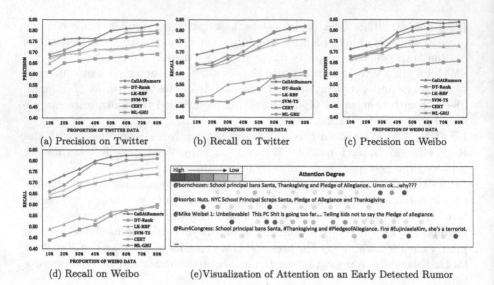

(a) Precision on Twitter   (b) Recall on Twitter   (c) Precision on Weibo

(d) Recall on Weibo   (e) Visualization of Attention on an Early Detected Rumor

Fig. 4. The charts in (a)–(d) reveal the performance for all methods with accumulative training data size. The effect of attention mechanism is visualized via (e). (Color figure online)

expressing users' doubting, esquiring and anger caused by the rumor. Despite the massive duplication from users' comments, by implementing textual attention mechanism, CallAtRumors is able to lay more emphasis on discriminative words, thus guaranteeing high performance in such case.

5 Conclusion

Rumor detection on social media is time-sensitive because it is hard to eliminate the vicious impact in its late period of diffusion as rumors can spread quickly and broadly. In this paper, we introduce CallAtRumors, a novel recurrent neural network model based on soft attention mechanism to automatically carry out early rumor detection by learning latent representations from the sequential social posts. We conducted experiments with five state-of-the-art rumor detection methods to illustrate that CallAtRumors is sensitive to distinguishable words, thus outperforming the competitors even when textual feature is sparse at the beginning stage of a rumor. In our future work, it would be appealing to investigate more complexed feature from opinion clustering results [29] and user behavior patterns [30] with our deep attention model to further improve the early detection performance.

References

1. Zimbra, D., Ghiassi, M., Lee, S.: Brand-related Twitter sentiment analysis using feature engineering and the dynamic architecture for artificial neural networks. In: 2016 49th Hawaii International Conference on System Sciences (HICSS), pp. 1930–1938. IEEE (2016)
2. Zafarani, R., Liu, H.: 10 bits of surprise: detecting malicious users with minimum information. In: CIKM, pp. 423–431. ACM (2015)
3. Wu, K., Yang, S., Zhu, K.Q.: False rumors detection on Sina Weibo by propagation structures. In: ICDE, pp. 651–662. IEEE (2015)
4. Ma, J., Gao, W., Wei, Z., Lu, Y., Wong, K.F.: Detect rumors using time series of social context information on microblogging websites. In: CIKM, pp. 1751–1754. ACM (2015)
5. Liu, X., Nourbakhsh, A., Li, Q., Fang, R., Shah, S.: Real-time rumor debunking on Twitter. In: CIKM, pp. 1867–1870. ACM (2015)
6. Rayana, S., Akoglu, L.: Collective opinion spam detection using active inference. In: Proceedings of the 2016 SIAM International Conference on Data Mining, pp. 630–638. SIAM (2016)
7. Rayana, S., Akoglu, L.: Collective opinion spam detection: bridging review networks and metadata. In: Proceedings of the 21th ACM SIGKDD International Conference on Knowledge Discovery and Data Mining, pp. 985–994. ACM (2015)
8. Zhao, Z., Resnick, P., Mei, Q.: Enquiring minds: early detection of rumors in social media from enquiry posts. In: Proceedings of the 24th International Conference on World Wide Web, pp. 1395–1405. ACM (2015)
9. Ma, J., et al.: Detecting rumors from microblogs with recurrent neural networks. In: Proceedings of IJCAI (2016)
10. Castillo, C., Mendoza, M., Poblete, B.: Information credibility on Twitter. In: Proceedings of the 20th International Conference on World Wide Web, pp. 675–684. ACM (2011)
11. Hu, X., Tang, J., Gao, H., Liu, H.: Social spammer detection with sentiment information. In: ICDM, pp. 180–189. IEEE (2014)
12. Sampson, J., Morstatter, F., Wu, L., Liu, H.: Leveraging the implicit structure within social media for emergent rumor detection. In: CIKM, pp. 2377–2382. ACM (2016)
13. Wu, L., Li, J., Hu, X., Liu, H.: Gleaning wisdom from the past: early detection of emerging rumors in social media. In: SDM (2016)
14. Yang, Z., Yang, D., Dyer, C., He, X., Smola, A., Hovy, E.: Hierarchical attention networks for document classification. In: Proceedings of NAACL-HLT, pp. 1480–1489 (2016)
15. Sutskever, I., Vinyals, O., Le, Q.V.: Sequence to sequence learning with neural networks. In: Advances in Neural Information Processing Systems, pp. 3104–3112 (2014)
16. Wu, L., Wang, Y., Li, X., Gao, J.: What-and-where to match: deep spatially multiplicative integration networks for person re-identification. Pattern Recognit. **76**, 727–738 (2018)
17. Wu, L., Wang, Y.: Where to focus: deep attention-based spatially recurrent bilinear networks for fine-grained visual recognition. arXiv preprint arXiv:1709.05769 (2017)
18. Wu, L., Wang, Y., Gao, J., Li, X.: Deep adaptive feature embedding with local sample distributions for person re-identification. Pattern Recognit. **73**, 275–288 (2018)

19. Bahdanau, D., Cho, K., Bengio, Y.: Neural machine translation by jointly learning to align and translate. arXiv preprint arXiv:1409.0473 (2014)
20. Vinyals, O., Kaiser, L., Koo, T., Petrov, S., Sutskever, I., Hinton, G.: Grammar as a foreign language. In: Advances in Neural Information Processing Systems, pp. 2773–2781 (2015)
21. Sharma, S., Kiros, R., Salakhutdinov, R.: Action recognition using visual attention. arXiv preprint arXiv:1511.04119 (2015)
22. Graves, A.: Generating sequences with recurrent neural networks. arXiv preprint arXiv:1308.0850 (2013)
23. Zaremba, W., Sutskever, I., Vinyals, O.: Recurrent neural network regularization. arXiv preprint arXiv:1409.2329 (2014)
24. Wu, L., Haynes, M., Smith, A., Chen, T., Li, X.: Generating life course trajectory sequences with recurrent neural networks and application to early detection of social disadvantage. In: Cong, G., Peng, W.-C., Zhang, W.E., Li, C., Sun, A. (eds.) ADMA 2017. LNCS (LNAI), vol. 10604, pp. 225–242. Springer, Cham (2017). https://doi.org/10.1007/978-3-319-69179-4_16
25. Chen, W., et al.: EEG-based motion intention recognition via multi-task RNNs. In: Proceedings of the 2018 SIAM International Conference on Data Mining. SIAM (2018)
26. Zhang, D., Yao, L., Zhang, X., Wang, S., Chen, W., Boots, R.: EEG-based intention recognition from spatio-temporal representations via cascade and parallel convolutional recurrent neural networks. arXiv preprint arXiv:1708.06578 (2017)
27. Collobert, R., Weston, J., Bottou, L., Karlen, M., Kavukcuoglu, K., Kuksa, P.: Natural language processing (almost) from scratch. J. Mach. Learn. Res. **12**(Aug), 2493–2537 (2011)
28. Kingma, D., Ba, J.: Adam: a method for stochastic optimization. arXiv preprint arXiv:1412.6980 (2014)
29. Chen, H., Yin, H., Li, X., Wang, M., Chen, W., Chen, T.: People opinion topic model: opinion based user clustering in social networks. In: WWW Companion, pp. 1353–1359. International World Wide Web Conferences Steering Committee (2017)
30. Yin, H., Chen, H., Sun, X., Wang, H., Wang, Y., Nguyen, Q.V.H.: SPTF: a scalable probabilistic tensor factorization model for semantic-aware behavior prediction. In: ICDM. IEEE (2017)

Impact of Indirect Contacts in Emerging Infectious Disease on Social Networks

Md Shahzamal[1,2]($\boxtimes$), Raja Jurdak[1,2], Bernard Mans[1], Ahmad El Shoghri[2,3], and Frank De Hoog[4]

[1] Macquarie University, Sydney, Australia
md.shahzamal@studentsmq.edu.au, bernard.mans@mq.edu.au
[2] Data61, CSIRO, Brisbane, Australia
raja.jurdak@data61.csiro.au
[3] University of New South Wales, Sydney, Australia
ahmad.elshoghri@students.unsw.edu.au
[4] Data61, CSIRO, Canberra, Australia
frank.dehoog@data61.csiro.au

Abstract. Interaction patterns among individuals play vital roles in spreading infectious diseases. Understanding these patterns and integrating their impact in modeling diffusion dynamics of infectious diseases are important for epidemiological studies. Current network-based diffusion models assume that diseases transmit through interactions where both infected and susceptible individuals are co-located at the same time. However, there are several infectious diseases that can transmit when a susceptible individual visits a location after an infected individual has left. Recently, we introduced a diffusion model called same place different time (SPDT) transmission to capture the indirect transmissions that happen when an infected individual leaves before a susceptible individual's arrival along with direct transmissions. In this paper, we demonstrate how these indirect transmission links significantly enhance the emergence of infectious diseases simulating airborne disease spreading on a synthetic social contact network. We denote individuals having indirect links but no direct links during their infectious periods as hidden spreaders. Our simulation shows that indirect links play similar roles of direct links and a single hidden spreader can cause large outbreak in the SPDT model which causes no infection in the current model based on direct link. Our work opens new direction in modeling infectious diseases.

Keywords: Social networks · Dynamic networks · Disease spreading

1 Introduction

Analysis of social contact networks is critical to understand and model diffusion processes within these networks. The contact patterns among individuals significantly impact spreading dynamics. Thus, a large body of work has attempted to reveal interactions among the network properties: e.g., temporal properties, burstiness and repetitive behaviors of contacts and the diffusion dynamics. Most

© Springer Nature Switzerland AG 2018
M. Ganji et al. (Eds.): PAKDD 2018, LNAI 11154, pp. 53–65, 2018.
https://doi.org/10.1007/978-3-030-04503-6_5

of these works assume that the interacting individuals are co-located in the same physical or virtual space at the same time to create a link [1,2]. However, this assumption does not hold in many diffusion processes such as airborne infectious disease spreading, vector borne disease spreading, and posted message diffusion in online social networks [3,4]. In these diffusion processes, the recipient may receive the spreading items from the sender without concurrent presence if the spreading item survives in the deposited location after its generation. This process can be explained with an example of airborne disease spreading where an infected individual deposits infectious particles at the locations where they visit. These particles can transmit to susceptible individuals who visit the locations even after the infected individual leaves as the airborne infectious particles suspend in the air for a long time [3,5]. Therefore, susceptible individuals do not need to be in the same place at the same time with the infected individual to contract disease.

Our recent work introduced a diffusion model called same place different time (SPDT) to capture such diffusion processes, e.g., airborne disease spreading. In the SPDT model, the transmission link is created between two individuals for visiting the same location within a specified time window [5]. For example, the infectious particles can transmit from the infected individual who arrived in a location to susceptible individuals who are present in that location or who arrive later on. Here, the created link is directional from the infected to the susceptible individual. We call these links SPDT links with components: (1) direct link when both individuals are present at the location; and/or (2) indirect link when the infected individual has left, but the susceptible individual is still present in the location or arrives later on. The transmission capability of SPDT links depends on the environmental conditions such as temperature and wind-flow etc. which determine the particle removal rate [3,5,6]. Thus, an SPDT link is characterized by the arrival and departure timings of infected and susceptible individuals and environmental conditions.

In the literature, previous works have aimed to characterize diffusion dynamics based on the interaction mechanisms among individuals [1,7,8]. These works have studied various aspects of human contact patterns from microscopic properties such as temporal behavior of contacts, burstiness, inter-event time and repetitiveness to the higher level structures such as clustering and community formation among individuals. The microscopic properties control the higher level structures of social contact networks and hence strongly influence the diffusion dynamics on it [9,10]. Thus, inclusion of indirect links in the SPDT model may modify the higher level structures that are present in the current same place same time (SPST) based individual to individual contact networks and influence the diffusion dynamic significantly. To the best of our knowledge, this work is the first to investigate the impact of indirect links occurring at the individual level through simulation of airborne disease spreading on social contact network.

To study the impact of indirect links, it is required to collect the sufficiently dense individual level interaction data with high spatial and temporal resolutions. However, it is quite difficult to gather such data for a population over a sufficient period of time due to privacy issues and the complexity of collection methods. Thus, we use synthetic traces generated by our SPDT network model that provides the SPDT links among the nodes present in the network. This model is fitted with the real SPDT contact network properties found among the users of social networking application Momo [11]. The generated SPDT links contain the timing information of nodes' interactions mimicking the arrival and departure of individuals in a location. Thus, SPDT links easily allow quantification of infection risk with environmental factors if neighbor nodes of a link are infected. The infection risk for concurrent interaction between susceptible and infected individuals is formulated in [12]. We improve this model to find infection risk for SPDT interaction. In our simulations, disease propagates on networks following the Susceptible-Exposed-Infected-Recovered (SEIR) epidemic model.

Our main goal is to determine the importance of indirect links for spreading infections. We first study diffusion dynamics on both the SPDT and the SPST networks selecting the seed nodes (initiating the spreading process) according to SPDT link components during their infection periods: having only indirect links, hidden spreaders, or both direct and indirect links. We gradually increase the proportion of hidden spreaders and find their impact on the diffusion, and highlight the inability of current models in capturing the effect of hidden spreaders on diffusion. We next explore how the changes in network properties that arise from the inclusion of indirect links impact the diffusion. We consider nodes creating small number of direct links when they are infected as seed nodes and look at the emergence of disease from each single seed node in both the SPST and the SPDT networks. Finally, we study the potential for disease emerge by a hidden spreader acting as a single seed node.

The rest of the paper is organized as follows. In Sect. 2, we present the improved infection risk assessment model for SPDT links. Section 3 describes our methodology. The experimental setup and analysis of results are presented in Sect. 4. Section 5 concludes our work and provides future research directions.

2 Infection Risk in SPDT

In this section, we present the methods of determining the infection risk for a susceptible individual that has SPDT links with the infected individuals. When an infected individual appears at a location L, he deposits infectious particles in the proximity of L. The number of infection particles n deposited (through coughing) per second by an infected individual into the proximity is

$$n = 0.2fvc \tag{1}$$

where f is the coughing frequency (coughs/second), v is the volume of each cough (m^3) and c is the concentration of infectious particles in the cough droplets (particles/m^3). If the particles are removed from the space of volume V with a rate r, the accumulation rate of particles in the proximity can be given by

$$V\frac{dN}{dt} = n - Nr$$

where N is the current number of particles at L and $r = (1 - b)(1 - g)$, b is the infectivity decay rate of particles and g is the air exchange rate from L [12]. The particle concentration at time t after the infected individual arrives at L is given as

$$\int_0^{N_t} \frac{dN}{n - Nr} = \frac{1}{V} \int_0^t dt$$

This leads to

$$N_t = \frac{n}{r} \left(1 - e^{-\frac{rt}{V}}\right) \tag{2}$$

If a susceptible individual stays within the proximity of L from t_c to $t_c + t_d$ for a period t_d when the infecter is concurrently present at L, the number of particles inhaled by the susceptible individual with pulmonary rate q for this direct link is

$$E_d = \frac{qn}{r} \int_{t_c}^{t_c + t_d} \left(1 - e^{-\frac{rt}{V}}\right) dt$$

$$= \frac{qn}{r} \left[\left(t_d + t_c + \frac{V}{r} e^{-\frac{r(t_c + t_d)}{V}}\right) - \left(t_c + \frac{V}{r} e^{-\frac{rt_c}{V}}\right)\right] \tag{3}$$

If the susceptible individual stays with the infected individual as well as after the latter leaves L, it will have both direct and indirect transmission links. The number of particles inhaled by the susceptible individual due to the direct link within the time t_c and t_a is given by

$$E_d = \frac{qn}{r} \left[\left(t_a + \frac{V}{r} e^{-\frac{rt_a}{V}}\right) - \left(t_c + \frac{V}{r} e^{-\frac{rt_c}{V}}\right)\right] \tag{4}$$

where t_a is the stay duration of infected individual. For the indirect link from time t_a to $t_c + t_d$, we need to compute the particle concentration during this period which decreases after the infected individual leaves. The particle concentration at time t_a can be given as

$$N_a = \frac{n}{r} \left(1 - e^{-\frac{rt_a}{V}}\right)$$

The particle concentration at time t after the susceptible leaves the proximity at time t_a is given by

$$V \frac{dN}{dt} = -Nr$$

Thus, the concentration at time t will be

$$\int_{N_a}^{N_t} \frac{dN}{N} = -\frac{r}{V} \int_{t_a}^t dt$$

Thus, we have

$$N_t = N_a e^{-\frac{r}{V}(t - t_a)} = \frac{n}{r} \left(1 - e^{-\frac{rt_a}{V}}\right) e^{-\frac{r}{V}(t - t_a)}$$

The susceptible individual inhales particles during the indirect period from t_a to $t_c + t_d$, quantified by

$$E_i = \int_{t_a}^{t_c+t_d} qN_t dt = -\frac{nqV}{r^2}\left(1 - e^{-\frac{rt_a}{V}}\right)\left[e^{-\frac{r}{V}(t_c+t_d-t_a)} - 1\right] \qquad (5)$$

If the susceptible individual is only present for the indirect period at the proximity, the number of inhaled particles for the period from t_c to $t_c + t_d$ is given

$$E_i = \int_{t_c}^{t_c+t_d} qN_t dt = -\frac{nqV}{r^2}\left(1 - e^{-\frac{rt_a}{V}}\right)\left[e^{-\frac{r}{V}(t_c+t_d-t_a)} - e^{-\frac{r}{V}(t_c-t_a)}\right] \qquad (6)$$

Thus, the total inhaled particles can be given for a susceptible individual by

$$E = E_d + E_i \qquad (7)$$

The equations determine the received exposure for one SPDT link with an infected individual, comprising both direct and indirect links. If a susceptible individual has m SPDT links during an observation period T, the total exposure is

$$E_T = \sum_{k=0}^{m} E_k$$

where E_k is the received exposure for k^{th} link. The probability of infection for causing disease can be determined by the dose-response relationship defined as

$$P(I = 1) = 1 - e^{-\sigma E_T} \qquad (8)$$

where σ is the probability that an infectious particle reaches to the respiratory tract and initiates infection [13]. It is assumed that inhaling one infectious particle has 50% chance of contracting the disease [14]. Therefore, we can calculate σ as

$$0.5 = 1 - e^{-\sigma}$$

$$\sigma = 0.693$$

In this risk formulation, σ is homogeneous for all susceptible individuals.

3   Methodology

For studying the impact of indirect contacts on the spreading process, individual level interaction data is required. However, it is difficult to collect such data with sufficiently high contact density, and with high spatial and temporal resolution. Thus, we generate synthetic individual to individual interaction data using our developed SPDT network model. This network model includes indirect links of disease transmission along with direct links in creating SPDT link between two nodes. Then, the SEIR epidemic model is simulated on this network with disease parameters of airborne diseases. In this section, we describe our methodology.

3.1 Contact Network

A network of M nodes is constructed following the approach of activity-driven temporal network generation where nodes switch between active periods and inactive periods over the simulation period of T discrete time steps. Active periods t_a mimic staying at a location and allow nodes to create SPDT links with other nodes while inactive periods t_w represent time windows when a node does not create links to other nodes, but receives SPDT links. The SPDT link is directed from the host node (as infected) to the neighbor node (susceptible node). The duration of active periods $t_a = \{1, 2, 3, \ldots\}$ are randomly drawn from $P_1(t_a) \sim geometric(\lambda)$ with the scaling parameter λ. The inactive period durations $t_w = \{1, 2, 3, \ldots\}$ are also drawn from $P_2(t_w) \sim geometric(\rho)$ with the scaling parameter ρ. The value of λ is constant for all nodes but ρ is assigned heterogeneously to model different frequencies of individuals in visiting locations. The heterogeneous activation potentiality ρ is drawn from a power law $F_1(\rho) \sim \rho^{-\alpha}$ with the scaling parameter α.

Corresponding to each active period, an indirect transmission period δ is added with t_a to capture the indirect links of disease transmission. During t_a and corresponding δ, a node creates a specific number of SPDT links d given by

$$Pr(d) = (1 - \mu)\mu^{d-1}$$

where μ is the propensity to access public places that is also drawn from the power law $F_2(\mu) \sim \mu^{-\beta}$ with the scaling parameter β. The variations in d capture spatio-temporal dynamics of social networks.

Each of these SPDT links has the timing characteristics: t_a representing the time duration node stays at the interacted location, t_c is the delay time of neighbor node arrival after the host node appeared, and t_d is the time duration neighbor node (end nodes of a link) stays at the interacted location. The link creation delay t_c is drawn from a truncated geometry distribution as follows as

$$Pr(t_c) = \frac{p_c(1 - p_c)^{t_c}}{1 - (1 - p_c)^{t_a + \delta}}$$

where p_c is the probability of creating a link with a neighbor. The value of t_d is drawn as $P_3(t_d) \sim geometric(p_b)$ with link breaking probability p_b. The value of p_c and p_b are constant for all nodes in the network.

Nodes maintain their social structure by applying their memory of previous contacts in selecting new neighbors with the probability $Pr(n_t + 1) = \mu_i\theta/(n_t + \mu_i\theta)$, where n_t is the number of nodes node i has contacted up to time t. With greater public accessibility μ, nodes will have higher contact set sizes. They also need to be selected as a neighbor by more nodes. Thus, the probability of being selected as a neighbor is $p(\mu_j) = \mu_j/M\langle\mu\rangle$. The neighbor selection mechanism also considers that if some nodes have included node j as neighbor, there is φ chance to select them as neighbors by node j.

The network model parameters are fitted with the real SPDT network constructed using the 2 million locations updates of 126K users collected over a week from the social networking application Momo. The location updates from Beijing city are applied to estimate the model parameters while the updates from Shanghai are used to validate the model. The capability of the model to simulate the SPDT diffusion process is also verified in detail. We generate synthetic SPDT networks to conduct our experiments using this model.

3.2 Epidemic Model

For propagating disease on the generated SPDT contact network, we consider a compartment-based Susceptible-Exposed-Infected-Recovered (SEIR) epidemic model. In this model, nodes remain in one of the four compartments, namely, Susceptible (S), Exposed (E), Infectious (I) and Recovered (R). If nodes in the susceptible compartment receive SPDT links from the nodes in the infectious compartment, the former will receive the infectious pathogens for both direct and indirect periods and may contract the disease. At the beginning of contraction, a susceptible node moves to the exposed (E) state where it cannot infect others. The exposed node becomes infectious (I) after a latent period. The infectious node continues to infect other nodes connecting through the SPDT links over its infection period until they enter the recovered state R [15]. It has been shown that the latent period is in the range over 1 to 2 days. Our simulations assume that if a node is infected in the current day of simulation, it starts infecting others on the next day of simulation. The infection period is shown to be in the range over 3 to 5 days for influenza-like diseases [16]. As the values can vary for each individual even for the same disease, we derive the parameters from a random uniform distribution within the observed empirical ranges.

4   Simulation and Analysis

We conduct various simulation experiments to understand the impact of indirect transmissions in shaping the epidemic dynamics on the social contact network. We consider a network of 300K nodes generated by the SPDT network model for a 32 day period. In this network, nodes create SPDT links that are inclusive of direct and indirect transmission links. Removing the indirect links from the above network results in an SPST network with only direct links. Therefore, the SPDT and SPST networks include the same nodes, but their connectivity differs due to the presence or absence of indirect links. If a node is infected and creates only indirect links during its infection period, we refer to this node as a 'hidden spreader'. The contribution of hidden spreaders to infections in the original SPST network is nil while they significantly contribute to spread in the SPDT network (and thus possibly promote new direct links). In this section, we explore the extent to which indirect links can impact the disease spread dynamics on contact networks through intensive simulations.

4.1 Simulation Setup

The generated synthetic SPDT links of 300K nodes provide the traces for running data-driven disease simulations over a period of $T = 32$ days. The disease on this network propagates according to the SEIR epidemic model. The simulation at $T = 0$ starts with some seed nodes that are randomly selected according to the requirements of experiments described below. The nodes' disease status are updated at the end of each simulation day. During each day of disease simulation, the received SPDT links for nodes are analyzed to find which nodes have received SPDT links from the infected neighbor nodes. Then, we calculate the particles inhaled by the susceptible neighbor node for each of these SPDT links according to Eq. 7 and sum them up to find the infection probability by Eq. 8. I order to keep the simulations simple, we assume that the inhaled infection particles can infect susceptible individual with the same rate over the day. The volume V of the proximity is fixed to a constant value assuming that the distance, within which a susceptible individual can inhale the infectious particles from an infected individual, is 40 m [3] and the particles will be available up to the ceiling height h of 3 m. We assign the other parameters as follows: cough frequency $f = 18/h$, total volume of the cough droplets $v = 6.7 \times 10^{-3}$ ml, pathogen concentration in the respiratory fluid $c = 3.7 \times 10^6$ pathogens/m^3, and pulmonary rate $q = 7.5$ l/min [17,18]. For each link, the infectivity decay rate b and air exchange rate g are selected randomly to find particles removal rate $r = (1 - b)(1 - g)$. The value of b is randomly drawn from the range $(0.005, 0.05)$ min^{-1} with a specified mean according to the experiments while g is randomly drawn from the range $(0.25, 5)$ h^{-1} with a specified median as the experiments require. The daily simulation outcomes are obtained for the epidemic parameters: the number of new infections, the disease prevalence as the number of current infections in the system and the cumulative number of infections. The statistics of these parameters provide the results of our experiments.

4.2   Results and Analysis

In the simulation of disease on the SPDT network, the infected nodes can be divided into two groups based on the SPDT links they form during their infection periods: (1) nodes with both direct and indirect links; and (2) nodes with only indirect links, i.e. hidden spreaders. A hidden spreader has zero infection force in the SPST network but can cause disease in the SPDT network. Thus, the increase in the proportion P of hidden spreaders in the seed nodes set will reduce spreading speed in the SPST network while spreading speed is sustained for SPDT according to the potentiality of the indirect links. In our first experiment, the simulations begin with 200 seed nodes where we randomly pick 200P seed nodes from the hidden spreader set and (1-P)200 seed nodes from the non-hidden spreader set. We vary the value of P from 0 to 1 with the step 0.1. The seed nodes start infecting at $T = 0$ and continue infecting for the period of days picked up randomly from the range (1, 5). With this set of seed nodes, we run simulations on the both SPST and SPDT networks and repeat 200 times for each value of P. We set the mean value of b to 0.01 min^{-1} and the median of g to 1 h^{-1}.

Figure 1 shows the changes in the disease spreading dynamics, averaged over the 200 simulation runs, for P = $\{0, 0.2, 0.4, 0.6, 0.8\}$. For the SPST network, both prevalence and cumulative infections shrink with increasing

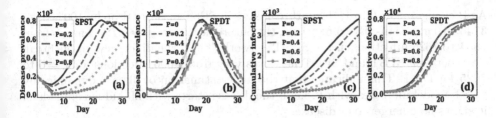

Fig. 1. Disease diffusion dynamics based on the disease prevalence (a, b) and cumulative infection (c, d) for different P in both SPST and SPDT networks

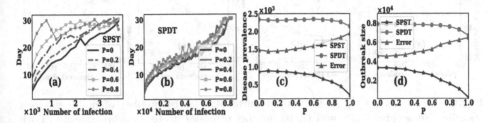

Fig. 2. Disease prediction performances for changes in P: (a, b) days requires to cause a specific number of infection since simulation starts and (c, d) error as differences between same metrics in the SPST and SPDT networks

P (see Fig. 1a and c), as the increased proportion of hidden spreaders as seed nodes reduces the likelihood of seed nodes to trigger spreading. We also observe that the rate of reduction in both prevalence and cumulative infections increases with P, where the shrinking set of non-hidden spreaders among seed nodes reduces the likelihood of spread more rapidly. The disease prevalences and outbreak sizes decrease significantly for increasing P and dropping rates increase at the higher P's. On the other hand, the disease spreading behaviors do not change much with P in the SPDT network (see Fig. 1b and d). There is a slight time shift in prevalence with increasing P, yet the size of the epidemic remains similar. Noting that changes in P result in different proportions of hidden spreaders, the SPDT results confirm that the potency of hidden spreaders is almost similar to non-hidden spreaders in determining diffusion outcomes.

We now explore disease prediction performance with changing P in Fig. 2 for delving deeper into the contribution of hidden spreaders. Figure 2(a) and (b) show the number of days required for causing a specific number of infections and how it is delayed with changing P. The SPST network fails to predict infection dynamics causing more delays to reach a specific number of cumulative infections as P increases (see Fig. 2(a)). The cumulative infections reaches 1000 by day 10 at P = 0 but it is delayed to day 30 when P is 0.8. However, the required days to reach a number of cumulative infections changes slightly for the SPDT network in Fig. 2b with changing P as the indirect links have similar impact as direct links. The differences in outbreak sizes and disease prevalences between SPST and SPDT networks with the changes in P are shown in Fig. 2(c) and (d).

As P changes from 0 to 1, the outbreak sizes in SPST network drop by 100% from 3745 to 0 infections. By comparison, the outbreak size drop by 2.7% changes from 7834 to 7625 infections in the SPDT network. Thus, the differences in outbreak sizes between SPST and SPDT networks vary from 4000 infections for P = 0 to 7625 for P = 1. This indicates that the underestimation of disease dynamics by SPST model increases with P, i.e. if the disease starts with all hidden spreaders, it shows no emergence of disease.

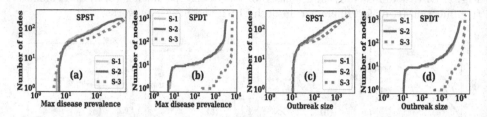

Fig. 3. Low connectivity nodes caused outbreaks: (a, b) number of nodes up to a specific disease prevalence and (c, d) number of nodes up to a outbreak size in both SPDT and SPST networks

The previous experiments show that the indirect links have strong impact on the spreading behaviors of diseases. Now, we study how these indirect links play vital roles at the individual level. To understand this, we investigate how the low direct connectivity nodes, that directly contact only one or two nodes during their infection periods through direct links, become important in SPDT network due to having the indirect links. We identify a low direct connection set of 10K nodes who have 1 or 2 neighbors in the SPST network over the first 5 days of our generated synthetic traces. Then, we run simulations by iterating through the nodes in this set to select each node as a seed node at $T = 0$ on both SPST and SPDT networks. The seed nodes are able to infect others for 5 days before recovering from the infectious state. We keep the same mean and median of previous experiments respectively for b and g. We also run simulations for 3 scenarios (changing b, g and σ) to understand how these nodes play a more significant role under certain conditions.

The results are presented in Fig. 3 for the nodes that cause outbreak sizes greater than 10. In the first scenario **S-1**, we set the parameters: mean of b is set to $0.01\,\mathrm{min}^{-1}$, median of g to $1\,\mathrm{h}^{-1}$ and $\sigma = 0.69$. With this configuration, we find only 206 nodes can trigger disease in the SPST network. Comparatively, 803 nodes become capable to trigger disease in the SPDT network and their outbreak size is twice the outbreak sizes in SPST. If we change the value of g to $0.5\,\mathrm{h}^{-1}$ in scenario **S-2**, the SPDT network allows 840 nodes to trigger disease while only one more node in the SPST network trigger disease. This is because the SPDT links are more pronounced for lower g while SPST links remain the same. If we make the scenario more favorable for spreading disease changing b to $0.005\,\mathrm{min}^{-1}$ and $\sigma = 0.80$ (scenario **S-3**), both networks offer more nodes to trigger disease where 307 nodes trigger disease in SPST network corresponding to 1649 nodes in the SPDT network. Under extreme conditions, the indirect links become significant, influencing the disease spreading strongly.

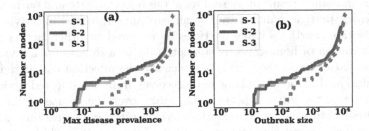

Fig. 4. Hidden spreaders (nodes with only indirect links during their infection periods) caused outbreaks: (a) number of nodes up to a disease prevalence and (b) number of nodes up to a specific outbreak size

In the third experiment, we investigate whether single seed nodes that are hidden spreaders can trigger significant outbreaks. Using the hidden spreader set from the first 5 days from our first experiment, we study emergence of disease through all nodes selecting each as seed node at time $T = 0$ of the simulations. The simulations are repeated 10 times for each node. We also explore whether the opportunities of emerging diseases are intensified for the favourable scenario of low b, g and high σ. The results are presented in Fig. 4. For the first scenario, we find that 324 nodes cause outbreak sizes greater than 10 with maximum 7656 infections over the 32 day period. The medium spreading scenario allows 53 (16.3%) more nodes to trigger diseases. If the diseases are more infectious having $\sigma = 0.80$ in scenario S-3, 1014 nodes among the 11K nodes (which have only indirect links during the first 5 days of selected network traces) are capable to trigger disease. In this favorable spread scenario, the outbreak sizes as well as maximum prevalence of disease increase with earlier prevalence peaks.

5   Discussion and Future Work

This paper has studied how the SPDT model enhances the opportunities of disease emergence due to indirect links. These outcomes can guide more robust policy for controlling the infectious diseases. Our simulations on hidden spreaders have shown that indirect links are equally important as the direct links in driving the spread process. The indirect links increase the connectivity of the network and make it favorable for disease spreading. Running disease simulations from the low connectivity nodes has shown that 5 times more nodes can contract the disease if indirect links are considered. Interestingly, we noticed that nodes having only indirect links, hidden spreaders, can cause outbreaks with as many as 10K infections. However, these nodes in the SPST model can not infect others as they do not have direct links. These outcomes require reconsideration of the disease spreading models of many infectious diseases.

Our work has some limitations. The infection risk is calculated assuming that the infection particles are distributed homogeneously in the space, which may not be true in reality, resulting in scaling of the infection rate and outbreak size. We only cover particle removal by the air exchange rate and infectivity decay rate of infectious particles while particles be removed by other factors as well. The consideration of homogeneous susceptibility to a disease for individuals is not fully realistic as well. Secondly, we calculate the infection risk at the end of each simulation day for all links a node receives from the infected nodes. The infectiousness of the inhaled particle may vary over the course of the day in reality. These factors can over estimate infection rates and spreading dynamics.

This work also opens some future research directions. It would be interesting to know why some nodes can initiate diseases. In our simulations, we observed that indirect links increase the connectivity among the nodes. This arises the question that the indirect links can lead nodes to become super-spreaders. We also observed favorable weather conditions caused larger outbreaks as the indirect links become stronger. Another direction can be to investigate how the SPDT model with its capability to capture human mobility as well as environmental conditions can aid to design optimal diffusion control strategies.

References

1. Holme, P.: Modern temporal network theory: a colloquium. Eur. Phys. J. B **88**(9), 234 (2015)
2. Pastor-Satorras, R., Castellano, C., Mieghem, P.V., Vespignani, A.: Epidemic processes in complex networks. Rev. Mod. Phys. **87**(3), 925 (2015)
3. Fernstrom, A., Goldblatt, M.: Aerobiology and its role in the transmission of infectious diseases. J. Pathog. (2013)
4. Gruhl, D., Guha, R., Liben-Nowell, D., Tomkins, A.: Information diffusion through blogspace. In: WWW Conference, pp. 491–501. ACM (2004)
5. Shahzamal, M., Jurdak, R., Arablouei, R., Kim, M., Thilakarathna, K., Mans, B.: Airborne disease propagation on large scale social contact networks. In: 2nd International Workshop on Social Sensing, pp. 35–40. ACM (2017)
6. Brenner, F., Marwan, N., Hoffmann, P.: Climate impact on spreading of airborne infectious diseases. Eur. Phys. J. Spec. Top. **226**(9), 1845–1856 (2017)
7. Zhang, N., Huang, H., Su, B., Ma, X., Li, Y.: A human behavior integrated hierarchical model of airborne disease transmission in a large city. Build. Environ. **127**, 211–220 (2018)
8. Meyer, S., Held, L.: Incorporating social contact data in spatio-temporal models for infectious disease spread. Biostatistics **18**(2), 338–351 (2017)
9. Shirley, M.D.F., Rushton, S.P.: The impacts of network topology on disease spread. Ecol. Complex. **2**(3), 287–299 (2005)
10. Min, B., Goh, K.-I., Vazquez, A.: Spreading dynamics following bursty human activity patterns. Phys. Rev. E **83**(3), 036102 (2011)
11. Chen, T., Kaafar, M.A., Boreli, R.: The where and when of finding new friends: analysis of a location-based social discovery network. In: ICWSM (2013)

12. Issarow, C.M., Mulder, N., Wood, R.: Modelling the risk of airborne infectious disease using exhaled air. J. Theor. Biol. **372**, 100–106 (2015)
13. Jones, R.M., Su, Y.-M.: Dose-response models for selected respiratory infectious agents: bordetella pertussis, group a streptococcus rhinovirus and respiratory syncytial virus. BMC Infect. Dis. **15**(1), 90 (2015)
14. Teunis, P.F., Brienen, N., Kretzschmar, M.E.: High infectivity and pathogenicity of influenza a virus via aerosol and droplet transmission. Epidemics **2**(4), 215–222 (2010)
15. Stehlé, J., et al.: Simulation of an seir infectious disease model on the dynamic contact network of conference attendees. BMC Med. **9**(1), 87 (2011)
16. Chowell, G., et al.: Characterizing the epidemiology of the 2009 influenza A/H1N1 pandemic in Mexico. PLOS Med. **8**(5), e1000436 (2011)
17. Loudon, R.G., Brown, L.C.: Cough frequency in patients with respiratory disease 1, 2. Am. Rev. Respir. Dis. **96**(6), 1137–1143 (1967)
18. Yin, S., Sze-To, G.N., Chao, C.Y.H.: Retrospective analysis of multi-drug resistant tuberculosis outbreak during a flight using computational fluid dynamics and infection risk assessment. Build. Environ. **47**, 50–57 (2012)

Interesting Recommendations Based on Hierarchical Visualizations of Medical Data

Ibrahim A. Ibrahim[1,2(✉)], Abdulqader M. Almars[1], Suresh Pokharel[1],
Xin Zhao[1], and Xue Li[1]

[1] The University of Queensland, Brisbane, Australia
{i.ibrahim,a.almars,s.pokharel,x.zhao,xueli}@uq.edu.au
[2] Minia University, Minya, Egypt
i.ibrahim@minia.edu.eg

Abstract. Due to the dramatic growth in Electronic Health Records (EHR), many opportunities are arising for discovering new medical knowledge. Those kinds of knowledge are useful for related stakeholders such as hospital planners, data analysts, doctors, insurance companies for better patients' management. However, many challenges need to be addressed while dealing with medical domain such as (1) how to measure the *interestingness* of information (2) how to visualize such *interestingness* (3) how to handle high dimensionality of medical data. To address these challenges, we present *MedVIS*, an interactive tool to visually explore the data space and finds interesting visualizations compared with another dataset. *MedVIS* structures visualizations into a hierarchical coherent tree to reveal *interestingness* of dimensions and other measures. We introduce a novel analysis work flow, and discuss various optimization mechanisms to effectively and efficiently explore the data space. Additionally, we discuss various approaches to mitigate the problem of high-dimensional medical data analysis and its visual exploration. In our experiments, we apply *MedVIS* to a real-world dataset and show promising visualization outcomes in terms of effectiveness and efficiency.

Keywords: Big data visualizations · Visual analytics
Knowledge discovery

1 Introduction

Recently, medical experts, biologists, and research scientists are confronted with fast growing large, complex, and high-dimensional data. This challenge will become immanent and more demanding with the ongoing trend towards personalized medicine [6,8]. One of the grand future challenges of biomedical informatics research is to gain knowledge from complex high-dimensional datasets [5]. Although, medical data is usually relevant and interesting structural and patterns but knowledge is often hidden and not accessible to domain experts.

© Springer Nature Switzerland AG 2018
M. Ganji et al. (Eds.): PAKDD 2018, LNAI 11154, pp. 66–79, 2018.
https://doi.org/10.1007/978-3-030-04503-6_6

Interactive Data Visualization Tools have interested the research community over the past few years, and it has presented a number of interactive data analytics tools such as ShowMe, Polaris, and Tableau [2,9,12,13,15]. Similar visualization specification tools have also been introduced by the database community, including Fusion Tables [3] and the Devise [14] toolkit.

Table 1. Relational table R of a patients admissions data

Patient id	Admission type	Admission location	Discharge location	Insurance	Admission count
P1	Emergency	Emergency admit	Long term hospital care	Private	2
P2	Urgent	Hospital transfer	Medical facility	Medicare	1
P3	Elective	Clinic referral	Home health care	Self pay	1
P4	Urgent	Transfer within facility	Long term hospital care	Medicaid	1
P5	Elective	Home referral/sick	Home health care	Private	2
P6	New born	Hospital transfer	Short term hospital care	Government	1
P7	Emergency	Clinic referral	Medical facility	Self pay	2
P8	New born	Emergency admit	Short term hospital care	Medicare	1
P9	Elective	Transfer within facility	Medical facility	Government	1

RtSEngine framework [9] has been proposed to automatically find a top (k) interesting visualizations in a given dataset based on the notion of deviation of data. The following example illustrates the *RtSEngine* analysis workflow to identify interesting visualizations from a real, large, and structured database Mimic-III which represents patient admission details. Figure 1, shows a snippet of the admission relation.

Example 1. Consider a hospital analytics team that is undertaking a study for a particular admission type: *ELECTIVE*. That admission type has poor performance and has received a lot of patients' complaints (Table 1).

Suppose that the team uses the *patients' admissions* database containing metrics such as number of admissions for each patient. Also, it has a large set of dimension attributes containing information such as admission locations, discharge locations, Insurance type, etc.

Given the large size of the database (millions of records), an analyst will overwhelmingly use a collection of visualization programs to gather insights into the behavior of admission types e.g. *ELECTIVE*. In a typical analysis workflow, an analyst would begin by using the program's GUI or a custom query language

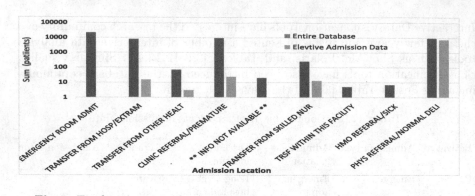

Fig. 1. Total patients numbers per admission locations (Color figure online)

to execute the equivalent of the following SQL query and pull all data from the database for *ELECTIVE* admission:

$$Q = \text{SELECT * FROM admissions WHERE admission type} = \text{"}ELECTIVE\text{"}$$

Next, the analyst would use an interactive GUI interface to manually visualize various metrics of the query result. For instance, the analyst may visualize average patients' numbers grouped by insurance type, discharge locations, maximum patients' numbers by admission locations, and so on. These visualization operations are essentially queries to the underlying data store and subsequent graphing of the results.   □

In our previous work *RtSEngine* [9], a bar chart shown in Fig. 1 is considered an interesting visualization as it shows a high deviation compared with entire dataset shown in blue bars. Although, *RtSEngine* framework successfully finds the top (k) interesting visualizations, but still has a main limitation. Firstly, the produced visualizations are based only on their deviation scores and it ignores the coherence of dimensions, aggregate functions, and other attributes that construct such visualizations. Thus, the analyst need to study all these visualizations in order to identify the impact of dimensions cause the deviation.

To tackle such limitations, in this paper we propose a technique *MedVis* to automatically produce a coherent tree of dimensions based on their *interestingness* and coherence. The hierarchical tree reflects the intention that the visualizations should be organized from left to right according to their interestingness and importance. To achieve that, we propose a new metric called *coherence* which used to structure the tree. Experiments on medical data show the effectiveness and efficiency of our framework.

Research Contribution: The main contributions of this paper are summarized as follows:

1. We propose a novel framework called *MedVis* which automatically produces a hierarchical tree of interesting visualizations.
2. We present a Coherence metric which measures the influence of dimensions that construct such visualizations.
3. We evaluate the proposed framework On real medical data. The result demonstrate the effectiveness and efficiency of our proposed framework.

The remainder of this paper is structured as follows: in the subsequent sections, we present related works and then we discuss a detailed description of the methodology in Sect. 3. A more discussion of the architecture in Sect. 4. Then, in Sect. 5, we describe the data sources being used, perform experiments, evaluate the work and finally concluding remarks.

2   Related Work

A few recent systems have attempted to automate some aspects of data analysis and visualization. Profiler is one such automated tool that allows analysts to detect anomalies in data [1,9,12]. Another related tool is VizDeck [13], in given a dataset, depicts all possible 2-D visualizations on a dashboard that the user can control by reordering or pinning visualizations. Given that VizDeck generates all visualizations however, it is only meant for small datasets; and VizDeck does not discuss techniques to speed-up the generation of these visualizations.

To support visual sense-making in medical diagnosis, INVISQUE [7,20] is an interactive visualization system proposed such as physical index cards on a two dimensional workspace. INVISQUE provides some features to support 'annotating, re-visiting, and merging two clusters. It discusses essential problems in designing medical diagnostic displays that can improving the review of a patient's medical history [7]. A recent work, SubVIS [8] is a visualization tool which assists the user to analyze and interactively explore computed subspaces to discover insights in highly dimensional and complex patient's datasets. SubVIS [8] introduces an analysis workflow to visually explore subspace clusters from various perspectives and it tackles some subspace clustering challenges such as difficulty of interpretation patient results, redundancy detection in subspaces and clusters, and multiple clustering results for different parameter settings.

OLAP: there has been some work on browsing data cubes, allowing analysts to variously find *explanations* for why two cube values were different, to find which neighboring cubes have similar properties to the cube under consideration, or get suggestions on what unexplored data cubes should be looked at next [10,16,17].

Database Visualization Work: Fusion tables [3] allow users to create visualizations layered on top of web databases; they do not consider the problem of automatic visualization generation. Devise [4] translated user-manipulated visualizations into database queries.

Although the aforementioned approaches provide assistance in query visualization, they lack the ability to automatically recommend interesting visualizations. SeeDB execution engine uses pruning optimizations to determine which aggregate views to discard. Specifically, partial results for each view based on the data processed so far are used to estimate utility and views with low utility are discarded. SeeDB execution engine supports two pruning schemes. The first uses confidence-interval techniques to bound utilities of views, while the second uses multi-armed bandit allocation strategies to find top utility views.

SeeDB pruning schemes experience some limitations, as they assume fixed data distribution [18, 19] for sampling to estimate the utility of views and require large samples for pruning low utility views with high guarantees. Moreover, aggregate functions MAX and MIN are not docile to sampling-based optimizations.

3   Methodology

3.1   Discovering Interesting Visualization

Background and Preliminaries. In this section, We start by explaining how a visualization (or a view) is constructed by an SPJ SQL query. Then, we define our scope of visualizations that our framework is focused on, and how to measure the interestingness of a visualization based on a model proposed by [19]. Finally, we formally present our problem statement.

3.2   Background and Scope

A visualization V_i is constructed by an SQL select-project-join query with a group-by clause over a database D. The attributes in a database table are classified into two sets: dimension attributes set $A = \{a_1, a_2, ...\}$, and measure attributes set $M = \{m_1, m_2, ...\}$. While the set $F = \{f_1, f_2, ...\}$ contains all aggregation functions. Hence, each visualization V_i is represented as a triple (a, m, f), where a is a group-by attribute applied to the aggregation function f on a measure attribute m.

As an example, $V_i(D)$ visualizes the results of grouping the data in D by a, and then aggregating the m values using f. This view is called the *reference view*. Consequently, $V_i(DQ)$ represents a similar visualization applied to the result set denoted as DQ for a given user query Q, and is called the *target view*. An example of a target view is shown in Fig. 1 where a is admission location, m is patients attribute, and f is the Sum aggregation function. Any combination of (a, m, f) represents a view. Accordingly, we can define the total number of possible views as follows:

$$\text{View Space}(SP) = 2 \times |A| \times |M| \times |F| \qquad (1)$$

Though, in the context of Big Data, SP is potentially a very large number. Hence, there is a need to automatically score all these SP views so that exploring them become efficient and practical.

3.3 Views Utility

Each view is associated with a *utility* value. The utility of a visualization is measured as its deviation from a reference dataset D_R. For instance, visualizations that show different trends in the query dataset (i.e. DQ) compared to a reference dataset D_R are supposed to have high utility. The reference dataset D_R may be defined as the entire underlying dataset D, the complement of $DQ(D - DQ)$ or data selected by any arbitrary query $Q'(DQ')$.

Given an aggregate view V_i and a probability distribution for a target view $P(V_i(DQ))$ and a reference view $P(V_i(D_R))$, the utility of V_i is the distance between these two normalized probability distributions. The higher the distance between the two distributions, the more likely the visualization is to be interesting and therefore higher utility value. Formally:

$$U(V_i) = S(P(V_i(DQ)), P(V_i(D)))\tag{2}$$

where S is a distance function (e.g., Euclidean distance, Earth Movers distance, etc.). In addition, S can be the Pearson's correlation coefficient to capture interesting trends in visualizations. The $RtSEngine$ framework outputs a sorted list $\mathcal{L}$ of visualizations $\mathcal{V}$ based on their utilities $\mathcal{U}$, where $\mathcal{L} = \{(V_1, u_1), (V_2, u_2), ..., (V_n, u_n)\}$. Now, we are in place to present our problem formulation for hierarchical construction.

3.4 Problem Formulation

Given a sorted list of interesting visualizations, our goal is to create a hierarchal tree where interesting visualizations with high coherent score appear on the left of the tree while the low score ones appear close to the right of the tree.

Definition 1. *A hierarchical tree $T = \{P_1, P_2, ...P_x\}$ is defined as a three level depth, where $P_t \in T$ is a path. Any interesting visulaistion $V = \{(a_i, m_j, f_x)\}$ is a path P from $a_i \to m_j \to f_x$ or vice versa.*

Definition 2. *Root node represents either the whole database D or a specific dataset D_Q. Any node is considered as a dimension attribute a_i, measure attribute m_i or aggregate function f_i associated with choherant score.*

Definition 3. *The coherent score is a measure to evaluate the interestingness of each node and can be computed using the utility score calculated in Sect. 1 plus the frequency score. The coherent score of a parent node is calculated by averaging the score of its children.*

In order to obtain the coherent score of a leaf node, we first calculate the utility score u_i and the frequency score f_i. The utility score is obtained by using the Eq. 2, while the frequency of leaf node is calculated based on the number of co-occurrence in V. To draw the final coherent score we use the equations below 3 and 4:

$$ch_{leaf} = (u) \times \alpha + (1 - \alpha) \times (f)\tag{3}$$

where u is the utility score of the view, f the frequency score of the view and alpah α is a hyperparameter to decide either utility score or frequancy control the structure of the tree. Thirdly, for non leaf elements m_i ,f_i, we compute the coherent score of it's children using the equation below:

$$ch_{non-leaf} = \frac{1}{N} \sum_{j=1} coherent_j, j \in children. \tag{4}$$

Given a sorted list $\mathcal{L} = \{(V_1, u_1), (V_2, u_2), ..., (V_n, u_n)\}$ of interesting visualizations, find a hierarchical tree $\mathcal{T}$ with $Max_{coherent(\mathcal{T})}$.

4   Architecture: *RtSEngine* Framework

The goal of *RtSEngine* is to recommend a set of aggregate views that are considered interesting because of their abnormal deviations. To achieve that, *RtSEngine* utilizes the following key idea: recommend views that are created from grouping high ranked dimension attributes A' within the set A. Conceptually, *RtSEngine* is designed as a recommendation plug-in that can be applied to any visualization engine, e.g., Tableau and Spotfire. However, in the previous work, we built *RtSEngine* as a standalone end-to-end system on top of SeeDB which allows users to pose arbitrary queries over data and obtain recommended visualizations. *RtSEngine* is comprised of two main modules (see Fig. 2):

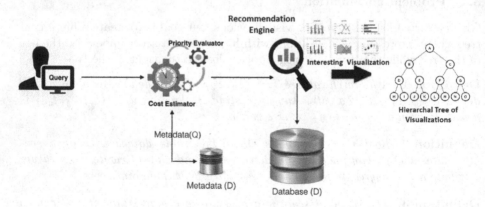

Fig. 2. *RtSEngine*: real time evaluation architecture for automatic recommendation

1. **Priority Evaluator:** An underlying module in front of any recommendation engine. Used to evaluate the dimension attributes that form visualizations according to a priority function Pr computed using our proposed techniques.
2. **Cost Estimator:** This module is supposed to run in parallel with the Priority Evaluator to estimate the retrieval and computation costs of each visualization using our estimation approaches. Estimating the visualization costs in

real-time improves the efficiency by discounting high costs and low priorities visualizations. Note that this module is an awareness cost approach which incorporates the estimated costs to assess visualizations based on their priorities and costs.

4.1 Priority Evaluator: Dimension Attributes Prioritizing

We illustrate the proposed approaches for prioritizing the dimension attributes in the both results set DQ and reference set (e.g. the entire dataset D) and suggest a set of visualizations that are likely to be interesting and score high deviation utilities in certain realtime limits such as maximum number of explored visualizations and execution time. In our previous work, $RtsEngine$ [9] applies the following three approaches to optimize the search space and reduce dimensions.

Ranking Dimension Attributes Based on Distinct Values. Scoring dimensions based on difference of distinct values is the first class of prioritizing algorithms. This approach is referred to as $Diff_D Val$, and it is based on the basic observation about the number of distinct values of the dimension attributes in the results set DQ and the entire database D. The $Diff_D Val$ algorithm scores the dimension attributes according to the difference between the normalized distinct values of attributes in the result set DQ and the entire database D. $Diff_D Val$ computes the required number of dimension attributes G that creates the limit number of views R, then it returns the set $\mathcal{H}$ of size G that contains a group of high priorities attributes.

Scoring Dimension Attributes Based on Selectivity. Our proposed approach $Sela$ utilizes the number of distinct values in the dimensions attributes and incorporates the query size to identify priorities of these dimensions by calculating a priority function for each dimension attribute. Then $Sela$ reorders the dimension attributes based on the priority. Using selectivity ratio and the number of distinct values for assessing visualizations in D and DQ gives closer insights about the data characteristics such as the size of aggregated views generated from group by attributes and the uniqueness degree of data in each dimension attribute.

However, capturing the change of both number in distinct values and the number of aggregated records in each dimension attribute in result set DQ and reference set D is essential to identify visualizations that produce high deviations among all possible visualizations.

Prioritizing Dimension Attributes Based on Histograms. We propose the $DimsHisto$ approach which attempts to capture data distribution inside the dimensions attributes by creating frequency histograms and directly measuring the distance among corresponding histograms to evaluate these dimensions attributes. $DimsHisto$ firstly generates frequency histograms for all dimensions attributes in each dataset. Then it computes the deviation in each dimension by calculating the normalized distances between each corresponding dimension attribute. And the computations of the distance metric.

All proposed algorithms $Diff_DVal$, $Sela$ and $DimsHisto$ have the same number of queries as the cost of retrieving data. While $DimsHisto$ has additional cost for distance computations, it shows high accuracy for most of the aggregate functions such as Sum, Avg, and Count, because these functions are relative to the data frequencies.

4.2 Coherent Hierarchical Representation

The problem that we address is how to construct a coherent tree-structured representation of the discovered interesting visualizations. It is important to mention that the hierarchical representation can help users easily navigate a huge database. Another advantages of the hierarchical tree is that it organizes the interesting visualizations based on internal score, which means it shows the most important and interesting visualizations close to the left, while less important ones appear nearer the right side of a tree representation as shown in Fig. 3. As a result, users can easily find the visualization with high effect on a dataset. To achieve that we propose a coherent matrix which measures the importance and effectiveness of each visualization in our dataset.

The input of the hierarchical algorithm is a set of interesting visualizations $\mathcal{L} = \{(V_1, u_1), (V_2, u_2), ..., (V_n, u_n)\}$. Each visualization (V_x, u_x) contains a view V_x with the associated score u_x. Each view V_x consists of three elements: a dimension attribute a, measure attribute m and aggregate function f. Our hierarchical function constructs a tree using three main steps. First, for all views in $\mathcal{L}$, we obtain the frequency of each element $Fr(a), Fr(m)$ and $Fr(f)$. The elements are sorted descending according to the occurrence in the whole list $\mathcal{L}$. Second, the proposed algorithm constructs a hierarchical tree in a bottom-up manner. For each view $V_x \in (V_x, u_x)$, we start from $a \in V_x$ as a leaf nodes and compute the coherent score using Eqs. 3 and 4.

Finally, our algorithm performs the same process again for each visualization to create a hierarchical tree. The final output is a sorted tree where the visualizations which have a strong effect on the dataset placed near the left.

5   Experiment and Evaluation

Before presenting our results, we describe the details of the conducted experiments including the used datasets, the proposed algorithms and the performance metrics which we use to measure the effectiveness and efficiency.

5.1   Evaluation Metric

We used the same SeeDB metrics [19] for evaluating the quality of aggregate views produced by proposed approaches.

Metrics: To evaluate the quality and correctness of the proposed algorithms, we used the following metric:

- Accuracy: if $\{VS\}$ is the set of aggregate views with the highest utility, and $\{VT\}$ is the set of aggregate views returned by SeeDB baseline, then the accuracy is defined as:

$$Accuracy = \frac{1}{|VT|} * \sum x \text{ where } \begin{cases} x = 1 & \text{if } VT_i = VS_i \\ x = 0 & otherwise \end{cases}$$

i.e., accuracy is the fraction of true positions in the aggregate views returned by SeeDB.

5.2 Datasets

MIMICIII: MIMIC III is freely available, deidentified **Intensive Care Unit** (ICU) medical dataset. It contains real time sensor data as well as history data. It has records of laboratory test results, medications, observations, mortality information etc. of 46,520 ICU patients between 2001 and 2012 [11]. Overall, it has 397 million records from different 26 tables such as admissions, chart events, input events, output events,..etc. The following are the tables that we have used in this research.

Admissions Table: This table contains 58,976 distinct hospital admissions records with 19 dimensions. This table records information related to patient admissions such as:admission, discharge, insurance, and mortality information.

Inputevents_cv Table: This table contains 17.5 million tuples with 22 dimensions using Philips CareVue system. The table include the information about any fluid given to patients for example tube feedings, oral, or intravenous solutions containing medications. It contains information about chart time, item, amount, rate for example e.g. 50 mL of normal saline etc.

5.3 Effectiveness Evaluation

For the interestingness visualization, we construct the hierarchical tree as follows: the first level shows aggregate functions, second is measure attributes and leaf nodes shows dimensions attributes of tree. The sub branch of each node are appear according to both interestingness of produced visualizations and frequency of related attributes to them. In another words, the most interesting and coherent attributes appear in the left and lest one in the right most position with siblings. The higher coherence value shows more the interesting attributes. For all experiment, we realized the hyper parameter α in Eq. 3, which controls the structure of tree to $\alpha = 0.6$ which means that the user is interesting in 60% of views utility and 40% of attributes coherence.

A Hierarchical Representation for Admission: the first experiment, we consider the following query for admission: *Q1: select * from admissions where* *admission location = "TRANSFER FROM OTHER HEALT"* to find interesting information of patients who transferred from other hospitals. Figure 3 shows a representative tree for admissions. The figure shows that the average paitents' numbers per diagnosis for result set of $Q1$ is the most interesting, while the total paitens' numbers SUM per admission type is the least one.

A Hierarchical Representation for Inputevents: In the second experiment, we consider the following query for inputevents: *select * from inputevents_cv where originalroute = "Intravenous Infusion"* to find interesting information related to *Intravenous Infusion* medications compared with other remaining methods in the database. As shown in Fig. 4, the minimum amounts per medicine is the most interesting dimension while the minimum amounts per patient dimension is less interesting. On the other side, the maximum amounts per medicine is the least interesting aggregate function. Since the coherence score of *MIN is 172.26* and *39.85 for MAX* function.

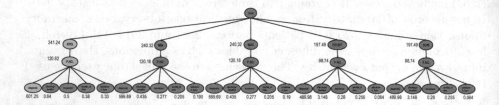

Fig. 3. Hierarchical representation for admission

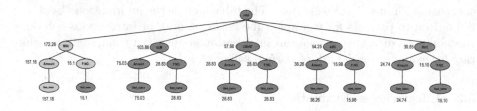

Fig. 4. Hierarchical representation for inputevents

5.4   Efficiency Evaluation

In this subsection we discuss the execution times of *RtSEngine* approaches with the original baseline of SeeDB to identify the improvements in the performance. As shown in Fig. 5a shows the total SeeDB and algorithms execution times compared with the original SeeDB baseline. This figure shows the improvements in the SeeDB performance with different space limits to find a top (K = 25) views. As shown, the improvements in the performance by using

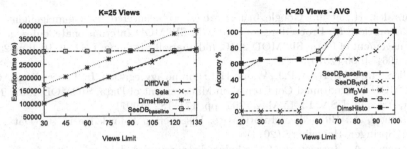

a) Execution times along different view b) Accuracy of algorithms along differ-
limits   ent view limit

Fig. 5. Results quality on varying view space sizes for the Algorithms $Sela$, $Diff_DVal$, $DimsHisto$, and $SeeDB_Rnd$

proposed algorithms are significant compared with the baseline furthermore, the execution costs increase linearly with the view space limit produced by algorithms. In summary, $Sela$ and $DimsHisto$ algorithms both produce results with accuracy $\leq 80\%$ for all aggregate functions and a variety of R Views Limits particularly when Views Limits $= 60$ views as shown in Fig. 5b. Moreover, they produce results with 100% accuracy after that limit. $Sela$ does slightly better than $DimsHisto$ as $Sela$ evaluates the recommended views by capturing the change of the selectivity ratios of dimension attributes.

6 Conclusion and Future Work

In this paper, we presented *MedVIS* an interactive tool for medical data exploration that finds interesting visualizations in a compared subset of data then, structures such visualizations in a hierarchical coherent tree. Furthermore, we discussed *RtSEngine* approaches to optimize the search space and tackle the dimensionality problem. Our experiments show high effectiveness and efficiency in recommending visualizations of a real dataset which will useful for medical decision makers to support better management. In our future work, as huge data are generating from different sensors and other devices, we will address streaming medical data recommendations.

References

1. Almars, A., Li, X., Zhao, X., Ibrahim, I.A., Yuan, W., Li, B.: Structured sentiment analysis. In: Cong, G., Peng, W.C., Zhang, W., Li, C., Sun, A. (eds.) ADMA 2017. LNCS, vol. 10604, pp. 695–707. Springer, Cham (2017). https://doi.org/10.1007/978-3-319-69179-4_49
2. Fisher, D.: Hotmap: looking at geographic attention. IEEE Trans. Vis. Comput. Graph. **13**(6), 1184–1191 (2007)

3. Gonzalez, H., et al.: Google fusion tables: web-centered data management and collaboration. In: Proceedings of the ACM SIGMOD International Conference on Management of Data, SIGMOD 2010, Indianapolis, Indiana, USA, 6–10 June 2010, pp. 1061–1066 (2010)

4. Hellerstein, J.M., Haas, P.J., Wang, H.J.: Online aggregation. In: Proceedings ACM SIGMOD International Conference on Management of Data, SIGMOD 1997, Tucson, Arizona, USA, 13–15 May 1997, pp. 171–182 (1997)

5. Holzinger, A.: Biomedical Informatics: Discovering Knowledge in Big Data, 1st edn. Springer, Heidelberg (2014)

6. Holzinger, A., Jurisica, I.: Knowledge discovery and data mining in biomedical informatics: the future is in integrative, interactive machine learning solutions. In: Holzinger, A., Jurisica, I. (eds.) Interactive Knowledge Discovery and Data Mining in Biomedical Informatics. LNCS, vol. 8401, pp. 1–18. Springer, Heidelberg (2014). https://doi.org/10.1007/978-3-662-43968-5_1

7. Holzinger, A., Simonic, K. (eds.): Information Quality in e-Health - 7th Conference of the Workgroup Human-Computer Interaction and Usability Engineering of the Austrian Computer Society. USAB 2011. LNCS, vol. 7058. Springer, Heidelberg (2011)

8. Hund, M., et al.: Visual analytics for concept exploration in subspaces of patient groups. Brain Inf. **3**, 1–15 (2016)

9. Ibrahim, I.A., Albarrak, A.M., Li, X.: Constrained recommendations for query visualizations. Knowl. Inf. Syst. **51**(2), 499–529 (2017)

10. Jagadish, H.V.: Review - explaining differences in multidimensional aggregates. ACM SIGMOD Digital Rev. **1** (1999)

11. Johnson, A.E., et al.: MIMIC-III, a freely accessible critical care database. Sci. Data **3**, 160035 (2016)

12. Kandel, S., Parikh, R., Paepcke, A., Hellerstein, J.M., Heer, J.: Profiler: integrated statistical analysis and visualization for data quality assessment. In: Proceedings of the International Working Conference on Advanced Visual Interfaces, pp. 547–554. ACM (2012)

13. Key, A., Howe, B., Perry, D., Aragon, C.R.: VizDeck: self-organizing dashboards for visual analytics. In: Proceedings of the ACM SIGMOD International Conference on Management of Data, SIGMOD 2012, Scottsdale, AZ, USA, 20–24 May 2012, pp. 681–684 (2012)

14. Livny, M., et al.: Devise: integrated querying and visualization of large datasets. In: Proceedings ACM SIGMOD International Conference on Management of Data, SIGMOD 1997, Tucson, Arizona, USA, 13–15 May 1997, pp. 301–312 (1997)

15. Mackinlay, J.D., Hanrahan, P., Stolte, C.: Show me: automatic presentation for visual analysis. IEEE Trans. Vis. Comput. Graph. **13**(6), 1137–1144 (2007)

16. Sarawagi, S.: User-adaptive exploration of multidimensional data. In: Proceedings of 26th International Conference on Very Large Data Bases, VLDB 2000, Cairo, Egypt, 10–14 September 2000, pp. 307–316 (2000)

17. Sathe, G., Sarawagi, S.: Intelligent rollups in multidimensional OLAP data. In: Proceedings of 27th International Conference on Very Large Data Bases, VLDB 2001, Roma, Italy, 11–14 September 2001, pp. 531–540 (2001)

18. Vartak, M., Madden, S., Parameswaran, A., Polyzotis, N.: SEEDB: towards automatic query result visualizations. Technical report, data-people. cs. illinois. edu/seedb-tr.pdf
19. Vartak, M., Madden, S., Parameswaran, A.G., Polyzotis, N.: SEEDB: automatically generating query visualizations. PVLDB 7(13), 1581–1584 (2014)
20. Wong, B.L.W., Chen, R., Kodagoda, N., Rooney, C., Xu, K.: INVISQUE: intuitive information exploration through interactive visualization. In: Proceedings of the International Conference on Human Factors in Computing Systems, CHI 2011, Extended Abstracts Volume, 7–12 May 2011, Vancouver, BC, Canada, pp. 311–316 (2011). http://doi.acm.org/10.1145/1979742.1979720

Efficient Top K Temporal Spatial Keyword Search

Chengyuan Zhang[1,2], Lei Zhu[1,2], Weiren Yu[3], Jun Long[1,2(✉)], Fang Huang[1], and Hongbo Zhao[4]

[1] School of Information Science, Central South University, Changsha, China
{cyzhang,leizhu,jlong,hfang}@csu.edu.cn
[2] Big Data and Knowledge Engineering Institute, Central South University, Changsha, China
[3] School of Engineering and Applied Science, Aston University, Birmingham, UK
w.yu3@aston.ac.uk
[4] School of Minerals Processing and Bioengineering, Central South University, Changsha, China
zhbalexander@csu.edu.cn

Abstract. Massive amount of data that are geo-tagged and associated with text information are being generated at an unprecedented scale in many emerging applications such as location based services and social networks. Due to their importance, a large body of work has focused on efficiently computing various spatial keyword queries. In this paper, we study the top-k temporal spatial keyword query which considers three important constraints during the search including time, spatial proximity and textual relevance. A novel index structure, namely SSG-tree, to efficiently insert/delete spatio-temporal web objects with high rates. Base on SSG-tree an efficient algorithm is developed to support top-k temporal spatial keyword query. We show via extensive experimentation with real spatial databases that our method has increased performance over alternate techniques.

1 Introduction

Due to the proliferation of user generated content and geo-equipped devices, massive amount of microblogs (e.g., tweets, Facebook comments, and Foursquare check-in's) that contain both text information [1] and geographical location information are being generated at an unprecedented scale on the Web. For instance, in the GPS navigation system, a POI (point of interest) is a geographically anchored pushpin that someone may find useful or interesting, which is usually annotated with textual information (e.g., descriptions and users' reviews). In social media (e.g., Flickr, Facebook, FourSquare, and Twitter), a large number of posts and photos are usually associated with a geo-position as well as a short text. In the above applications, a large volume of spatio-textual objects may continuously arrive with high speed. For instance, it is reported that there are about 30 million people sending geo-tagged data out into the Twitterverse, and

© Springer Nature Switzerland AG 2018
M. Ganji et al. (Eds.): PAKDD 2018, LNAI 11154, pp. 80–92, 2018.
https://doi.org/10.1007/978-3-030-04503-6_7

2.2% of the global tweets (about 4.4 million tweets a day) provide location data together with the text of their posts[1].

In this paper, we aim to take advantage of the combination of geo-tagged information within microblogs to support temporal spatial keyword search queries on microblogs, where users are interested in getting a set of recent microblogs each of which contains all keywords and closet to user's location. Due to the large numbers of microblogs that can satisfy the given constraints, we limit the query answer to k microblogs, deemed most relevant to the querying user based on a ranking function f_{st} that combines the time recency and the spatial proximity of each microblog to the querying user.

Challenges. There are three key challenges in efficiently processing temporal spatial keyword queries over temporal spatial keyword microblogs streams. Firstly, a massive number of microblogs, typically in the order of millions, are posted in many applications, and hence even a small increase in efficiency results in significant savings. Secondly, the streaming temporal spatial keyword microblogs may continuously arrive in a rapid rate which also calls for high throughput performance for better user satisfaction. Thirdly, the above challenges call for relying on **only** in-memory data structures to index and query incoming microblogs, where memory is a scarce resource.

Based on the above challenges, we first discuss what kinds of techniques should be adopted from different angles. Then, we propose a novel index technique, namely the Segment Signature Grid trees (SSG-Trees for short), to effectively organize continuous temporal spatial keyword microblogs. In a nutshell, SSG-Trees is essentially a set of Signature Grid-Trees, each node of which is enriched with the reference to a frequency signature file for the objects contained in the sub-tree rooted at the node. Then, an efficient temporal spatial keyword search algorithm is designed to facilitate the online top-k temporal spatial keyword search. Extensive experiments show that our SSG-Trees based **TSK** algorithm achieves very substantial improvements over the nature extensions of existing techniques due to strong filtering power.

The rest of this paper is organized as follows. Section 2 formally defines the problem of top k temporal spatial keyword search. We introduce the techniques should be adopted in Sect. 3. Section 4 presents the framework of SSG-Trees and algorithm. Extensive experiments are reported in Sect. 5.

2   Preliminaires

In this section, we present problem definition and necessary preliminaries of top k temporal spatial keyword Search. Table 1 below summarizes the mathematical notations used throughout this section.

In this section, $\mathcal{O}$ denotes a sequence of incoming stream geo-textual objects. A **geo-textual object** is a textual message with geo-location and timestamp,

---

[1] http://www.futurity.org/tweets-give-info-location.

Table 1. Notations

Notation	Definition
$o(q)$	s geo-textual object (query)
$o.\psi(q.\psi)$	a set of keywords (terms) used to describe o (query q)
$o.loc(q.loc)$	location of the object o (query q)
$o.t(q.t)$	timestamp of the object o (query q)
$\mathcal{V}$	vocabulary
w	a keyword (term) in $\mathcal{V}$
l	the number of query keywords in $q.\psi$
α	the preference parameter to balance the spatial proximity and temporal recency

such as geo-tagged and check-in tweets. Formally, a geo-textual object o is modeled as $o = <\psi, loc, t>$, where $o.\psi$ denotes a set of distinct keywords (terms) from a vocabulary set $\mathcal{V}$, $o.loc$ represents a geo-location with latitude and longitude, and $o.t$ represented the timestamp of object.

Definition 1 (Top-k *Temporal Spatial-Keyword* (TSK) Query). *A top-k temporal spatial keyword query q is defined as $q = <\psi, loc, t, k>$, where $q.\psi$ is a set of distinct user-specified keywords (terms), $q.loc$ is the query location, $q.t$ is the user submitted timestamp, k is the number of the result user expected.*

Definition 2 (Spatial Proximity $f_s(o.loc, q.loc)$). *Let δ_{max} denote the maximal distance in the space, the spatial relevance between the object o and the query q, denoted by $f_s(o.loc, q.loc)$, is defined as $\frac{\delta(q.loc, o.loc)}{\delta_{max}}$.*

Definition 3 (Temporal Recency $f_t(o.t, q.t)$). *Let λ_{max} denote the maximal time difference in the timestamp, the temporal recency between the object o and the query q, denoted by $f_t(o.t, q.t)$, is defined as $\frac{\lambda(o.t, q.t)}{\lambda_{max}}$.*

Based on the spatial proximity and temporal recency between the query and the object, the **Spatial-temporal Ranking Score** of an object o regarding the query q can be defined as follows.

Definition 4 (Spatial-temporal Ranking Score $f_{st}(o, q)$). *Let α $(0 \leq \alpha \leq 1)^2$ be the preference parameter specifying the trade-off between the spatial proximity and temporal recency, we have*

$$f_{st}(o, q) = \alpha * f_s(o.loc, q.loc) + (1 - \alpha)f_t(o.t, q.t). \tag{1}$$

*Note that the objects with the **small score values** are preferred (i,e., ranked higher).*

---

[2] $\alpha = 1$ indicates that the user cares only about the spatial proximity of geo-textual objects, $\alpha = 0$ gives the k most recent geo-textual objects in dataset.

Definition 5 (Temporal Spatial Keyword Search). *Given a set of geo-textual objects $\mathcal{O}$ and a temporal spatial keyword query q, we aim to find the top k geo-textual objects with **smallest** spatial-temporal score, and each of which contains **all** of the query keywords.*

3   Related Work

3.1   Efficient Textual Retrieval

Existing textual retrieval indexes, which can effectively combine with other spatial or temporal indexes, are mainly falling into one of two categories: inverted index [2–4] and signature file [5–8]. Owing to only the indexes of related keywords have been extracted in inverted index, inverted index excels in query processing efficiency while compared with signature. However, signature has faster insertion speed and utilizes significantly less storage overhead. According to [9], inverted index requires much larger space overhead than signature file ($\approx$10 times), and demands expensive updates of the index when insert a new document, due to many terms of inverted index needs to store more than once and frequently undergo re-organization triggers under intensive information insertion/updating procedures. Furthermore, the inverted index is also reported to perform poorly for multiple terms queries in [10]. Obviously, taking the properties of fast update online system into consideration, signature file seems a better choice.

3.2   Efficient Spatial Partition

To support high arrival rates of incoming objects, space-partitioning index (e.g., Quadtree [11–15], Pyramid [16–18], and Grid structure [2,19,20] is more famous than object-partitioning index (e.g., R-tree). As stressed in [21], space-partitioning index is more suitable to high update system because of its disjoint space decomposition policy, while the shape of object-partitioning index is highly affected by the rate and order of incoming data, which may trigger a large number of node splitting and merging. Motivated by this, we should adopt space-partitioning index as our proposed spatial partition index.

3.3   Efficient Temporal Partition

Regarding the temporal partition techniques, which can combine with other textual or spatial index, are mainly divided into two categories: Log Structure [22] and Sliding Window [23]. Log Structure partitions the data into a sequence of indexes with exponentially increasing size, while Sliding Window partitions the data into a sequence of indexes with equal size or with equal time range. Obviously, the performance of Log Structure is better than Sliding Window if top-k results can be find in most recent data. However, a sharp drop will be met if the top-k results can be find in most recent data, due to its exponent increase size partition. Furthermore, the insertion cost of Log Structure will significant increase while combining with other index.

4   SSG-Tree Framework

In this section, we present a segment signature grid trees (SSG-Trees for short) that supports update at high arrive rate and provides the following required functions for geo-textual object search and ranking.

4.1   SSG-Tree Structure

In order to support efficient geo-textual object search, the SSG-Trees clusters a set of geo-textual objects into a series of continual signature Grid-Tree, which cluster the objects into disjointed subsets of nodes and abstracts them in various granularities. By doing so, it capacitates the pruning of those (textually, spatially or temporally) irrelevant subsets or trees. The efficiency of SSG-Trees depends on its pruning power is highly related to the effectiveness of object clustering and the search algorithms. Our SSG-Trees clusters spatially and temporally close objects together and carries textual information in its node signatures.

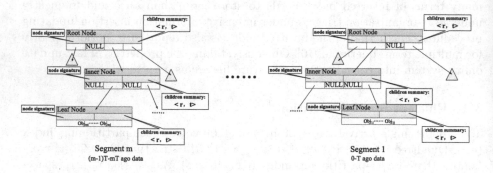

Fig. 1. SSG-tree structure

SSG-Trees is essentially a set of Signature Grid-Trees, each node of which is enriched with the reference to a signature file for the objects contained in the sub-tree rooted at the node. In particular, each node of an SSG-Trees contains all spatial, temporal, and keyword information; the first is in the form of a rectangle, the second is in the form of timestamp, and the last is in the form of a signature.

More formally, the leaf node of SSG-Trees has the form $(nSig, r, t, oSig)$. $oSig$ refers to a set of signatures created by the objects of current node, $nSig$ is the **OR-ing** of all signature in $gSig$, r is the area covered by current node, and t is the latest timestamp aggregated from the objects. An inner node has the form $(nSig, r, t, cp)$. cp are the address of the children nodes, $nSig$ is the **OR-ing** of all the signatures of its children, r is the area covered by current node, and t is the latest timestamp aggregated from its children nodes. To simplify the following presentation we degrade the Grid-Tree to its special case Quadtree in the example. Figure 1 depicts an SSG-Trees indexing structure.

4.2   Processing of TSK Queries

We proceed to present an important metric, the minimum spatial-temporal distance $MIND_{st}$, which will be used in the query processing. Given a query q and a node N in the SSG-Trees, the metric $MIND_{st}$ offers a lower bound on the actual spatial-temporal distance between query q and the objects enclosed in the rectangle of node N. This bound can be used to order and efficiently prune the paths of the search space in the SSG-Trees.

Definition 6 ($MIND_{st}(q, N)$). *The distance of a query point q from a node N in the SSG-Trees, denoted as $MIND_{st}(q, N)$, is defined as follows:*

$$MIND_{st}(q, N) = \alpha * \frac{MIND_s(q.loc, N.r)}{\delta_{max}}$$
$$+ (1 - \alpha) * \frac{MIND_t(q.t, N.t)}{\lambda_{max}} \qquad (2)$$

where α, δ_{max}, and λ_{max} are the same as in Eq. 1; $MIND_s(q.loc, N.r)$ is the minimum Euclidian distance between $q.loc$ and $N.r$, $MIND_t(q.t, N.t)$ is the minimum time difference between $q.t$ and $N.t$.

A salient feature of the proposed SSG-Trees structure is that it inherits the nice properties of the Quadtree for query processing. Given a query point q, a node N, and a set of objects $\mathcal{O}$ in node N, for any $o \in \mathcal{O}$, we have $f_{st}(q, N) \leq DIST_{st}(q, o)$.

When searching the SSG-Trees for the k objects nearest to a query q, one must decide at each visited node of the SSG-Trees which entry to search first. Metric $MIND_{ST}$ offers an approximation of the spatial-temporal ranking score to every entry in the node and, therefore, can be used to direct the search. Note that only node satisfied the constraint of query keywords need to be loaded into memory and compute $MIND_{ST}$.

To process **TSK** queries with SSG-Trees framework, we exploit the best-first traversal algorithm for retrieving the top-k objects. With the best-first traversal algorithm, a priority queue is used to keep track of the nodes and objects that have yet to be visited. The values of f_{st} and $MIND_{st}$ are used as the keys of objects and nodes, respectively.

When deciding which node to visit next, the algorithm picks the node N with the smallest $MIND_{st}(q, N)$ value in the set of all nodes that have yet to be visited. The algorithm terminates when k nearest objects (ranked according to Eq. 1) have been found.

Algorithm 1 illustrates the details of the SSG-Trees based **TSK** query. A minimum heap $\mathcal{H}$ is employed to keep the Grid-Tree's nodes where the key of a node is its minimal spatial-temporal ranking score. For the input query, we calculate its frequency signature in Line 3. In Line 4, we find out the root node of current time interval, calculate the minimal spatial-temporal ranking score for the root node, and then pushed the root node into the $\mathcal{H}$. The algorithm executes the while loop (Line 5–21) until the top-k results are ultimately reported in Line 23.

Algorithm 1. TSK Search(q, k, $\mathcal{I}$)

Input: q : the spatial-keyword temporal query, k : the number of object return, $\mathcal{I}$: current SSG-Trees index

Output: $\mathcal{R}$: top-k query result results

1:   $\mathcal{R} := \emptyset$; $\mathcal{H} = \emptyset$, $\lambda_{max} = \infty$
2:   $\mathcal{H} \leftarrow$ new a min first heap
3:   build frequency signature for query
4:   $\mathcal{H}$.Enqueue($\mathcal{I}.root, MIND_{st}(q, \mathcal{I}.root)$)
5:   **while** $\mathcal{H} \neq \emptyset$ **do**
6:     $e \leftarrow$ the node popped from $\mathcal{H}$
7:     **if** e is a leaf node **then**
8:       **for** each object o in node e **do**
9:         **if** o passed the signature test **AND** $f_{st}(q, o) \leq \lambda_{max}$ **then**
10:           $\lambda_{max} \leftarrow f_{st}(q, o)$
11:           update $\mathcal{R}$ by $(o, f_{st}(q, o))$
12:         **end if**
13:       **end for**
14:     **else**
15:       **for** each child e' in node e **do**
16:         **if** e' passed the signature test **AND** $MIND_{st}(q, e') \leq \lambda_{max}$ **then**
17:           $\mathcal{H}$.Enqueue($e', MIND_{st}(q, e')$)
18:         **end if**
19:       **end for**
20:     **end if**
21:    process the root node of next SSG-Trees
22: **end while**
23: **return** $\mathcal{R}$

In each iteration, the top entry e with minimum spatial-temporal ranking score is popped from $\mathcal{H}$. When the popped node e is a leaf node (Line 7), for each signature in node e, we will iterator extract the objects that satisfy query constraint and check whether its spatial-temporal ranking score is less than λ_{max}. If its score is not larger than λ_{max}, we push o into result set and add update λ_{max}. When the popped node e is a non-leaf node (Line 15), a child node e' of e will be pushed to $\mathcal{H}$ if it can pass the query signature test and the minimal spatial-temporal ranking score between e' and q, denoted by $MIND_{st}(q, e')$, is not larger than λ_{max} (Line 15–17). We process the root node of next interval in Line 21. The algorithm terminates when $\mathcal{H}$ is empty and the results are kept in $\mathcal{R}$.

5   Experiments

5.1   Baseline Algorithms

(I) Inverted File plus Quadtree (**IFQ**). IFQ first employs Quadtree to partition objects into leaf cells according to their location information. Then, the objects inside each cell are stored in a reversed chronological order. Finally, for the objects in each leaf cell, we build inverted file for keyword filtering proposed and recursively construct the inverted file for its ancestral cells.

(II) Segment Based Inverted File plus Quadtree (**SIFQ**). SIFQ is an enhanced version of IFQ. The major difference between them is that IFQ organizes all the objects in an single quadtrees, but SIFQ partitions all the incoming objects into a set of quadtrees or segments by time unit.

5.2 Experiment Setup

In this section, we implement and evaluate following algorithms.

- **IFQ.** IFQ based **TSK** algorithm proposed in baseline algorithm.
- **SIFQ.** Enhanced IFQ by partitioning the objects into a set of time unit in baseline algorithm.
- **SSG.** SSG-Trees based **TSK** algorithm proposed in Sect. 4.

Dataset. All experiments are based on a real-life dataset **TWEETS**. TWEETS is a real-life dataset collected from Twitter [5], containing 13 million geo-textual tweets from May 2012 to August 2012. The statistics of **TWEETS** are summarized in Table 2.

Table 2. Dataset details

Property	# of objects	Vocabulary	Avg. # of term per obj.
TWEETS	13.3M	6.89M	10.05

Workload. The workload for the **TSK** query consists of 1000 queries, and the average query response time are employed to evaluate the performance of the algorithms. The number of query keywords (l) varies from 1 to 5, the number of results (k) grows from 10 to 50, and the preference parameter α changes from 0.1 to 0.9. By default, l, k, and α are set to **3**, **10**, and **0.5** respectively. All experiments are implemented in C++. The experiments are conducted on a PC with 2.9 GHz Intel Xeon 2 cores CPU and 32 GB memory running Red Hat Enterprise Linux.

5.3 Index Maintenance

In this subsection, we evaluate the insertion time, deletion time, storage overhead of all the algorithms[3]. Since all the objects are indexed in one single quadtree in IFQ, the insertion and deletion time are longer than the other algorithms. What's worse, the time to process 2000 objects for IFQ is large than one second. Thus, we ignore IFQ's insertion time and deletion time in comparison.

---

[3] We ignore IFQ's insertion time and deletion time in comparison, due to it cannot meet the current arrive rate.

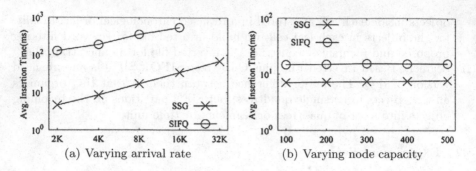

    (a) Varying arrival rate   (b) Varying node capacity

Fig. 2. Index insertion time

Evaluation on Insertion Time. Figure 2(a) gives the performance when varying the arrival rate from 2K to 32K microblogs/second. The figure presents the time of bulk insertion every 1 s, i.e., all microblogs that have arrived in the last second are inserted in bulk. Both SSG perform much better than SIFQ. While SIFQ is able to digest only 16K microblogs/second, SSG are able to bulk insert 32K microblogs in less than 0.12 s. For the current Twitter rate (4,600 microblogs/second), all above algorithms are able to insert all the incoming 4,600 items in less than 56 ms. Figure 2(b) gives the same experiment with varying node capacity from 100 to 500. The performance is stable for all alternatives with 177, and 7.5 ms for SIFQ, and SSG, respectively.

Evaluation on Deletion Time. Figure 3(a) gives the performance when varying the delete rate from 2K to 32K microblogs/second. While SIFQ is able to erase only 32K microblogs/second, SSG are able to bulk delete 32K microblogs in less than 10 ms. Obviously, all the algorithms can satisfied current deletion requirement. Figure 3(b) gives the same experiment with varying node capacity from 100 to 500. The performance of all algorithms meets a slight improvement. As expected, SSG always has the best performance when node capacity changes.

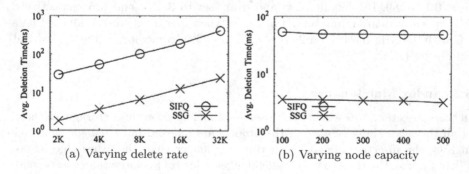

    (a) Varying delete rate   (b) Varying node capacity

Fig. 3. Index deletion time

5.4   Time Evaluation

In this subsection, we evaluate the query response time of all the algorithms.

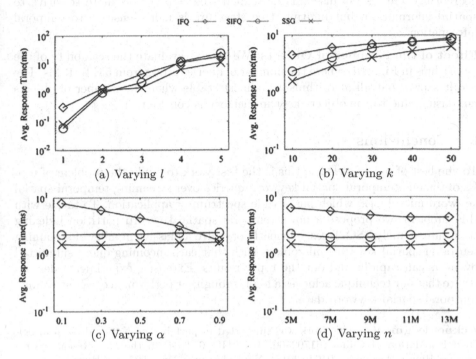

Fig. 4. Response time

Effect of the Number of Query Keywords. Figure 4(a) evaluates the response time of four algorithms against TWEETS where the number of query keywords varies from 1 to 5. Not surprisingly, the performance of three algorithms degrades with the growth of the number of query keywords, due to the growth of search region. When there is only one query keyword, the performance of SIFQ is better than the other algorithm, due to there is not false positive probability in inverted index and SIFQ stores the objects in a set of trees instead of individual quadtree. However, when the number of query keywords is larger than 1, SSG performs the best. Moreover, the margin becomes significant when there are more than one query keyword.

Effect of the Number of Result. We evaluate response time of four algorithms against TWEETS where the number of result varies from 10 to 50 in Fig. 4(b). It is shown that SSG demonstrates superior performance in comparison with other algorithms, due to the frequency signature can reduce interference from low frequency terms. Compared with SIFQ and SSG, IFQ becomes more and more efficient as the number of result increases.

Effect of the Value of Preference Parameter. Figure 4(c) the effect of varying α from 0.1 to 0.9 on response time. It is interesting that when α grows, the performance of IFQ improves, while the other algorithms meet slightly decreasing. This is because this individual tree structure is more sensitive to spatial information, but multiple tree structure is more sensitive to temporal information.

Effect of the Number of Objects. We turn to evaluate the scalability of the algorithms in Fig. 4(d) where the number of queries grows from 5M to 13Ms. The result shows that all algorithms are almost stable when the number of objects increasing, due to the objects kept in memory is constant.

6   Conclusions

To the best of our knowledge, this is the first work to study the problem of top-k continuous temporal spatial keyword queries over streaming temporal spatial keyword microblogs, which has a wide spectrum of application. To tackle with this problem, we propose a novel temporal spatial keyword partition indexing structure, namely SSG-Trees, efficiently organize a massive number of streaming temporal spatial keyword microblogs such that each incoming query submitted by users can rapidly find out the top-k results. Extensive experiments demonstrate that our technique achieves a high throughput performance over streaming temporal spatial keyword data.

Acknowledgments. This work was supported in part by the National Natural Science Foundation of China (61702560, 61379110, 61472450), the Key Research Program of Hunan Province (2016JC2018), Natural Science Foundation of Hunan Province (2018JJ3691), and Science and Technology Plan of Hunan Province (2016JC2011).

References

1. Wang, Y., Lin, X., Wu, L., Zhang, W., Zhang, Q.: Exploiting correlation consensus: towards subspace clustering for multi-modal data. In: Proceedings of the ACM International Conference on Multimedia, MM 2014, Orlando, FL, USA, 03–07 November 2014, pp. 981–984 (2014)
2. Wang, Y., Lin, X., Wu, L., Zhang, W.: Effective multi-query expansions: collaborative deep networks for robust landmark retrieval. IEEE Trans. Image Process. **26**(3), 1393–1404 (2017)
3. Wang, Y., Lin, X., Wu, L., Zhang, W., Zhang, Q., Huang, X.: Robust subspace clustering for multi-view data by exploiting correlation consensus. IEEE Trans. Image Process. **24**(11), 3939–3949 (2015)
4. Wang, Y., Zhang, W., Wu, L., Lin, X., Zhao, X.: Unsupervised metric fusion over multiview data by graph random walk-based cross-view diffusion. IEEE Trans. Neural Netw. Learning Syst. **28**(1), 57–70 (2017)
5. Wang, Y., Lin, X., Wu, L., Zhang, W., Zhang, Q.: LBMCH: learning bridging mapping for cross-modal hashing. In: Proceedings of the 38th International ACM SIGIR Conference on Research and Development in Information Retrieval, Santiago, Chile, 9–13 August 2015, pp. 999–1002 (2015)

6. Wu, L., Wang, Y., Li, X., Gao, J.: What-and-where to match: deep spatially multiplicative integration networks for person re-identification. Pattern Recogn. **76**, 727–738 (2018)
7. Wu, L., Wang, Y., Ge, Z., Hu, Q., Li, X.: Structured deep hashing with convolutional neural networks for fast person re-identification. Comput. Vis. Image Underst. **167**, 63–73 (2018)
8. Zhang, C., Zhang, Y., Zhang, W., Lin, X., Cheema, M.A., Wang, X.: Diversified spatial keyword search on road networks. In: Proceedings of the 17th International Conference on Extending Database Technology, EDBT 2014, Athens, Greece, 24–28 March 2014, pp. 367–378 (2014)
9. Christodoulakis, S., Faloutsos, C.: Design considerations for a message file server. IEEE Trans. Softw. Eng. **10**(2), 201–210 (1984)
10. Faloutsos, C., Jagadish, H.V.: Hybrid index organizations for text databases. In: Pirotte, A., Delobel, C., Gottlob, G. (eds.) EDBT 1992. LNCS, vol. 580, pp. 310–327. Springer, Heidelberg (1992). https://doi.org/10.1007/BFb0032439
11. Gargantini, I.: An effective way to represent quadtrees. Commun. ACM **25**(12), 905–910 (1982)
12. Wang, Y., Lin, X., Zhang, Q., Wu, L.: Shifting hypergraphs by probabilistic voting. In: Tseng, V.S., Ho, T.B., Zhou, Z.-H., Chen, A.L.P., Kao, H.-Y. (eds.) PAKDD 2014. LNCS (LNAI), vol. 8444, pp. 234–246. Springer, Cham (2014). https://doi.org/10.1007/978-3-319-06605-9_20
13. Wang, Y., Wu, L.: Beyond low-rank representations: orthogonal clustering basis reconstruction with optimized graph structure for multi-view spectral clustering. Neural Netw. **103**, 1–8 (2018)
14. Zhang, C., Zhang, Y., Zhang, W., Lin, X.: Inverted linear quadtree: efficient top k spatial keyword search. In: 29th IEEE International Conference on Data Engineering, ICDE 2013, Brisbane, Australia, 8–12 April 2013, pp. 901–912 (2013)
15. Zhang, C., Zhang, Y., Zhang, W., Lin, X.: Inverted linear quadtree: efficient top k spatial keyword search. IEEE Trans. Knowl. Data Eng. **28**(7), 1706–1721 (2016)
16. Aref, W.G., Samet, H.: Efficient processing of window queries in the pyramid data structure. In: Proceedings of the Ninth ACM SIGACT-SIGMOD-SIGART Symposium on Principles of Database Systems, 2–4 April 1990, Nashville, Tennessee, USA, pp. 265–272 (1990)
17. Wang, Y., Zhang, W., Wu, L., Lin, X., Fang, M., Pan, S.: Iterative views agreement: an iterative low-rank based structured optimization method to multi-view spectral clustering. In: Proceedings of the Twenty-Fifth International Joint Conference on Artificial Intelligence, IJCAI 2016, New York, NY, USA, 9–15 July 2016, pp. 2153–2159 (2016)
18. Wu, L., Wang, Y., Gao, J., Li, X.: Deep adaptive feature embedding with local sample distributions for person re-identification. Pattern Recogn. **73**, 275–288 (2018)
19. Mouratidis, K., Hadjieleftheriou, M., Papadias, D.: Conceptual partitioning: an efficient method for continuous nearest neighbor monitoring. In: Proceedings of the ACM SIGMOD International Conference on Management of Data, Baltimore, Maryland, USA, 14–16 June 2005, pp. 634–645 (2005)
20. Wang, Y., Lin, X., Zhang, Q.: Towards metric fusion on multi-view data: a crossview based graph random walk approach. In: 22nd ACM International Conference on Information and Knowledge Management, CIKM 2013, San Francisco, CA, USA, 27 October–1 November 2013, pp. 805–810 (2013)

21. Magdy, A., Mokbel, M.F., Elnikety, S., Nath, S., He, Y.: Mercury: a memory-constrained spatio-temporal real-time search on microblogs. In: IEEE 30th International Conference on Data Engineering, Chicago, ICDE 2014, IL, USA, 31 March–4 April 2014, pp. 172–183 (2014)
22. O'Neil, P.E., Cheng, E., Gawlick, D., O'Neil, E.J.: The log-structured merge-tree (LSM-tree). Acta Inf. **33**(4), 351–385 (1996)
23. Magdy, A., et al.: Taghreed: a system for querying, analyzing, and visualizing geotagged microblogs. In: SIGSPATIAL (2014)

HOC-Tree: A Novel Index for Efficient Spatio-Temporal Range Search

Jun Long[1,2], Lei Zhu[1,2], Chengyuan Zhang[1,2(✉)], Shuangqiao Lin[1,2],
Zhan Yang[1,2], and Xinpan Yuan[3]

[1] School of Information Science and Engineering, Central South University,
Changsha, People's Republic of China
{jlong,leizhu,cyzhang,linshq,zyang22}@csu.edu.cn
[2] Big Data and Knowledge Engineering Institute, Central South University,
Changsha, People's Republic of China
[3] School of Computer, Hunan University of Technology, Zhuzhou, China
xpyuan@hut.edu.cn

Abstract. With the rapid development of mobile computing and Web
services, a huge amount of data with spatial and temporal information
have been collected everyday by smart mobile terminals, in which an
object is described by its spatial information and temporal information. Motivated by the significance of spatio-temporal range search and
the lack of efficient search algorithm, in this paper, we study the problem of spatio-temporal range search (STRS), a novel index structure is
proposed, called HOC-Tree, which is based on Hilbert curve and OC-
Tree, and takes both spatial and temporal information into consideration. Based on HOC-Tree, we develop an efficient algorithm to solve the
problem of spatio-temporal range search. Comprehensive experiments on
real and synthetic data demonstrate that our method is more efficient
than the state-of-the-art technique.

Keywords: Hilbert curve · Spatio-temporal · Range search
HOC-Tree

1 Introduction

With the rapid development of mobile computing and Web services, a huge
amount of data [8,9,11,15] with spatial and temporal information have been
collected everyday by smart mobile terminals, such as smart phones, tablets,
wearable devices etc., or devices of Iot which are equiped with GPS or wireless modules. In addition, Location Based Services (LBS) and social network
services provide users with location-dependent information and services in dai-
lylife. Everyday a vast number of pictures [10,17,19] and texts with geotags [21–
23]and timestamps are posted to Fackbook or Instagram. Foursquare supports
more than 45 million users who have checked-in more than 5 billion times at over
1.6 million businesses. Users can search any interested information by specified
time interval and geolocation.

© Springer Nature Switzerland AG 2018
M. Ganji et al. (Eds.): PAKDD 2018, LNAI 11154, pp. 93–107, 2018.
https://doi.org/10.1007/978-3-030-04503-6_8

In this paper, we study an important search problem in spatio-temporal data query area, named spatio-temporal range search (STRS for short). Spatio-temporal range search aims to retrieval all spatio-temporal objects whose location is within a specific geographical region during a time region. In many application scenarios, it plays an important role for data management and geo-social networks. For example, in location-based social networks platforms, such as Facebook, Twitter, Weibo, etc. Users prefer to make friends with the people who usually do daily activities in the same geographic region and same time range, because same daily activities like shopping, doing outdoor exercises, going to cinema, etc. are important factors to establish relationships. Thus according to the posts with spatio-temporal data, they can find the users who have the same hobbies within a given area and given time interval, shown in Fig. 1. The red square is the geographical range of search for daily activities. Likewise, location based services like Facebook's Nearby and Foursquare's Radar return the friends that recently checked-in at close proximity to a user's current location [1,8,14,16]. In the big data age, as swift growth of the amount of spatio-temporal data, spatio-temporal range search has become a hot issue in data searching and management area.

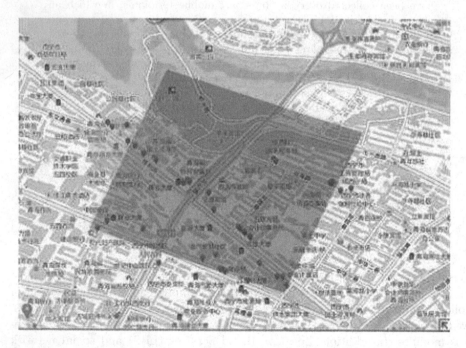

Fig. 1. An example of spatio-temporal search in ocation-based social network services

Motivation. The challenges for the problem of spatio-temporal range search are two-fold. (i) Due to the massive amount of spatio-temporal objects in lots of important applications, large-scale heterogeneous social networks with spatial and temporal information have been constructed. How to efficient management and access geo-social data is a core problem. (ii) For the various application requirements in social network services, high efficient search algorithms need to be developed by combining spatial and temporal features of social data.

Motivated by the significance of spatio-temporal range search and the lack of efficient search algorithm, we propose a novel spatio-temporal index structure, named HOC-Tree based on Hilbert curve and OC-Tree. Besides, we develop an efficient range search algorithm for STRS problem. OC-Tree is an important index structure in spatial database area. It is most often used to partition a three-dimensional space by recursively subdividing it into eight octants. HOC-Tree is a nature extension of OC-Tree, but it not only inherits the valuable properties in 3-dimensional partition, such as the data that close in space and close in time are partition into same cell, but also provides an efficient 3D Morton code generation mechanism, which can easily and effectively combine the spatial and temporal information together to support spatio-temporal search.

Contributions. To summarize, our key contributions in this paper are summarized as follows:

(1) We propose a novel spatio-temporal index based on Hilbert curve and OC-Tree named HOC-Tree to solve the problem of spatio-temporal range search. To the best of our knowledge, this study is the first time to design a novel spatio-temporal indexing mechanism for efficient spatio-temporal range search.
(2) We develop an efficient spatio-temporal range search algorithm based on HOC-Tree.
(3) We conduct comprehensive experiments on real and synthetic datasets. The results show that our method can solve spatio-temporal range search effectively and efficiently, and it outperforms the state-of-the-art approaches.

Roadmap. The rest of the paper is organized as follows: We present the related work in Sect. 2. Section 3 formally defines the problem and describes the index structure. We elaborate the search algorithm in Sect. 4 and extensive experiments are presented in Sect. 5. Finally, we offer conclusions in Sect. 6.

2 Related Work

In this section, we review geo-social networks queries and collective spatial queries, which are two kinds of techniques related to our works.

Geo-Social Networks Queries. A typical geo-social network [9,13,18] combines social networks techniques [12,16] and spatial data queries techniques. Many research findings from academia and techniques applying to industry have been proposed. In industrial circle, the most famous social networks platform, Facebook, provided a location-based social network service named *Nearby* [1] which aims to find the friends who are in the neighborhood of a user currently. Geoloqi is another analogous platform for building location aware applications. It provides the service which notifies users when their friends get into a certain geographical region. Uber is a advanced mobile Internet platform for texi service based on geo-location information of texi drivers and riders. The riders can search the near drivers around them and send messages for request. These applications just only focus on the spatial attributes of data on social networks or cloud platforms for range search. In academic circles, the problem of spatio-temporal search is concerned by lots of researchers. In [1], armenatzoglou et al. proposed a general framework that offers flexible data management and algorithmic design. The nearest star group query contained in the framework returns the k nearest subgraphs of m users. In [4], Liu et al. proposed propose the k-Geo-Social Circle of Friend Query which aims to finds the group g of $k + 1$ users, which (i) is connected, (ii) contains u, and (iii) minimizes the maximum distance between any two of its members. In [6], Scellato et al. proposed three more geo-social networks metrics: (i) average distance (ii) distance strength, and (iii) average triangle length. In [20], Yang et al. developed a hybrid index named Social R-Tree to solve the problem of socio-spatial group query. The studies mentioned above did not combine the spatial and temporal attributes of objects in database for searching.

Geo-Social Networks Queries. Collective spatial query is another important problem. In [24], Zhang et al. presented a novel spatial keyword query problem called the m-closest keywords (mCK) query, which aims to aims to find the spatially closest tuples which match m user-specified keywords. They proposed a new index named bR*-tree extended from the R*-tree to address this problem. The R*-tree designed by Beckmann et al., which incorporates a combined optimization of area, margin and overlap of each enclosing rectangle in the directory [7]. In [3], Guo et al. proved that answering mCK query is NP-hard and designed two approximation algorithms called *SKEC*a and *SKEC*a+. In [2], Deng et al. proposed a generic version of closest keywords search named best keyword cover which considers inter-objects distance as well as the keyword rating of objects. In [5], Long et al. studied two types of the CoSKQ problems, MaxSum-CoSKQ and Dia-CoSKQ. These studies aim to solve the problem of spatial keyword queries to find a set of objects. They did not develop efficient index structure and search algorithms for range search in a specific geographical area and time interval.

3 Model and Structure

This section first presents a Definition of problem, then describes the proposed data structure, named HOC-Tree, which based on Hilbert curve and OC-Tree. Table 1 below summarizes the symbols used frequently throughout the paper.

Table 1. The summary of notations

Notation	Definition
D	A given data set of spatio-temporal data
o	A spatio-temporal object
x_i	The longitude of a spatio-temporal data
y_i	The latitude of a spatio-temporal data
t_i	The timestamp of a spatio-temporal data
h_n	The value of Morton order in Hilbert curve
L	The deepest level of HOC-Tree
ψ	The division threshold value for a node in HOC-Tree
v	The Morton order of a leaf node in HOC-Tree
q	A spatio-temporal range search ($[x_{min}, x_{max}], [y_{min}, y_{max}], [t_{start}, t_{end}]$)
d	The Euclidean distance between two points in spatial space

3.1 Problem Definition

Definition 1 (Spatio-temporal Object Set). *A spatio-temporal object set can be defined as $D = \{o_1, o_2, \ldots, o_n\}$. Each spatio-temporal object o is associated with a spatial location $o.(x_i, y_i)$ and the timestamp $o.t_i$.*

Definition 2 (Spatio-temporal Range Search (STRS)). *Given a spatio-temporal objects data set D, a range query is defined as $q([x_{min}, x_{max}], [y_{min}, y_{max}], [t_{start}, t_{end}])$ where $([x_{min}, x_{max}], [y_{min}, y_{max}])$ is the query spatial region and $([t_{start}, t_{end}])$ is the query temporal interval, this work aims to select all the records which satisfy the query q from D.*

3.2 Index Structure

In this section, we introduce a novel spatio-temoral index, named HOC-Tree, which is based on OC-Tree and Hilbert curve. This data structure is the key technique of this work.

As it will be shown in Subsect. 4.1, the more subspaces overlapping with range query q, the more time will be consumed when searching HOC-Tree. To solve this problem, a *MBRsign* tag data structure is used to reduce non-promising nodes access, which can avoid unnecessary I/O costs. For each subspace, the spatial locations of all the points in it can be associated with a minimum bounding rectangle (MBR), so a *MBRsign* tag is maintained for each non-empty leaf node to keep the MBR information. For a given range query q, the covering non-empty leaf nodes which don't satisfy the spatial constraint will not be accessed in searching process with the help of tags. HOC-Tree keeps two end points information of the MBR, which only require 16 bytes for each non-empty leaf node. The more detail of using the tags will be described particularly in Subsect. 4.1, where elucidates the search algorithms. Figure 2 illustrates the structure of HOC-Tree with *MBRsign* tags.

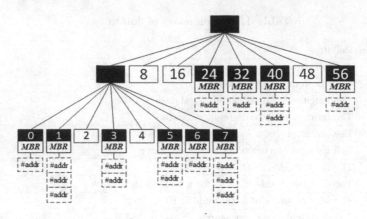

Fig. 2. An example of HOC-Tree with *MBRsign* tags

The black blocks represent non-empty nodes which contains a list of spatio-temporal data locations while the white blocks represent empty nodes. Each leaf node is labeled by its Morton order value according to our approach as mentioned above, and tags are kept for them to maintain the MBR.

4   Spatio-Temporal Query Algorithms

This section gives exhaustive description of spatio-temporal range search based on HOC-Tree.

4.1   Range Search Algorithm

Range query q is an essential function in spatio-temporal data processing. In our algorithms as shown in Algorithm 1, this work is done in several stages. The input query $q = ([x_{min}, x_{max}], [y_{min}, y_{max}], [t_{start}, t_{end}])$ is in three-dimensional space, where $[x_{min}, x_{max}]$ gives the range of longitude, $[y_{min}, y_{max}]$ gives the range of latitude and $[t_{start}, t_{end}]$ gives the time interval. The output S is a set of entries inside spatio-temporal query q. This algorithm only accesses the optimized nodes when searching HOC-Tree. A prune process is executed to check the entries whether they satisfy the query range or not and remove false positives to refine results.

Mapping Hilbert Curve Values: For a given range query q, the Hilbert curve values of covering spatial spaces can be calculated immediately according to the region $([x_{min}, x_{max}], [y_{min}, y_{max}])$ of q. The function *getHilbertValues()* maps the rectangle region into a set of one-dimensional values in line 2.

Finding Spatio-Temporal Covering Cubes: Before searching HOC-Tree in corresponding regions locally, the function *getOverlappingCubes()* line in 6 computes the covering nodes which overlap with three-dimensional query range.

Algorithm 1. Spatio-Temporal Range Search

Input: q : $x_{min}, x_{max}, y_{min}, y_{max}, t_{start}, t_{end}$.

Output: S :set of entries inside query range q.

1: $S \leftarrow \emptyset; CoveringSpacialSpaces \leftarrow \emptyset; RegionSet \leftarrow \emptyset$;
2: $CoveringSpacialSpaces \leftarrow getHilbertValues(x_{min}, x_{max}, y_{min}, y_{max})$;
3: $RegionSet \leftarrow getRegions(CoveringSpacialSpaces)$;
4: $CoveringNodes \leftarrow \emptyset; Q.L^f \leftarrow \emptyset; Q.L^p \leftarrow \emptyset$;
5: **for** each region $\in RegionSet$ **do**
6:     $CoveringNodes \leftarrow getOverlappingCubes(x_{min}, x_{max}, y_{min}, y_{max}, t_{start}, t_{end})$;
7:     $N^f, N^p \leftarrow Identify(CoveringNodes)$;
8:     **for** each Node $v \in N^p$ **do**
9:         **MBRCheck**($MBRsign_v, x_{min}, x_{max}, y_{min}, y_{max}$);
10:     **end for**
11:     **for** each Node $v \in N^p$ that *survive from* **MBRCheck** *process* **do**
12:         $Q.L^p \leftarrow getEntriesList(v)$;
13:     **end for**
14:     **for** each Node $v \in N^f$ **do**
15:         $Q.L^f \leftarrow getEntriesList(v)$;
16:     **end for**
17:     $S \leftarrow Q.L^f + Prune(Q.L^p)$;
18: **end for**
19: **return** S.

The covering cubes can be partial or full. The left part of Fig. 3 shows a spatio-temporal range query q (the shaded cube) which would overlap multiple sub-spaces. For simplicity, the partition of space does not present here. The cubes overlapping with query range in spatial dimension is illustrated in the right part of Fig. 3, where the deepest level L of the HOC-Tree is 4. We can see that cube A has a full spatial overlap while the rest cubes have partial spatial overlap.

For each covering node, it needs to be searched HOC-Tree to get the list of addresses refer to the locations of data point. All the points in fully spatio-temporal overlapping cubes will satisfy the spatio-temporal range search which do not need to do an additional refinement step. **Algorithm 2 Identify** distinguishes these two kinds of covering nodes by N^f and N^p, where N^f denotes the set of fully covering nodes and N^p denotes the set of partially covering nodes. The identification of full overlaps helps to reduce the computation time, which can avoid unnecessary CPU checking overhead in the refinement step.

Confirming Non-empty Covering Nodes: The benefit of coupling the spatial and temporal information in our index will be more clear in this stage. As shown in Fig. 3 (right part), the overlap of cube B with query's spatial dimensional area is very small w.r.t cube A, which has a full overlap. If searching the index without any spatial discrimination, then a very small overlap (i.e., the cube B) will need the same I/O costs with that of a full overlap (i.e., the cube A). As a result, many false positive results will be collected, which have to be later pruned through the spatial criteria. Especially when the data is skew, there might be a lot of empty partial covering nodes. This case can happen because the points in

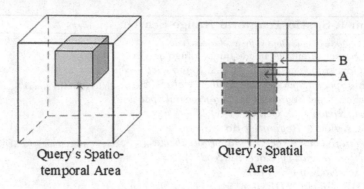

Query's Spatio-
temporal Area

Query's Spatial
Area

Fig. 3. Spatio-temporal range search

Algorithm 2. Identify(CoveringNodes)

Input: *CoveringNodes.*
Output: N^p, N^f
 1: $N^f \leftarrow \emptyset; N^p \leftarrow \emptyset;$
 2: **for** each region $\in CoveringNodes$ **do**
 3:     **if** v is a full overlapping leaf node **then**
 4:         add v into N^f;
 5:     **end if**
 6:     **if** v is a partial overlapping leaf node **then**
 7:         add v into N^p;
 8:     **end if**
 9: **end for**
10: **return** N^p, N^f.

---

that cube do not satisfy with the spatial criteria of q. In line 8 to 10, the information kept in *MBRsign* tag is used to check whether the MBR overlap with the spatial criteria or not. The checking is just needed in partial covering nodes because the points in full overlaps will all satisfy the spatio-temporal criteria. The confirmation of non-empty spatial covering nodes can efficiently reduce the number of false positive results in region search. As shown in Fig. 4, the Morton values of overlapping nodes (i.e. the nodes in the rectangle marked with dotted lines) is obtained by the function *getOverlappingCubes*(), which overlap with the range query given in Fig. 3. For simplicity, the further division of the Node v_3 as shown is omitted here. Then with the help of *MBRsign* tag, it can further confirm the non-empty covering nodes (i.e. the Node v_3, the Node v_7 and the Node v_{24}), which need to be searched in HOC-Tree.

Searching HOC-Tree: Spatio-temporal adjacent nodes will be stored nearby each other by this encoding in HOC-Tree. Identifying full and partial covering nodes helps to reduce the computation time, while confirming non-empty spatial covering nodes can reduce the number of I/O during searching HOC-Tree. Furthermore, the property of Hilbert curve can ensure that the generated Morton value of query range will contain all the valid points, which discussed in Subsect. 3.2. According to the stages described above, the algorithm can get a set of full covering nodes N^f and a set of non-empty partial covering nodes N^p which need to be searched in HOC-Tree. Line 11 to 16 give the search result by $Q.L^f$ and $Q.L^p$, where the notation $Q.L^f$ and $Q.L^p$ denote the sets of entries in full and partial covering nodes respectively.

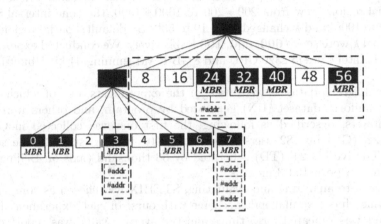

Fig. 4. Query overlapping nodes and non-empty covering nodes

Refining Results: The entries in $Q.L^p$ which have partial overlap need further refinement. There might be some points in partial overlapping nodes that not satisfy with the spatio-temporal query range. **Algorithm 3 Prune** checks each entry in $Q.L^p$ whether it is inside query range or not and removes unrelated results immediately.

Algorithm 3. Prune($Q.L^p$)

Input: $Q.L^p$
Output: *results*
1: *Results* $\leftarrow \emptyset$;
2: **while** $Q.L^p \neq \emptyset$ **do**
3:    **for** each entry $e \in Q.L^p$ **do**
4:       **if** query range contains e **then**
5:          add e into results;
6:       **end if**
7:    **end for**
8: **end while**
9: **return** *results*.

---

5 Experiments

5.1 Experiment Settings and Datasets

With the implementation of HOC-Tree, a comprehensive experimental evaluation is conducted to verify the performance of the scheme in a real cloud environment. The locations of all datasets were scaled to the two-dimensional space $[0, 10000]^2$, and the timestamp of all datasets were scaled to $[0, 5000]$. In addition, the spatial region grew from $200 * 200$ to $1000 * 1000$, the time interval varied from 200 to 1000, and k changed from 10 to 500. By default, spatial region, time interval and k were set to **600, 600, 100** respectively. We conducted experiments on a 3 GHz Intel Core i5 2320 CPU and 8 GB RAM running 64-bit Ubuntu 16.04 LTS.

Three different datasets were used in the experiments, one of which was a synthetic uniform dataset (UN) generated by program, and others were real-world datasets, described as following: the first one was collected in Geolife project [22] (GL) by 182 users from April 2007 to August 2012, the second one was T-Drive [23,24] (TD) generated by 33 thousand taxis on Beijing road network over a period of 3 months.

For accurate analyzing and evaluating, STEHIX was chosen as comparative object, which has a similar index scheme with ours. In each experimental case, the process was repeated for 5 times and the average value was reported. For the HOC-Tree in all the tests, the deepest level L was set to 16 and the division threshold value ψ was set to 200.

5.2 Performance Evaluation

Evaluation on Different Datasets. A series of evaluation was performed on index construction time, index size and data query performance separately against three datasets GL, TD and UN, where other parameters had default settings.

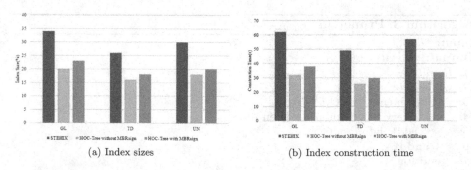

(a) Index sizes   (b) Index construction time

Fig. 5. Evaluation on different datasets

Figure 5(a) depicts the rate of space occupying by the index sizes. STEHIX requires more space due to the two kinds of indices (called *s-index* and *t-index*) kept for all the entries. The storage cost of the STEHIX increases faster in larger datasets. In contrast, an index record is maintained for each entry only once so that our index saves more space in memory. Particularly, HOC-Tree with *MBRsign* tag occupies a very small index size compared with HOC-Tree without *MBRsign* tag. Figure 5(b) shows the difference of construction time between HOC-Tree and STEHIX. Due to the simple split and code algorithms of an HOC-Tree, our method has a shorter constructing time as compared to STEHIX, which need to traverse two indices during the construction.

The experiment results of range query on different datasets are shown in Fig. 6, where the spatial region and time interval were both set 600. The query performance was measured by computing the duration time between when the regions started searching and when client received all accurate results.

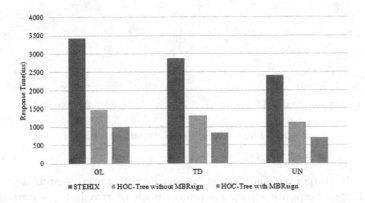

Fig. 6. Data query performance on various datasets

The HOC-Tree demonstrates superior performance in comparison with STE-HIX. Our analysis is as follows, STEHIX calculates the number of addresses in *s-index* and *t-index* separately to choose the high-selectivity list for further retrieval. In sense that, each query will decompose into two processes to collect results in temporal dimension and spatial dimension, which will provoke more I/O costs. On the other hand, STEHIX uses a period time T to divide all entries in temporal dimension because of the periodicity in timestamps, which means if let $T = 24$ (a period of 24 h is a cycle) and divide T into several segments such as 8 segments, then all the entries will map into the 8 segments by their temporal information. For a given temporal range query, all results returned by STEHIX are confused by modulo value, which have same time intervals but different in dates. Therefore, it will take more time to remove false positive results, which delays the response time in queries.

HOC-Tree with *MBRsign* tag improves a little efficiency comparing with HOC-Tree without *MBRsign* tag for uniform data (UN), because the benefit of MBR information maintained in *MBRsign* is apparently in skew data such as GL and TD.

Evaluation on the Effect of Varies Extent in STRS. A series experiments was conducted to investigate the effect of spatial region and temporal interval respectively. These experiments purposed to present the benefits of coupling spatio-temporal information in HOC-Tree and maintaining the *MBRsign* tag.

In order to show the trend of performance change in different spatial region, the temporal interval was set as the default value 600. Two representative datasets were used in the experiments, one was real-world dataset (TD) and another was uniform synthetic dataset (UN). As shown in Fig. 7(a) and (b), the time cost on different datasets is plotted.

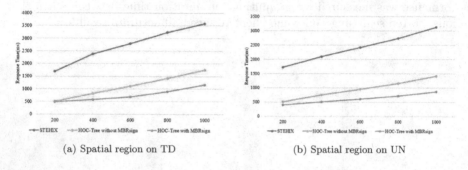

(a) Spatial region on TD   (b) Spatial region on UN

Fig. 7. The effect of varies extent in STRS

Apparently, larger spatial region means larger spatial search area, which results in longer response time. Therefore, both of two indexes perform better when the spatial region is small. On the other hand, the performance of range query in uniform dataset is better than real-world dataset, which mainly because that real-world dataset is a skew data. As is evident from the experiments, HOC-Tree shows an improvement over STEHIX especially when the spatial region is large. Larger spatial query range leads to much more unrelated entries identified as candidates in STEHIX. It spends much more running time and CPU cost because of the high computation for refinement step. However, our index performs better due to the non-fully-decoupling spatial and temporal properties so that all the points are placed by their spatio-temporal proximity in HOC-Tree which can help to reduce I/O load when searching trees. For a given three-dimensional query, HOC-Tree can immediately locate the covering nodes and explore the corresponding HOC-Tree which is owing to the efficient nodes' pruning and the use of Morton value. As pointed out earlier, STRS identifies as more full covering nodes as possible during executing query operation, which helps to reduce the CPU cost for checking fully satisfied entries.

To evaluate the effect of temporal interval on response time of HOC-Tree and STEHIX, experiments were conducted in the same manner with the previous one and spatial region was set as the default value 600. The experimental results are demonstrated in Fig. 8(a) and (b).

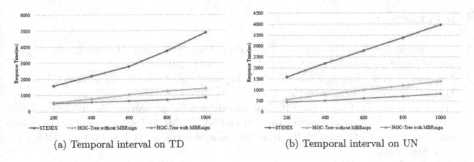

(a) Temporal interval on TD   (b) Temporal interval on UN

Fig. 8. Performance effected by temporal interval

A large temporal query range would cover a lot of partial overlapping nodes, which fully satisfy with temporal restriction but non-fully satisfy with spatial restriction. As the temporal query range becomes larger in STEHIX, all these nodes have to be accessed by *s-index* and *t-index*, which increases the number of I/Os obviously and there are much more candidates to check in the refinement step while HOC-Tree has removed a lot earlier. Because, a *MBRsign* tag data structure is designed in HOC-Tree reduce non-promising nodes access so that these nodes can be removed earlier to avoid unnecessary I/O costs. Figure 8(a) and (b) demonstrate the running time of HOC-Tree with and without the tag, and our index performs better especially for skewed data. In such a scenario, the *MBRsign* tag makes full use of non-fully-decoupled spatial and temporal information to confirm non-empty spatial covering nodes and thus many unnecessary I/O load can be avoided. For uniform dataset (UN), tag is still helpful when there are large number of empty partial overlapping nodes.

6   Conclusions

The problem of spatio-temporal search is very significant due to the increasing amount of spatio-temporal data collected in widely applications. The proposed HOC-Tree is based on Hilbert curve and OC-Tree. Based on HOC-Tree, we design an efficient algorithm to solve the problem of spatio-temporal range search. The results of our experiments on real and synthetic data demonstrate that HOC-Tree is able to achieve a reduction of the processing time by 60–80% compared with prior state-of-the-art methods.

Acknowledgments. This work was supported in part by the National Natural Science Foundation of China (61702560, 61379110, 61472450), the Key Research Program of Hunan Province (2016JC2018), Natural Science Foundation of Hunan Province (2018JJ3691), and Science and Technology Plan of Hunan Province (2016JC2011).

References

1. Armenatzoglou, N., Papadopoulos, S., Papadias, D.: A general framework for geo-social query processing. PVLDB **6**(10), 913–924 (2013)
2. Deng, K., Li, X., Lu, J., Zhou, X.: Best keyword cover search. IEEE Trans. Knowl. Data Eng. **27**(1), 61–73 (2015)
3. Guo, T., Cao, X., Cong, G.: Efficient algorithms for answering the m-closest keywords query. In: Proceedings of the 2015 ACM SIGMOD International Conference on Management of Data, Melbourne, Victoria, Australia, 31 May–4 June 2015, pp. 405–418 (2015)
4. Liu, W., Sun, W., Chen, C., Huang, Y., Jing, Y., Chen, K.: Circle of friend query in geo-social networks. In: Lee, S., Peng, Z., Zhou, X., Moon, Y.-S., Unland, R., Yoo, J. (eds.) DASFAA 2012. LNCS, vol. 7239, pp. 126–137. Springer, Heidelberg (2012). https://doi.org/10.1007/978-3-642-29035-0_9
5. Long, C., Wong, R.C., Wang, K., Fu, A.W.: Collective spatial keyword queries: a distance owner-driven approach. In: Proceedings of the ACM SIGMOD International Conference on Management of Data, SIGMOD 2013, New York, NY, USA, 22–27 June 2013, pp. 689–700 (2013)
6. Scellato, S., Noulas, A., Lambiotte, R., Mascolo, C.: Socio-spatial properties of online location-based social networks. In: Proceedings of the Fifth International Conference on Weblogs and Social Media, Barcelona, Catalonia, Spain, 17–21 July 2011 (2011)
7. Beckmann, N., Kriegel, H.P., Schneider, R., et al.: The R*-tree: an efficient and robust access method for points and rectangles. ACM SIGMOD Rec. **19**(2), 322–331 (1990)
8. Wang, Y., Lin, X., Wu, L., Zhang, W.: Effective multi-query expansions: collaborative deep networks for robust landmark retrieval. IEEE Trans. Image Process. **26**(3), 1393–1404 (2017)
9. Wang, Y., Lin, X., Wu, L., Zhang, W., Zhang, Q.: Exploiting correlation consensus: towards subspace clustering for multi-modal data. In: Proceedings of the ACM International Conference on Multimedia, MM 2014, Orlando, FL, USA, 03–07 November 2014, pp. 981–984 (2014)
10. Wang, Y., Lin, X., Wu, L., Zhang, W., Zhang, Q.: LBMCH: learning bridging mapping for cross-modal hashing. In: Proceedings of the 38th International ACM SIGIR Conference on Research and Development in Information Retrieval, Santiago, Chile, 9–13 August 2015, pp. 999–1002 (2015)
11. Wang, Y., Lin, X., Wu, L., Zhang, W., Zhang, Q., Huang, X.: Robust subspace clustering for multi-view data by exploiting correlation consensus. IEEE Trans. Image Process. **24**(11), 3939–3949 (2015)
12. Wang, Y., Lin, X., Zhang, Q., Wu, L.: Shifting hypergraphs by probabilistic voting. In: Tseng, V.S., Ho, T.B., Zhou, Z.-H., Chen, A.L.P., Kao, H.-Y. (eds.) PAKDD 2014. LNCS (LNAI), vol. 8444, pp. 234–246. Springer, Cham (2014). https://doi.org/10.1007/978-3-319-06605-9_20
13. Wang, Y., Wu, L.: Beyond low-rank representations: orthogonal clustering basis reconstruction with optimized graph structure for multi-view spectral clustering. Neural Netw. **103**, 1–8 (2018)
14. Wang, Y., Wu, L., Lin, X., Gao, J.: Multiview spectral clustering via structured low-rank matrix factorization. IEEE Trans. Neural Netw. Learn. Syst. **29**, 4833–4843 (2018)

15. Wang, Y., Zhang, W., Wu, L., Lin, X., Zhao, X.: Unsupervised metric fusion over multiview data by graph random walk-based cross-view diffusion. IEEE Trans. Neural Netw. Learn. Syst. **28**(1), 57–70 (2017)
16. Wu, L., Wang, Y., Gao, J., Li, X.: Deep adaptive feature embedding with local sample distributions for person re-identification. Pattern Recognit. **73**, 275–288 (2018)
17. Wu, L., Wang, Y., Ge, Z., Hu, Q., Li, X.: Structured deep hashing with convolutional neural networks for fast person re-identification. Comput. Vis. Image Underst. **167**, 63–73 (2018)
18. Wu, L., Wang, Y., Li, X., et al.: Deep attention-based spatially recursive networks for fine-grained visual recognition. IEEE Trans. Cybern. **PP**(99), 1–12 (2018)
19. Wu, L., Wang, Y., Li, X., Gao, J.: What-and-where to match: deep spatially multiplicative integration networks for person re-identification. Pattern Recognit. **76**, 727–738 (2018)
20. Yang, D., Shen, C., Lee, W., Chen, M.: On socio-spatial group query for location-based social networks. In: 18th ACM SIGKDD International Conference on Knowledge Discovery and Data Mining, KDD 2012, Beijing, China, 12–16 August 2012, pp. 949–957 (2012)
21. Zhang, C., Zhang, Y., Zhang, W., Lin, X.: Inverted linear quadtree: efficient top K spatial keyword search. In: 29th IEEE International Conference on Data Engineering, ICDE 2013, Brisbane, Australia, 8–12 April 2013, pp. 901–912 (2013)
22. Zhang, C., Zhang, Y., Zhang, W., Lin, X.: Inverted linear quadtree: efficient top K spatial keyword search. IEEE Trans. Knowl. Data Eng. **28**(7), 1706–1721 (2016)
23. Zhang, C., Zhang, Y., Zhang, W., Lin, X., Cheema, M.A., Wang, X.: Diversified spatial keyword search on road networks. In: Proceedings of the 17th International Conference on Extending Database Technology, EDBT 2014, Athens, Greece, 24–28 March 2014, pp. 367–378 (2014)
24. Zhang, D., Chee, Y.M., Mondal, A., Tung, A.K.H., Kitsuregawa, M.: Keyword search in spatial databases: towards searching by document. In: Proceedings of the 25th International Conference on Data Engineering, ICDE 2009, 29 March–2 April 2009, Shanghai, China, pp. 688–699 (2009)

A Hybrid Index Model for Efficient Spatio-Temporal Search in HBase

Chengyuan Zhang[1,2], Lei Zhu[1,2(✉)], Jun Long[1,2], Shuangqiao Lin[1,2],
Zhan Yang[1,2], and Wenti Huang[1,2]

[1] School of Information Science, Central South University,
Changsha, People's Republic of China
{cyzhang,leizhu,jlong,linshq,zyang22,174601025}@csu.edu.cn
[2] Big Data and Knowledge Engineering Institute, Central South University,
Changsha, People's Republic of China

Abstract. With advances in geo-positioning technologies and geo-location services, there are a rapidly growing massive amount of spatio-tempor-al data collected in many applications such as location-aware devices and wireless communication, in which an object is described by its spatial location and its timestamp. Consequently, the study of spatio-temporal search which explores both geo-location information and temporal information of the data has attracted significant concern from research organizations and commercial communities. This work study the problem of spatio-temporal k-nearest neighbors search (STkNNS), which is fundamental in the spatial temporal queries. Based on HBase, a novel index structure is proposed, called **H**ybrid Spatio-**T**emporal HBase **I**ndex (**HSTI** for short), which is carefully designed and takes both spatial and temporal information into consideration to effectively reduce the search space. Based on HSTI, an efficient algorithm is developed to deal with spatio-temporal k-nearest neighbors search. Comprehensive experiments on real and synthetic data clearly show that HSTI is three to five times faster than the state-of-the-art technique.

1 Introduction

Massive amount of data, which include both geo-location and temporal information, are being generated at an unparalleled scale on the Web. For example, more than 3.2 Billion comments have been posted to Facebook every day [1], while more than 400 million daily tweets containing texts and images [1–4] have been generated by 140 million twitter active users [4]. Combined with the advances in location-aware devices [5–7] (GPS-enabled devices, RFIDs, etc.) and wireless communication, spatio-temporal data storage and processing have entered a new age. There is an increasing demand to manage spatio-temporal data in many applications, such as wireless sensor networks (WSNs) [8,9], spatio-temporal multimedia retrieval [10–12]. Consequently, the study of spatio-temporal search which explores both geo-location information and temporal information of the

M. Ganji et al. (Eds.): PAKDD 2018, LNAI 11154, pp. 108–120, 2018.
https://doi.org/10.1007/978-3-030-04503-6_9

data has attracted significant concern from research organizations and commercial communities.

This work investigates the problem of spatio-temporal k-nearest neighbors search (STkNNS), which is applied in a variety of applications, such as spatio-temporal database management systems (STDBM), location based web service, spatio-temporal information based recommending system and industrial detection system. For example, there is a pipe leakage occurs in residential area, to confirm the leakage location and leakage reason as soon as possible, the water inspector has to query the spatio-temporal data for a given region and given time interval. However, there are lots of data satisfy the given spatio-temporal constraint sometimes. Thus, the water inspector might wish to limit the query answer to k nearest spatio-temporal data.

Challenges. There are three key challenges in spatio-temporal k-NN query. Firstly, vast amount of data, typically in the order of TB scale or even PB scale, are uploaded to the service. Thus, the cloud storage systems, which can exploit a distributed hash table (DHT) approach to index data, should be adopted. Secondly, current cloud storage systems, such as HBase, provide a key-value store system, but they cannot maturely extend to support multi-attribute queries, which greatly restrict their application. Hence, it is important to design a transformation mechanism to convert multi-dimension value to one-dimension value. Thirdly, novel techniques need to be created to design spatio-temporal indexing scheme that supports spatial pruning and temporal pruning synchronously.

Based on the above observation, a novel index technique is proposed, namely **H**ybrid **S**patio-**T**emporal HBase **I**ndex (**HSTI** for short), to effectively organize spatio-temporal data. In brief, HSTI is a two-layered structure which follows the retrieval mechanism of HBase. In the first layer, the whole space is partitioned into equal-size cells, and a space filling curve technique, Z-order, is employed to map these two-dimensional spaces to one-dimensional sequence number. In the second layer, to effectively partition the spatio-temporal data, a three-dimensional tree structure is designed, named Z-Octree.

Contributions. The principle contributions of this work are summarized as follows.

- A novel Hybrid Spatio-Temporal Hbase Index is devised to deal with the problem of spatio-temporal query. As far as we know, this work is the first spatio-temporal indexing mechanism which integrates the spatial and temporal information during index construction.
- Based on HSTI, an efficient spatio-temporal k-nearest neighbors query algorithm is developed.
- Comprehensive experiments on real and synthetic datasets demonstrate that our new index achieve substantial improvements over the state-of-the-art technique.

Roadmap. The rest of the paper is organized as follows. Section 2 introduces related work. Section 3 describes the data model, and the index structure. Section 4 presents algorithms and optimization strategies for search and refinement. Extensive experiments are depicted in Sect. 5. Finally, Sect. 6 concludes the paper.

2 Related Work

With the emergence of the era of big data [13,14], relational DBMSs are incompetent with the increasing volume of data because of low insertion rate and insufficient scalability. Therefore, to manage and process multidimensional spatial data efficiently, tree-structured spatial indices, such as R-Tree [15], R*-Tree [16], Quad-Tree [9,17], Kd-Tree [18], are widely used in traditional DBMSs.

Recently, spatial data with temporal attribute becomes one of the largest volumes of data collected by web services. Traditional relational DBMSs can no longer handle the quantity, and thus some researchers study on many NoSQL database implementations for scaling datastores horizontally. Fox et al. [19] presented a spatio-temporal index built on top of Accumulo to store and search spatio-temporal data sets efficiently.

To processing large volumes of data, SpatialHadoop [20,21] and Hadoop-GIS [22,23], which are based on MapReduce, are widely used. These systems can efficiently support high-performance spatial queries. However, they cannot directly be used in real-time system, as they do not take the temporal constraints into consideration.

Spatial indices have also been extended to NoSQL-based solutions. To partition the space on top of HBase, Nishimura et al. [24] built a multi-dimensional index layer, called MD-HBase, by using multidimensional index structures (Kd Tree and Quad-Tree). Linearization techniques (Z-ordering [25]) are used to convert multidimensional data to one dimension. However, MD-HBase does not index inner storage structure of slave nodes, and only provides an index layer in the META table. Hence, full scan operations are executed in each slave node, which reduces its efficiency.

Hsu et al. [26] presented a novel key formulation scheme for spatial index in HBase, called KR+-tree. R+-tree is used to divide the data into disjoint rectangles, while gird is used for further division. Then, Hilbert-curve is exploited to encode the grid cells. During processing, KR+-tree first searches the rectangle cells, which satisfied the query constraint, in the KeyTable. Then, it finds the corresponding data according to the rectangles. However, the scan operations still need to be executed in slave nodes, because the lookup mechanism of HBase is not considered.

Zhang et al. [27] proposed scalable spatial data storage based on HBase, called HBaseSpatial. Compared with MongoDB and MySQL, HBase has better performance while searching complex spatial data, especially searching MultiLingString and LingString data types. But this storage model can only support spatial queries in HBase.

To the best of our knowledge, [28] is the only work, which takes both spatial information and temporal information into consideration. In [28], Chen et al. proposed a spatio-temporal index scheme, called STEHIX (Spatio-TEmporal HBase IndeX) based on HBase. However, the spatial information and temporal information are not considered as an entirety.

3   Model and Structure

This section first presents a description of problem, then gives an overview of storage model based on HBase, last the proposed index structure HSTI, which based on the two-level lookup mechanism of HBase, is introduced. Table 1 summarizes the mathematical notations used throughout this paper to facilitate the discussion of our study.

Table 1. The summary of notations

Notation	Definition
O	A given data set of spatio-temporal data
o_i	A spatio-temporal data object
o_{id}	The identification of an object
x_i	The longitude of a spatio-temporal data
y_i	The latitude of a spatio-temporal data
t_i	The timestamp of a spatio-temporal data
p	The location point of an object o_i
z_n	The value of Morton order in Z-order curve
L	The deepest level of Z-Octree
ξ	The division threshold value for a node in Z-Octree
v	The Morton order of a leaf node in Z-Octree
q	A spatio-temporal k-NN search
k	The nearest neighbors' number of a k-NN search
δ_e	The Euclidean distance between two points in spatial space

3.1   Problem Description

Definition 1 (Spatio-temporal object). *A spatio-temporal object can be represented as the form o_i (o_id, x_i, y_i, t_i), where o_{id} is the identification of the object that has a spatial location p, including longitude and latitude (x_i, y_i) along x and y dimensions at the timestamp t_i.*

Definition 2 (Spatio-temporal k-nearest neighbors search). *Given a set O of spatio-temporal objects, a spatio-temporal k-nearest neighbors query is denoted as $q(x_q, y_q, [t_{start}, t_{end}], k)$, where (x_q, y_q) is the query spatial location and $([t_{start}, t_{end}])$ is the query temporal interval. This work aims to find a set $O(q) \subseteq O \cap |O(q)| = k$, and for each location point of $o \in O(q)$, $\delta_e(p, (x_q, y_q)) \leq \delta_e(p', (x_q, y_q)), \forall p \in O(q), p' \in O\varnothing(q)$, and p.t, p'.t $\in [t_{start}, t_{end}]$, where the δ_e is the Euclidean distance.*

3.2   Index Structure

This section introduces the novel two-layer index structure, namely HSTI, which is based on the two-level internal lookup mechanism in HBase. The first layer index in HSTI can achieve high efficiency of servers routing in HBase cluster. With the help of the second layer index in HSTI, the distributed retrievals in involved RegionServers can be completed immediately. Figure 1 shows the overall structure of HSTI. The details of index design are described as below.

The First Layer: *META* Table Design. This subsection induces the first level of HSTI. Firstly, the whole spatial space is divided into non-overlapping cells, in which spatio-temporal data points are distributed based on their spatial locations. The linearization technique Z-order curve [25] is used to map data points from multi-dimension to one-dimension. As will be shown in Sect. 4, the reason why choosing Z-order curve is that the proposed index should have the special property: given any two-dimensional rectangle, the bottom leftmost point and the top rightmost point should be mapped as the minimum value and the maximum value respectively. All the points within the specified rectangle are distributed in this region.

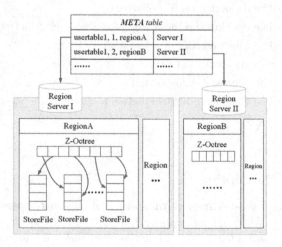

Fig. 1. The overall structure of HSTI

Given a data point with spatio-temporal information, the spatial Z-order curve value of the point can be calculated by its longitude and latitude (x_i, y_i). According to the hash principle of HBase, the prefix of row key should be different so that it can improve the insertion rate over different RegionServers and increase the chance of load balancing. Meanwhile, the data points that are spatially close should share the same prefix in row key, thus these points can be stored adjacently by their spatial proximity. Therefore, as the first layer to index spatio-temporal data, Z-order curve value Zn is used as row key in the *META* table, and then each data point can be placed into the corresponding RegionServer.

The Secound Layer: Z-Octree Structure. For searching local data efficiently, a novel indexing structure, namely Z-order Octree (Z-Octree), is proposed as the second layer in HSTI. Z-Octree is an in-memory list-like structure, kept in each RegionServer. Each entry in Z-Octree records a list of addresses pointing to spatio-temporal data in StoreFiles.

In Z-Octree, the spatio-temporal space is considered as a three-dimensional space, in which all the data points are mapped by their spatial and temporal information. In the three-dimensional data processing, octree structure is an efficient storage method so that octree is adopted here to store the code values. In octree, each non-leaf nodes have 8 sub-nodes as show in Fig. 5(a). In [18], a three-dimensional space can be recursively partitioned into uniform 8^{L-1} subspaces where L is the level of the partition. Each subspace is assigned a Morton value based on its visiting order. Nevertheless, unlikely with the traditional process, a more applicable strategy is proposed to construct an octree (called Z-Octree).

To keep the Z-Octree adaptive, it is constructed based on the data dense and skewness. An inner subspace would divide into 8 subspaces only when there are sufficient points in it. Hence in our Z-Octree structure, the leaf nodes can exist on different levels which is different with traditional octree. When construct the Z-Octree, the whole spatio-temporal space is firstly processed as a root node, and then if a space contains more than ξ points, it will be recursively divided into 8 subspaces. As shown in Fig. 2(a), the whole space is divided into eight sub-nodes, and only one of them satisfies the division threshold value ξ. From this way, Z-Octree can easily handle the skew data and eliminate the hotspot efficiently.

The generation of Morton order value for each leaf node is described as follows. For the given deepest level L of the octree, we assume that the whole spatio-temporal space can be divided into 8^{L-1} virtual subspaces and the Morton order value of each virtual subspace is calculated [25]. In Z-Octree, it does not have to implement each virtual subspace in sense that, if the number of a node does not reach the division threshold value ξ, this node will not be divided. The minimum of virtual subspace, which covers the not divided leaf node, is used as the Morton value v of this specified node. Figure 2(b) shows an example of Z-Octree built from Fig. 2(a) which has a deepest level 3. The Morton value of each leaf node is denoted in the corresponding circle.

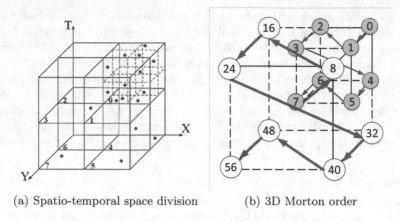

(a) Spatio-temporal space division   (b) 3D Morton order

Fig. 2. An example of space division and 3D Morton order

4   *K*-Nearest Neighbors Search Algorithm

A spatio-temporal *k*-NN query q is defined in Subsect. 3.1. Usually, searching the *k*-nearest neighbors in spatio-temporal dimension is more difficult than the search of spatial *k*-NN. Because the restrictions on both spatial and temporal information should be considered.

A *k*-NN query algorithm based on HSTI is proposed. For a given spatial location (x_q, y_q), the Z-ordering space Zn containing (x_q, y_q) is computed in first layer of HSTI. Then the Z-Octree in the corresponding RegionServer is utilized to retrieve a list of data points. Meanwhile, the adjacent spaces of Z_n are also computed. A priority queue Q is maintained for all the points and adjacent spaces, where priority metric is the distance from location (x_q, y_q) to point or Z-ordering space. The element in priority queue is constantly dequeued, either being processed to result list or being retrieved to obtain adjacent points to be enqueued, until *k*-nearest neighbors are found.

Algorithm 1 is the *k*-NN query pseudocode. The input is query $q_k = (x_q, y_q, [t_{start}, t_{end}], k)$, and the output is the *k*-nearest points of the spatial location (x_q, y_q) during $[t_{start}, t_{end}]$. In line 2, a priority queue Q is initialized to order the elements which are enqueued by the distance. The computed Z-ordering space z_n is first enqueued to Q (line 4), while the adjacent spaces of Z_n are also found and enqueued where the priority metric is the distance MINDIST [29]. Then elements in Q are constantly dequeued and processed in line 6. If the element is a Z-ordering space, RegionSearch (line 12) would be executed in each involved RegionServer.

Since Z-Octree is used for space partition in each RegionServer, the query of *k*-nearest neighbors will be transformed into the search of Z-Octree node's neighbors. Each leaf node in the Z-Octree corresponds to a sub-cube in the

---

Algorithm 1. k Nearest Neighbors Search

Input: $x_q, y_q, t_{start}, t_{end}, k$.
Output: S :list of k-nearest neighbors.
 1: $S \leftarrow \emptyset; Z_n \leftarrow \emptyset; CoveringCubes \leftarrow \emptyset; AdjacentSpaces \leftarrow \emptyset; AdjacentCubes \leftarrow \emptyset;$
    $e \leftarrow \emptyset; AS \leftarrow \emptyset;$
 2: $Q \leftarrow CreatePriorityQueue();$
 3: $Z_n \leftarrow getZorderingSpace(x_q, y_q);$
 4: Enqueue(Z_n, MINDIST(($x_q, y_q), Z_n$), Q);
 5: **while** $Q \neq \emptyset$ **do**
 6:   $e \leftarrow Dequeue(Q);$
 7:   **if** e is typeof Zordering space **then**
 8:     $AdjacentSpaces \leftarrow getAdjacentSpaces(e);$
 9:     **for** each $AdjacentSpace AS \in AdjacentSpaces$ **do**
10:       Enqueue(AS, MINDIST(($x_q, y_q), AS$), Q);
11:     **end for**
12:     RegionSearch($e, x_q, y_q, t_{start}, t_{end}, k$);
13:   **end if**
14:   **if** e is typeof cube **then**
15:     $PointsSet \leftarrow getPoints(e);$
16:     **for** each $point \in PointsSet$ **do**
17:       Enqueue($point$, $\delta_e((x_q, y_q), point)$, Q);
18:     **end for**
19:   **end if**
20:   **if** e is typeof point **then**
21:     Add e into S;
22:     **if** $S.length = k$ **then**
23:       **return** S;
24:     **end if**
25:   **end if**
26: **end while**

---

spatio-temporal space. If cubes are adjacent, then the nodes in octree are neighbors with each other. The k-nearest neighbors of a spatial location (x_q, y_q) are searched not only in the covering cube, but also in all the cubes adjacent to the cube within $[t_{start}, t_{end}]$. Figure 3 illustrates all cubes that need to be searched in Z-Octree.

As shown in Algorithm 1, if the element e dequeued from Q is a cube, the algorithm keeps enqueue the points in the cube by the distance line in 15 to 18. Otherwise, if e is a point, which means the element is a result, it would be added into the result list S (line 21). The algorithm described above is looped until the length of list S reaches k. The results of query q_k are obtained in S.

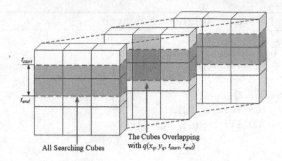

Fig. 3. The searching cubes in Z-Octree

5   Experiments

Experiment Settings. In the experiments, the Zookepper was responsible for coordination and synchronization in HBase cluster, the locations of all datasets were scaled to the two-dimensional space [0, 10000][0, 10000], and the timestamp of all datasets were scaled to [0, 5000]. In addition, the top k value changed from 10 to 500, and cluster size varied from 2 to 8. By default, top k value and cluster size were set to 100, 4 respectively. For accurate analyzing and evaluating, STEHIX was chosen as baseline.

Datasets. Three different datasets were used in the experiments, one of which was a synthetic uniform dataset (UN for short) generated by program, and others were real-world datasets, geolife project [30] (GL for short) and T-Drive [31,32] (TD for short).

Evaluation on Different Datasets. We investigate the query response time, index construction time and index size of 2 algorithms against three datasets Tigers, GL, TD and UN, where other parameters are set to default values. Figure 4(a) depicts the rate of space occupying by the index sizes. The baseline requires more space due to the two kinds of indices (called *s-index* and *t-index*) kept for all the entries and the storage cost of it increases faster in larger datasets. In contrast, an index record is maintained for each entry only once so that our index saves more space in memory. Figure 4(b) shows the difference of construction time between HSTI and the baseline. Due to the simple split and code algorithms of an Z-Octree, HSTI has a shorter constructing time as compared to the baseline, which need to traverse two indices during the construction. In Fig. 4(c), HSTI demonstrates superior performance in comparison with baseline in all datasets (Table 2).

Evaluation on the Effect of the Number of Results k. Figure 5 report the query response time of the algorithms as a function of k on dataset UN. As expected, the performance of all algorithms degrades regarding the increase of k (i.e., the larger search region). And the performance of HSTI always outperforms

Table 2. Dataset statistics

Property	UN	GL	TD
Number of records (millions)	20	25.84	17.76
Size of dataset (GB)	0.74	1.54	0.71

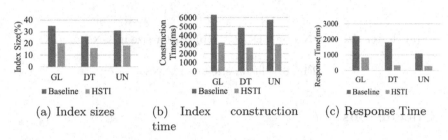

(a) Index sizes   (b) Index construction   (c) Response Time
time

Fig. 4. Performance on various datasets

the baseline for both high-density and low-density points. The reasons are as follow. First of all, the retrieval procedures of baseline have been divide into two parts: *s-index* and *t-index*. Thus, each information extraction will decompose into two processes to collect results in temporal dimension and spatial dimension, which will provoke more I/O costs. On the other hand, the periodicity of *t-index* gives rise to lots of discrete timestamps are mixed together.

Evaluation on the Effect of Cluster Size. As shown in Fig. 6(a) and (b), with cluster size increased, running time decreased gradually. More clusters means less data on each sever. Obviously, the processing time will decline. Meanwhile, we also observe that the performance of k-NN query on uniform dataset UN is better than that of real-world dataset TD. This is because the data distribution of TD might be inhomogenous, and there may be some hotpot in TD.

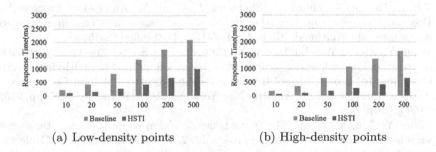

(a) Low-density points   (b) High-density points

Fig. 5. Performance effected by k on UN

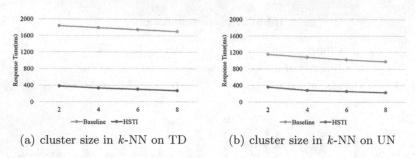

(a) cluster size in k-NN on TD   (b) cluster size in k-NN on UN

Fig. 6. Cluster size in STkNNS

6 Conclusions

The problem of spatio-temporal k-NN search is important due to the increasing amount of spatio-temporal data collected in a wide spectrum of application. The proposed hybrid spatio-temporal index scheme is based on the two-level internal lookup mechanism in HBase. Based on HSTI, an efficient algorithm is developed to support spatio-temporal k-NN search. Our comprehensive experiments on real and synthetic data clearly show that HSTI is able to achieve a reduction of the processing time by 60–80% compared with prior state-of-the-art methods.

Acknowledgments. This work was supported in part by the National Natural Science Foundation of China (61702560, 61379110, 61472450), the Key Research Program of Hunan Province (2016JC2018), Natural Science Foundation of Hunan Province (2018JJ3691), and Science and Technology Plan of Hunan Province (2016JC2011).

References

1. Wang, Y., Lin, X., Wu, L., Zhang, W.: Effective multi-query expansions: collaborative deep networks for robust landmark retrieval. IEEE Trans. Image Process. **26**(3), 1393–1404 (2017)
2. Wu, L., Huang, X., Zhang, C., Shepherd, J., Wang, Y.: An efficient framework of Bregman divergence optimization for co-ranking images and tags in a heterogeneous network. Multimed. Tools Appl. **74**(15), 5635–5660 (2015)
3. Wang, Y., Zhang, W., Wu, L., Lin, X., Fang, M., Pan, S.: Iterative views agreement: an iterative low-rank based structured optimization method to multi-view spectral clustering. In: Proceedings of the Twenty-Fifth International Joint Conference on Artificial Intelligence, IJCAI 2016, New York, NY, USA, 9–15 July 2016, pp. 2153–2159 (2016)
4. Wang, Y., Lin, X., Zhang, Q.: Towards metric fusion on multi-view data: a cross-view based graph random walk approach. In: 22nd ACM International Conference on Information and Knowledge Management, CIKM 2013, San Francisco, CA, USA, 27 October–1 November 2013, pp. 805–810 (2013)
5. Zhang, C., Zhang, Y., Zhang, W., Lin, X.: Inverted linear quadtree: efficient top K spatial keyword search. IEEE Trans. Knowl. Data Eng. **28**(7), 1706–1721 (2016)

6. Wu, L., Wang, Y., Li, X., Gao, J.: What-and-where to match: deep spatially multiplicative integration networks for person re-identification. Pattern Recognit. **76**, 727–738 (2018)

7. Wu, L., Wang, Y., Ge, Z., Hu, Q., Li, X.: Structured deep hashing with convolutional neural networks for fast person re-identification. Comput. Vis. Image Underst. **167**, 63–73 (2018)

8. Liu, A., Liu, X., Long, J.: A trust-based adaptive probability marking and storage traceback scheme for wsns. Sensors **16**(4), 451 (2016)

9. Wu, L., Wang, Y., Gao, J., Li, X.: Deep adaptive feature embedding with local sample distributions for person re-identification. Pattern Recognit. **73**, 275–288 (2018)

10. Wang, Y., Zhang, W., Wu, L., Lin, X., Zhao, X.: Unsupervised metric fusion over multiview data by graph random walk-based cross-view diffusion. IEEE Trans. Neural Netw. Learn. Syst. **28**(1), 57–70 (2017)

11. Wang, Y., Lin, X., Wu, L., Zhang, W., Zhang, Q., Huang, X.: Robust subspace clustering for multi-view data by exploiting correlation consensus. IEEE Trans. Image Process. **24**(11), 3939–3949 (2015)

12. Wang, Y., Wu, L., Lin, X., Gao, J.: Multiview spectral clustering via structured low-rank matrix factorization. IEEE Trans. Neural Netw. Learn. Syst. **29**(10), 4833–4843 (2018)

13. Wang, Y., Wu, L.: Beyond low-rank representations: orthogonal clustering basis reconstruction with optimized graph structure for multi-view spectral clustering. Neural Netw. **103**, 1–8 (2018)

14. Wu, L., Wang, Y., Li, X., et al.: Deep attention-based spatially recursive networks for fine-grained visual recognition. IEEE Trans. Cybern. **PP**(99), 1–12 (2018)

15. Guttman, A.: R-trees: a dynamic index structure for spatial searching. In: SIGMOD 1984, Proceedings of Annual Meeting, Boston, Massachusetts, 18–21 June 1984, pp. 47–57 (1984)

16. Beckmann, N., Kriegel, H., Schneider, R., Seeger, B.: The R*-tree: an efficient and robust access method for points and rectangles. In: Proceedings of the 1990 ACM SIGMOD International Conference on Management of Data, Atlantic City, NJ, 23–25 May 1990, pp. 322–331 (1990)

17. Finkel, R.A., Bentley, J.L.: Quad trees: a data structure for retrieval on composite keys. Acta Inf. **4**, 1–9 (1974)

18. Brown, R.A.: Building a balanced k-d tree in O(kn log n) time. CoRR abs/1410.5420 (2014)

19. Fox, A.D., Eichelberger, C.N., Hughes, J.N., Lyon, S.: Spatio-temporal indexing in non-relational distributed databases. In: Proceedings of the 2013 IEEE International Conference on Big Data, Santa Clara, CA, USA, 6–9 October 2013, pp. 291–299 (2013)

20. Eldawy, A., Mokbel, M.F.: A demonstration of spatialhadoop: an efficient mapreduce framework for spatial data. PVLDB **6**(12), 1230–1233 (2013)

21. Wang, Y., Lin, X., Wu, L., Zhang, W., Zhang, Q.: Exploiting correlation consensus: towards subspace clustering for multi-modal data. In: Proceedings of the ACM International Conference on Multimedia, MM 2014, Orlando, FL, USA, 03–07 November 2014, pp. 981–984 (2014)

22. Aji, A., et al.: Hadoop-GIS: a high performance spatial data warehousing system over mapreduce. PVLDB **6**(11), 1009–1020 (2013)

23. Wang, Y., Lin, X., Zhang, Q., Wu, L.: Shifting hypergraphs by probabilistic voting. In: Tseng, V.S., Ho, T.B., Zhou, Z.-H., Chen, A.L.P., Kao, H.-Y. (eds.) PAKDD 2014. LNCS (LNAI), vol. 8444, pp. 234–246. Springer, Cham (2014). https://doi.org/10.1007/978-3-319-06605-9_20

24. Nishimura, S., Das, S., Agrawal, D., El Abbadi, A.: MD-HBase: a scalable multidimensional data infrastructure for location aware services. In: 12th IEEE International Conference on Mobile Data Management, MDM 2011, Luleå, Sweden, 6–9 June 2011, vol. 1, pp. 7–16 (2011)

25. Morton, M.G.: A Computer Oriented Geodetic Data Base and a New Technique in File Sequencing. International Business Machines Company, New York (1966)

26. Hsu, Y., Pan, Y., Wei, L., Peng, W., Lee, W.: Key formulation schemes for spatial index in cloud data managements. In: 13th IEEE International Conference on Mobile Data Management, MDM 2012, Bengaluru, India, 23–26 July 2012, pp. 21–26 (2012)

27. Zhang, N., Zheng, G., Chen, H., Chen, J., Chen, X.: HBaseSpatial: a scalable spatial data storage based on HBase. In: 13th IEEE International Conference on Trust, Security and Privacy in Computing and Communications, TrustCom 2014, Beijing, China, 24–26 September 2014, pp. 644–651 (2014)

28. Chen, X.Y., Zhang, C., Ge, B., Xiao, W.D.: Efficient historical query in HBase for spatio-temporal decision support. Int. J. Comput., Commun. Control. 11(5), 613–630 (2016)

29. Roussopoulos, N., Kelley, S., Vincent, F.: Nearest neighbor queries. In: Proceedings of the 1995 ACM SIGMOD International Conference on Management of Data, San Jose, California, 22–25 May 1995, pp. 71–79 (1995)

30. Zheng, Y., Xie, X., Ma, W.: GeoLife: a collaborative social networking service among user, location and trajectory. IEEE Data Eng. Bull. 33(2), 32–39 (2010)

31. Yuan, J., Zheng, Y., Xie, X., Sun, G.: Driving with knowledge from the physical world. In: Proceedings of the 17th ACM SIGKDD International Conference on Knowledge Discovery and Data Mining, San Diego, CA, USA, 21–24 August 2011, pp. 316–324 (2011)

32. Yuan, J., et al.: T-Drive: driving directions based on taxi trajectories. In: Proceedings of 18th ACM SIGSPATIAL International Symposium on Advances in Geographic Information Systems, ACM-GIS 2010, San Jose, CA, USA, 3–5 November 2010, pp. 99–108 (2010)

Rumor Detection via Recurrent Neural Networks: A Case Study on Adaptivity with Varied Data Compositions

Tong Chen[✉], Hongxu Chen, and Xue Li

The University of Queensland, Brisbane, Australia
{tong.chen,hongxu.chen}@uq.edu.au, xueli@itee.uq.edu.au

Abstract. Rumor detection is a meaningful research problem due to its significance in preventing potential threats to cyber security and social stability. With the recent popularity of recurrent neural networks (RNNs), the application of RNNs in rumor detection has resulted in promising results as RNNs can naturally blend into the task of language processing and sequential data modelling. However, since deep learning models require large data scale for training in order to extract sufficient distinctive patterns, their adaptivity with varied data compositions can become a challenge for real-life application scenarios where rumors are always the minority (outlier) in the streaming data. In this paper, we present a case study to investigate how the ratio of rumors in training data affects the calssification performance of RNN based rumor detection models and successfully address some issues on the model adaptivity.

Keywords: Recurrent neural networks · Rumor detection

1 Introduction

Originated from the research on information credibility on Twitter [1], rumor detection on social media has been a significant yet unsolved problem. With the rapid development of contemporary social media platforms in recent years, the increasing range and speed of information diffusion on social media further allows the widespread of rumors. As a typical instance of rumors' threat to cyber security and social stability, on April 23rd 2013, massive social panic was triggered by a tweet posted by a hacked Twitter account named Associated Press. The rumor claimed that two explosions happened in the White House and Barack Obama got injured. As demonstrated in Fig. 1, despite the quick clarification reaction from both the White House and Associated Press, the fast diffusion to millions of users had resulted in a huge loss of $136.5 billion in the stock market[1]. This real case reflects both the difficulty and necessity in accurate rumor detection.

---

[1] http://www.dailymail.co.uk/news/article-2313652/AP-Twitter-hackers-break-news-White-House-explosions-injured-Obama.html.

© Springer Nature Switzerland AG 2018
M. Ganji et al. (Eds.): PAKDD 2018, LNAI 11154, pp. 121–127, 2018.
https://doi.org/10.1007/978-3-030-04503-6_10

Rumor detection is commonly formulated as binary classification tasks in related researches using features extracted from textual comments, user profiles and information propagation patterns. Among these rumor detection approaches, text based detection techniques are most studied. The main reasons are: (1) textual features are relatively stable in contrast to user profiles [2] and are free from sensitive data; (2) feature extraction from user posts is more efficient compared with propagation graph construction and users' post sequences also contain supplementary information of rumor dissemination. In regards to text analysis and time series processing, recurrent neural networks (RNNs) have showcased considerable superiority [3,4] due to its capability of handling sequential inputs as well as bypassing the sophisticated feature engineering stage [5].

In this paper, we aim to examine the adaptivity of RNN based rumor detection models in **real-life application scenario**. According to [6], anomalous information only takes very low proportions on social media (usually around 5%), thus can be regarded as outliers. However, previous studies tend to balance the data for both model training, thus being impractical in industrial applications as the streaming data collected in real time is apparently imbalanced. Thus, in this case study, we set some empirical benchmarks for RNN based rumor detection with varied rumor-to-non-rumor ratio within the training dataset.

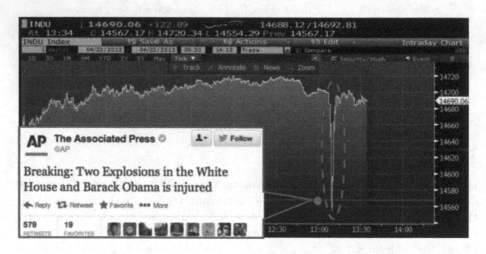

Fig. 1. A real case of 'White House Explosion' rumor and its impact.

2   Related Work

Our work is inspired by recent works focusing on text based rumor detection. The extraction and selection of discriminative features significantly affects the performance of the classifier. Hu *et al.* first conducted a study to analyze the sentiment differences between spammers and normal users and then presented an optimization formulation that incorporates sentiment information into a novel social spammer detection framework [7]. Similarly, in [8], some very rare but informative enquiry phrases related to a specific topic are identified using pre-defined rules, and such features increase the detection performance by a noticeable margin. More recently, the dynamics of rumor propagation has been emphasised in order to detect viral rumors in their formative stages to allow early action [9,10]. For example, a time-series based feature structure [11] is introduced to seize context variation over time. More recently, recurrent neural network was first introduced to rumor detection by Ma *et al.* [5], utilizing sequential user-generated texts to timely capture temporal textual characteristics during information diffusion. In [12], textual attention mechanism is embedded into the RNN model, which helps detecting rumor earlier with higher accuracy.

However, similar to opinion based community detection [13], since the acquisition of ground truth relies on professional human verification, when being involved in the real-life application without abundant labelled data for rumors, the performance of these methods may vary significantly due to the lack of distinguishable and large-scale patterns of rumor posts, especially RNNs. In the following chapters, we conduct an empirical case study on RNN model's adaptivity with varied data compositions in the training phase.

3   Ground Truth and Network Structure

In this section, we present instructions on the experimental dataset and the network structure we use in the experiments.

3.1   Dataset

As we aim to determine whether a trending event (topic) is a rumor or not by investigating related posts published by social media users in a time series, we use the dataset condensed from the public Twitter dataset in [5]. The data collection and verification process is demonstrated in Fig. 2. In the Twitter dataset, 498 rumors are collected using the keywords extracted from verified fake news published on Snopes[2], a real-time rumor debunking website. It also contains 494 normal events from Snopes and two public datasets [1,14]. For one event, because we treat each post as the input for every time step, we limited the maximum number of posts in each event by 500 (all posts are sorted and truncated chronologically). The statistics of the dataset is listed in Table 1.

Table 1. Statistical details of the dataset. PPE stands for posts per event.

Total posts	Total events	Rumors	Non-rumors	Avg. PPE	Min. PPE	Max. PPE
175,783	992	498	494	177	8	500

---

[2] www.snopes.com.

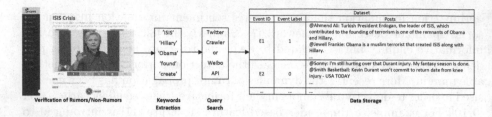

Dataset		
Event ID	Event Label	Posts
E1	1	@Ahmend Ali: Turkish President Erdogan, the leader of ISIS, which contributed to the founding of terrorism is one of the remnants of Obama and Hillary. @Jewell Frankie: Obama is a muslim terrorist that created ISIS along with Hillary.
E2	0	@Sonny: I'm still hurting over that Durant injury. My fantasy season is done. @Smith Basketball: Kevin Durant won't commit to return date from knee injury - USA TODAY
...	...	...

Fig. 2. Data collection. For each event, its authenticity is verified through official news verification services. Then, suitable keywords for each event are determined to ensure a precise search result of relevant posts. After that, posts are crawled with query search and store our collected data using the data storage structure shown in the table. Rumor events are labelled as 1 and normal events are labelled as 0.

3.2 Network Structure

We employ long short-term memory (LSTM), a variation of RNN for better long-range dependency learning [15] to learn latent representations for rumors:

$$
\begin{aligned}
i_t &= \sigma(U_i h_{t-1} + W_i x_t + V_i c_{t-1} + b_i), \\
f_t &= \sigma(U_f h_{t-1} + W_f x_t + V_f c_{t-1} + b_f), \\
c_t &= f_t c_{t-1} + i_t \tanh(U_c h_{t-1} + W_c x_t + b_c), \\
o_t &= \sigma(U_o h_{t-1} + W_o x_t + V_o c_t + b_o), \\
h_t &= o_t \tanh(c_t),
\end{aligned}
\tag{1}
$$

where $\sigma(\cdot)$ is the logistic sigmoid function, and i_t, f_t, o_t, c_t are the input gate, forget gate, output gate and cell input activation vector, respectively. In each of them, there are corresponding input-to-hidden, hidden-to-output, and hidden-to-hidden matrices: U, V, W and the bias vector b. We used both **2-layer** (with 512 and 128 hidden states) and **3-layer** (with 1,024, 512 and 128 hidden states) LSTM stacks for the experiment. Similar to [12], for each time step t, we encode the sentence into a *tf-idf* vector with 10,000 vocabulary dimensions as the input feature vector x_t. Also, we employ cross-entropy loss with $l2$ regularization with **2 different settings**:

$$
\begin{aligned}
\mathcal{L} &= -\sum_{c=1}^{C} y_{t,c} \log \hat{y}_{t,c} + \gamma \phi^2, \\
\mathcal{L}' &= -\sum_{n=1}^{N} \left(w \sum_{c=1}^{C} y_{t,c,n} \log \hat{y}_{t,c,n} + \gamma \phi^2 \right),
\end{aligned}
\tag{2}
$$

where y_t is the one hot label represented by 0 and 1, $\hat{y}_t$ is the predicted binary class probabilities at the last time step t, $C = 2$ is the number of output classes, γ is the weight decay coefficient, and ϕ represents all the model parameters. To test the performance gain from skew data training strategy following [16], in $\mathcal{L}'$ we set w to be the inverse of class ratio for the ground truth y_t and N to be the total amount of samples in the training batch.

4   Empirical Evaluation and Conclusion

We evaluate the rumor detection performance using *Precision*, *Recall* and *F1 Score*. We randomly split the dataset with the ratio of 70%, 10% and 20% for training, validation and test respectively. In the training phase, we utilize all the non-rumor events and tested 4 increasing proportions of rumor events, specifically 5%, 20%, 35% and 50% among all the training data. Results for our empirical evaluation on 4 different RNN settings are shown in Fig. 3.

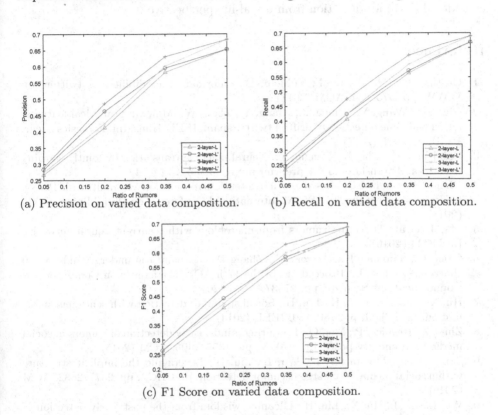

(a) Precision on varied data composition.   (b) Recall on varied data composition.

(c) F1 Score on varied data composition.

Fig. 3. Results for empirical evaluation.

Apparently, the performance gradually grows as the RNN models have more rumor samples for training and the highest Precision, Recall and F1 Score appears with the most complex RNN model with 3-layer LSTM stack and weighted loss. We can draw the observation from the results that RNN models fail to extract enough distinctive patterns for rumor detection when the rumors in training set is significantly less than non-rumors (5% and 20%). Also, adding hidden layers and applying weighted loss in $\mathcal{L}'$ both contributed to about 2.5% performance gain in the starting period, but the improvement is gradually weakened with the increase of the rumor samples in the training data.

With the performance of RNN based rumor detection model tested on varied data compositions, we notice that when facing real-life scenario, the lack of rumors in the training data may trigger significant drop in these models. The adaptivity on low rumor ratio is not satisfying, but once the amount of labelled rumors grows, continuous model accuracy and sensitivity can be acquired from RNNs. In summarize, we conclude that in order to ensure the adaptability of RNN models for applicable rumor detection, either more complex textual features or deeper neural network structures (e.g., attention mechanism) should be considered in rumor detection from a real-life perspective.

References

1. Castillo, C., Mendoza, M., Poblete, B.: Information credibility on twitter. In: WWW, pp. 675–684. ACM (2011)
2. Chen, C., Wang, Y., Zhang, J., Xiang, Y., Zhou, W., Min, G.: Statistical features-based real-time detection of drifted twitter spam. IEEE Trans. Inf. Forensics Secur. **12**(4), 914–925 (2017)
3. Bahdanau, D., Cho, K., Bengio, Y.: Neural machine translation by jointly learning to align and translate. arXiv preprint arXiv:1409.0473 (2014)
4. Chen, W., et al.: EEG-based motion intention recognition via multi-task RNNs. In: Proceedings of the 2018 SIAM International Conference on Data Mining, SIAM (2018)
5. Ma, J., et al.: Detecting rumors from microblogs with recurrent neural networks. In: IJCAI (2016)
6. Grier, C., Thomas, K., Paxson, V., Zhang, M.: @spam: the underground on 140 characters or less. In: Proceedings of the 17th ACM Conference on Computer and Communications Security, pp. 27–37. ACM (2010)
7. Hu, X., Tang, J., Gao, H., Liu, H.: Social spammer detection with sentiment information. In: ICDM, pp. 180–189. IEEE (2014)
8. Zhao, Z., Resnick, P., Mei, Q.: Enquiring minds: early detection of rumors in social media from enquiry posts. In: WWW, pp. 1395–1405. ACM (2015)
9. Sampson, J., Morstatter, F., Wu, L., Liu, H.: Leveraging the implicit structure within social media for emergent rumor detection. In: CIKM, pp. 2377–2382. ACM (2016)
10. Wu, L., Li, J., Hu, X., Liu, H.: Gleaning wisdom from the past: early detection of emerging rumors in social media. In: SDM (2016)
11. Ma, J., Gao, W., Wei, Z., Lu, Y., Wong, K.F.: Detect rumors using time series of social context information on microblogging websites. In: CIKM, pp. 1751–1754. ACM (2015)
12. Chen, T., Wu, L., Li, X., Zhang, J., Yin, H., Wang, Y.: Call attention to rumors: deep attention based recurrent neural networks for early rumor detection. arXiv preprint arXiv:1704.05973 (2017)
13. Chen, H., Yin, H., Li, X., Wang, M., Chen, W., Chen, T.: People opinion topic model: opinion based user clustering in social networks. In: WWW Companion, International World Wide Web Conferences Steering Committee, pp. 1353–1359 (2017)
14. Kwon, S., Cha, M., Jung, K., Chen, W., Wang, Y.: Prominent features of rumor propagation in online social media. In: ICDM, pp. 1103–1108. IEEE (2013)

15. Greff, K., Srivastava, R.K., Koutník, J., Steunebrink, B.R., Schmidhuber, J.: LSTM: a search space odyssey. IEEE Trans. Neural Netw. Learn. Syst. **28**, 2222–2232 (2016)
16. Rosenberg, A.: Classifying skewed data: importance weighting to optimize average recall. In: Thirteenth Annual Conference of the International Speech Communication Association (2012)

TwiTracker: Detecting and Extracting Events from Twitter for Entity Tracking

Meng Xu[✉], Jiajun Cheng, Lixiang Guo, Pei Li, Xin Zhang,
and Hui Wang

Science and Technology on Information Systems Engineering Laboratory,
College of Systems Engineering, National University of Defense Technology,
Changsha 410073, People's Republic of China
mengxu_simone@foxmail.com, {jiajun.cheng,
guolixiang10,peili,huiwang}@nudt.edu.cn,
ijunzhanggm@gmail.com

Abstract. Among the existing social platforms, Twitter plays a more and more important role in social sensing due to its real-time nature. In particular, it reports various types of events occurred in the real world, which provides us the possibility of tracking entities of interest (e.g., celebrities, organizations) in real time via event analysis. Hence, this paper presents TwiTracker, a system for obtaining the timelines of entities on Twitter. The system uses Twitter API and keyword search to collect tweets containing the entities of interest, and combines event detection and extraction together to extract elements including activities, time, location and participants. Online incremental clustering is further applied to fuse extraction results from different tweets to remove redundant information and enhance accuracy. Echarts is used to visualize the dynamic trajectory of each entity under tracking. For evaluation, we take Golden State Warriors, a famous NBA team, as well as the stars in the team as the experimental objects to compute their timelines, and compare the experimental results with the ground truth data hunted from the Internet, which demonstrates TwiTracker is effective for tracking entities and can provide information that is not covered by newswires.

Keywords: Twitter · TwiTracker · Event extraction · Event fusion

1 Introduction

Social media's popularity changes people's life enormously. Especially, it makes ordinary netizens play new roles of social sensors that combine the information producer, deliverer and consumer in a body. Among numerous social platforms, Twitter [1] has a distinct function of social sensing: global users update their status on Twitter frequently and most news portals register their own accounts and report events promptly on Twitter. Twitter's social sensing function has been fully demonstrated in major events such as the earthquakes in Japan [2], the floods in the Red River in the United States [3] and the Arab Spring Campaign [4]. All these lead Twitter to be a globally distributed social sensing network, which can be used to not only conduct

© Springer Nature Switzerland AG 2018
M. Ganji et al. (Eds.): PAKDD 2018, LNAI 11154, pp. 128–134, 2018.
https://doi.org/10.1007/978-3-030-04503-6_11

situation awareness but also track status of important entities (e.g., celebrities, organizations).

The application of conducting situation awareness has been extensively studied, while to track important entities via Twitter has not been well explored. Marcus et al. [5] create the Twitinfo system for visualizing and summarizing events on Twitter to build timelines of entities, but not provide a structured description of events. Popescu et al. [6] detect events involving known entities from Twitter and extract events, entities, actions and opinions, which nearly describe the event clearly. Li et al. [7] generate timelines for individuals from tweets published by themselves on Twitter. Hence, this paper presents the TwiTracker system focusing on the task of utilizing Twitter tracking entities of interest. To summarize, we make the following main contributions in this paper:

- We present the TwiTracker system that aggregates the data collection, event extraction, event fusion and visualization. It is a platform that provides timelines of entities in real time.
- We combine event detection and event extraction. This paper utilizes the identification of activity elements to detect events. Such combination improves the efficiency of algorithm and makes event detection more accurate.

The rest of the paper is organized as follows: We start with the overview and architecture of the TwiTracker system and our algorithm in detail. Then we carry on an experiment, discuss the result, and draw the conclusion.

2   The TwiTracker System

In this section, we first show the architecture of the TwiTracker framework, and then introduce the event extraction and event fusion methods serially. Finally, we use Echarts [8] to visualize the trajectory.

2.1   Overview and the Architecture

There are four modules in the architecture (Fig. 1) including tweets collection, event extraction, event fusion and visualization.

TwiTracker uses Twitter official API to search tweets containing one of keywords in the keyword list we have made for entities on Twitter. Then the simhash algorithm is used to remove the duplicates and the left English tweets are turned to the event extraction step. In the event extraction step, we extract activity, time, location and participant elements from collected data. After that, to remove the redundant information about an event and get more detailed description, event fusion is processed. Finally, the timelines of entities that contain the four elements at each node are visualized for users.

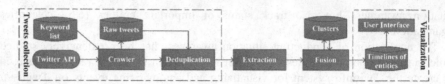

Fig. 1. The architecture of TwiTracker.

2.2 Event Extraction

The definition of events is defined firstly in this section. Then every step of event extraction is introduced, including the following four aspects: activity recognition, time recognition, location extraction and participant extraction.

In this paper, an event is defined as an activity or action that entities play a key role, which may have a specific location or time, so we use a quaternion: *Event = (activity, time, location, participant)* to describe events in a structured manner. The activity points at what entities do, e.g. play, visit, lead. Time elements are when events happen, which may be when tweets are published or illustrated in tweets. Location elements are where events happen. Participant elements are the entities and other objects that participate in the activities.

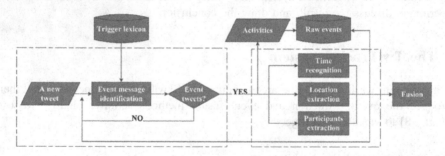

Fig. 2. The processing pipeline of event extraction.

The goal of event extraction is extracting the quaternion of events efficiently. As shown in Fig. 2, it is the first for event extraction to determine whether the tweet is an event tweet, the keys of which are triggers. Trigger words are notional verbs related to entities and tweets that contain them are thought as event tweets. When a new tweet comes, matching it with trigger words by regular expression, if successful, it is regarded as an event tweet. The activity matched is recognized and stored in raw event database containing the four elements at the same time. Then time, location and participant elements are extracted serially via Stanford NLP and stored in the database. After that, the extracted events (i.e., all the extracted elements) will go into the fusion step. If matched unsuccessfully, the tweet will be treated as a rubbish tweet and thrown away and TwiTracker will turn to the next tweet.

2.3   Event Fusion

The raw events are acquired after event extraction. Then we choose Single Pass algorithm [9] to assemble those about one event into a cluster, and different into various ones (Fig. 3). TwiTracker uses the extracted event elements to calculate event similarity. Let E_i and E_j denote two of the events and $Sim_{txt}(E_i, E_j)$ denote the similarity between two event texts. The similarity of time, location and participant elements between two events is denoted as $Sim_{tim}(E_i, E_j)$, $Sim_{loc}(E_i, E_j)$ and $Sim_{par}(E_i, E_j)$ separately. Let ω_k be the weight of each elements satisfying $\sum_{i=1}^{4} \omega_k = 1$. Then the event similarity is defined as follows: $EventSim(E_i, E_j) = \omega_1 Sim_{txt}(E_i, E_j) + \omega_2 Sim_{tim}(E_i, E_j) + \omega_3 Sim_{loc}(E_i, E_j) + \omega_4 Sim_{par}(E_i, E_j)$, and the parameters i and j must obey the constraint $0 < i, j \leq n$ (n is the number of events). The similarity of event texts is calculated by cosine similarity measure method of eigenvectors. Time similarity is decided by the threshold value we set in advance, if the difference between two time expressions exceeds the threshold, the two events are regarded as different ones, and vice versa. TwiTracker calculates the difference between locations' latitude and longitude to get location similarity. The participant elements are involved in the entity dictionary. If they belong to an entity's name, we consider the similarity is 1, or else it is 0.

Various classes of events are obtained after clustering. However, each cluster may have different descriptions about one event, so it is necessary to merge the elements that reflect an event's different sides to get more detailed information. TwiTracker applies a rule-based event fusion algorithm containing three regulations: normalization (turn extracted elements into a normal style), reliable high frequency (regard the normalized elements referred mostly as the result), and level decreasing regulation (the lowest level of expression is regarded as the result).

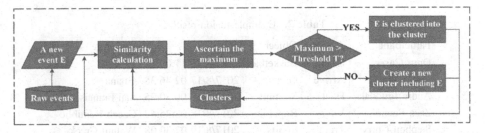

Fig. 3. The schematic diagram of Single Pass clustering in event fusion.

2.4   Visualization

After event extraction and event fusion, we get a detailed structured description of events, which is shown in a table style. It is very clear, but not intuitional enough. It is also difficult to look up in the table to find entities' timelines. To solve these problems and improve the efficiency, a map containing the four elements is created.

TwiTracker uses Echarts developed by Baidu front-end technology department to visualize the trajectory, namely the dynamic curve that links chronologically the places where the entity has reached on the map. The dynamic map displays every places they pass and each node involves the activity, time and participant elements, which is easy to understand and discover useful information.

3 Experiments

To the best of our knowledge, Golden State Warriors are an American professional basketball team based in Oakland, California. Consequently, they have a lot of fans and there are plenty of users discussing about them on Twitter. Tweets about them are numerous and it turns out to be significant to study their behaviors. Therefore, we take Golden State Warriors and the stars in them as the research objects and make use of TwiTracker to get the timelines of them.

This paper collects a total of 55875 tweets from August 2017 to September 2017. After deduplication, the left 19708 English tweets are selected, then event message identification and event extraction are combined to identify 3487 event tweets. Through event fusion, there are 829 clusters at last (Table 1). Parts of the event fusion results are shown in Table 2. The dynamic trajectory is visualized in Fig. 4, which displays five elements at each node including participants, activities, objects, time and location separately from top to bottom.

Table 1. The overall results.

Collected data	55875	Event tweets	3487
English tweets after deduplication	19708	Clusters after fusion	829

Table 2. Example fusion results.

Participant	Activity	Object	Time	Location
Omri Casspi	Lead	Basketball	2017/8/15 17:09:10	Israel
David West	Arrive	Ghana	2017/8/17 02:46:38	Ghana
Andre Iguodala	Host	Summit	2017/8/18 00:30:05	San Francisco
Zaza Pachulia	Vs	Republic	2017/8/18 08:25:52	Czech Republic
Stephen Curry	Train	Sports	2017/8/19 07:30:08	Walnut Creek
Klay Thompson	Play	life	2017/8/19 15:54:29	China

From the trajectory, we can find that the experiment results correspond with the reality basically. According to the news between August and September sin 2017, NBA China Games began selling tickets to Warriors vs. Timberwolves on August 18; David West hosted basketball training camps for young people in Ghana on August 20; Stephen Curry stayed at his home in Walnut Creek and so on, which were nearly the

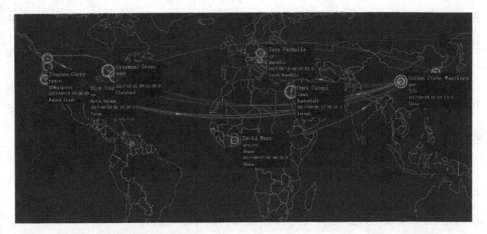

Fig. 4. The obtained trajectory of Golden State Warriors during Aug. to Sept. in 2017.

same as Fig. 4. In addition, the trajectory also shows information that has not been reported officially. The experiment results suggest that TwiTracker is useful and efficient for tracking entities.

4 Conclusions

Social media, especially Twitter, play a more and more important role in detecting, delivering and processing information in the rapidly evolving society. If we can take full advantage of this, it can solve problems of getting information in many fields. This paper presents TwiTracker, a system for making the activity trajectory of entities by extracting elements from Twitter. TwiTracker combines Stanford NLP algorithms and regular expressions to extract elements, uses Single Pass clustering algorithm for event fusion and ECharts to visualize the trajectory. We gather the tweets about Golden State Warriors on Twitter from August 2017 to September 2017 in a total of 55875 tweets and draw their activity route map during the two months, which shows their trajectory intuitively and displays information that has not been reported.

References

1. Li, R., Lei, K.H., Khadiwala, R., et al.: TEDAS: a Twitter-based event detection and analysis system. In: IEEE International Conference on Data Engineering (ICDE), 1–5 April 2012
2. Sakaki, T., Okazaki, M., Matsuo, Y.: Earthquake shakes Twitter users: real-time event detection by social sensors. In: The International Conference on World Wide Web (WWW), Raleigh, North Carolina, USA (2010)
3. Vieweg, S., Hughes, A. L., Starbird, K., Palen, L.: Microblogging during two natural hazards events: what twitter may contribute to situational awareness. In: The SIGHI Conference on Human Factors in Computing Systems, Atlanta, Georgia, USA (2010)

4. González-Bailón, S., Borge-Holthoefer, J., Rivero, A., Moreno, Y.: The dynamics of protest recruitment through an online network. Sci. Rep. **1**, 197 (2011)
5. Marcus, A., Bernstein, M.S., Badar, O., et al.: TwitInfo: aggregating and visualizing microblogs for event exploration. In: The SIGCHI Conference on Human Factors in Computing Systems, Vancouver, BC, Canada, ACM (2011). https://doi.org/10.1145/1978942.1978975
6. Popescu, A.-M., Pennacchiotti, M., Paranjpe, D.: Extracting events and event descriptions from Twitter. In: The International Conference on World Wide Web (WWW), Hyderabad, India. ACM (2011). https://doi.org/10.1145/1963192.1963246
7. Li, J., Cardie, C.: Timeline generation: tracking individuals on Twitter. In: The International Conference on World Wide Web (WWW), Seoul, Korea. ACM (2014). https://doi.org/10.1145/2566486.2567969
8. Echarts Homepage. http://echarts.baidu.com/. Accessed 27 Feb 2018
9. Yin, F.J., Xiao, W.D., Ge, B., et al.: Incremental algorithm for clustering texts in internet-oriented topic detection. Appl. Res. Comput. **28**(1), 54–57 (2011)

Australasian Workshop on Machine Learning for Cyber-Security (ML4Cyber)

Rapid Anomaly Detection Using Integrated Prudence Analysis (IPA)

Omaru Maruatona[1(✉)], Peter Vamplew[2], Richard Dazeley[2],
and Paul A. Watters[3]

[1] PwC, Melbourne, Australia
omaru.maruatona@pwc.com
[2] Federation University, Ballarat, Australia
[3] La Trobe University, Melbourne, Australia

Abstract. Integrated Prudence Analysis has been proposed as a method to maximize the accuracy of rule based systems. The paper presents evaluation results of the three Prudence methods on public datasets which demonstrate that combining attribute-based and structural Prudence produces a net improvement in Prudence Accuracy.

Keywords: Integrated prudence · Expert systems · Prudence analysis

1 Introduction

We present results of evaluations of three types of prudence methods: structural, attribute-based and integrated prudence methods, represented by three systems: Rated MCRDR, RDMs (RDMs) and Integrated Prudence Analysis (IPA) respectively [1]. These evaluations on public datasets extend the preliminary tests conducted by Maruatona et al. [2]. Prudence is the ability of a KBS to indicate to a system administrator when it issues a conclusion it is unsure of [3]. RDR publications usually refer to Prudence as a warning mechanism triggered when the system reaches the limits of its knowledge [4]. Prudence has been trialed in a number of applications including in Internet banking fraud detection, network traffic classification and in eHealth document monitoring [2, 5, 6]. Other fields, including cyber incident response [7], sentiment analysis [8], and identifying problem gamblers [9] could benefit from more prudent classifications.

2 Methods

Seven different public datasets were used in testing and evaluating the three prudent RDR algorithms; Rated MCRDR, RDMs and IPA. Each dataset was randomised before every test. All datasets included were extracted from the UCI Machine Learning repository [10]. A simulated expert was developed for each of these.

© Springer Nature Switzerland AG 2018
M. Ganji et al. (Eds.): PAKDD 2018, LNAI 11154, pp. 137–141, 2018.
https://doi.org/10.1007/978-3-030-04503-6_12

3 Results

This section presents a number of results and analysis of classification and prudence accuracies of Rated MCRDR, RDMs and IPA on categorical and numerical public datasets.

3.1 Simple Accuracy

Three public numerical datasets of up to 5,000 cases and with simulated expert decision tree sizes of up to 100 rules were tested on the MCRDR system. The MCRDR system represents the Rated MCRDR, RDMs and IPA systems without their prudence components. Table 1 presents the base simple accuracy of an MCRDR classifier. According to Table 1, MCRDR's relative accuracy across the numerical datasets is 81%. This means that the system has correctly learnt at least 80% of all the knowledge from the numerical datasets. The system's average simple accuracy across numerical datasets is 69%.

Table 1. MCRDR's simple and relative accuracy on numerical datasets

Dataset	TC %	FC %	Acc %	RA %
Physical	59.4	40.6	59.4	60.6
Poker	52.0	48.0	52.0	86.7
Iris	97.3	2.7	97.3	97.3

In Table 2, MCRDR's simple accuracy results on the Car Evaluation, Tic tac toe, Garvan and the Adult datasets are presented. On categorical data, MCRDR has an average relative accuracy of 76%, which is comparable to the system's average Relative Accuracy in numeric datasets. The average classification accuracy in categorical datasets is 71%, also proximal to the system's average simple accuracy of 69% in numerical datasets. In general, results from Tables 2 and 3 show that MCRDR recorded consistent results in both categorical and numerical datasets. Prudence results for Rated MCRDR, RDMs and IPA are discussed in the next section.

Table 2. MCRDR's simple and relative accuracy on categorical datasets

Dataset	TC %	FC %	Acc %	RA %
Car	65.0	35.0	65.0	69.1
Tic tac toe	68.9	31.1	68.9	68.9
Garvan	89.0	19.0	89.0	90.8
Adult	62.5	37.5	62.5	73.5

3.2 Prudence Accuracy

Table 3 presents the Specificity (Sp), Sensitivity (Se) and Prudence Accuracy results of Rated MCRDR, RDMs and IPA in numerical datasets. The student t-test was used to assess the statistical significance of the observed differences in performance between the different prudent systems. On average, IPA appears to have an advantage over Rated MCRDR and RDMs with an average Prudence Accuracy of 80.9% across the numerical datasets. Both IPA and Rated MCRDR were significantly better ($p < 0.001$) than RDMs on the Poker dataset, while IPA was also significantly better than Rated MCRDR on this dataset. All three approaches performed identically on the Iris dataset, and were not significantly different on the Physical dataset.

Table 3. Rated MCRDR, RDMs and IPA prudence accuracy on datasets

	TP %	TN %	FP %	FN %	PA %
Physical					
RDMs	27.0	43.8	15.3	13.9	70.1
Rated MCRDR	27.6	43.0	16.4	13.0	70.2
IPA	29.7	39.9	19.5	10.9	70.2
Poker					
RDMs	29.0	39.2	12.8	19.0	67.9
Rated MCRDR	29.1	42.1	10.4	18.4	70.7
IPA	29.1	45.2	7.3	18.4	73.7
Iris					
RDMs	2.7	95.2	2.1	0.0	98.9
Rated MCRDR	2.7	95.2	2.1	0.0	98.9
IPA	2.7	95.2	2.1	0.0	98.9

Table 4 presents the systems' Prudence Accuracy results on categorical datasets. Again IPA performed well, yielding statistically significant improvements over Rated MCRDR and RDMs on the Tic-tac-toe, Adult and Garvan datasets.

Table 4. Rated MCRDR, RDMs and IPA prudence accuracy on categorical datasets

	TP %	TN %	FP %	FN %	PA %
Cars					
RDMs	21.1	42.7	23.4	12.6	63.61
Rated MCRDR	21.2	33.0	32	13.8	55.67
IPA	22.5	37.3	28.4	11.9	61.09
Tic tac toe					
RDMs	23.5	51.1	17.8	6.9	75.73
Rated MCRDR	20.8	50.8	18.0	10.0	70.68
IPA	24.8	51.2	17.7	6.3	77.03
Garvan					
RDMs	7.8	84.7	4.4	3.1	83.31
Rated MCRDR	7.8	81.0	8.1	3.1	81.23
IPA	8.1	84.3	4.8	2.8	84.46
Adult					
RDMs	29.4	45.7	18.8	6.1	76.83
Rated MCRDR	31.1	42.2	20.3	6.4	75.23
IPA	32.8	43.4	19.1	4.7	78.45

3.3   Impact of Prudence on Simple Accuracy

The MCRDR simple accuracy was recalculated on each prudent system to determine the impact of prudence on a system's simple accuracy. Similar tests had earlier been conducted by Dazeley and Kang [3] to investigate the effect of missed warnings in a prudent classifier. It was concluded that generally, there was relatively an insignificant impact on classifier accuracy from missed warnings. The tests results presented in Table 5 will further clarify whether a prudent system has a better classification accuracy and vice versa for a less prudent system.

Table 5 shows the systems' simple accuracies before and after prudence on both categorical and numerical datasets. The net change shows the effective impact on a system's simple accuracy after prudence. There does not appear to be a significant improvement or drop in classification accuracy after prudence. Each of the three systems recorded either; an equal classification accuracy after prudence as the original MCRDR, a small improvement of not more than 1%, or an accuracy drop of not more than 0.5%. On average, the three systems' combined accuracy improvement across the seven datasets is 0.15%. This validates Dazeley & Kang's [3] earlier proposal that the effect of missed warnings in a prudent classifier are relatively minor.

Table 5. Simple accuracy before and after prudence

Dataset	System	Before (%)	After (%)	Net change
Physical	RDMs	59.4	59.1	−0.3
	Rated MCRDR	59.4	59.4	0.0
	IPA	59.4	59.4	0.0
Poker	RDMs	52.0	52.0	0.0
	Rated MCRDR	52.0	52.5	0.0
	IPA	52.0	52.5	0.5
Iris	RDMs	97.3	97.3	0.0
	Rated MCRDR	97.3	97.3	0.0
	IPA	97.3	97.3	0.0
Car	RDMs	65.0	66.1	0.1
	Rated MCRDR	65.0	65.0	0.0
	IPA	65.0	65.7	0.7
Tic tac.	RDMs	68.9	68.9	0.0
	Rated MCRDR	68.9	68.8	−0.1
	IPA	68.9	68.9	0.0
Garvan	RDMs	89.0	89.1	0.1
	Rated MCRDR	89.0	89.1	0.1
	IPA	89.0	89.1	0.1
Adult	RDMs	62.5	64.5	2.0
	Rated MCRDR	62.5	62.5	0.0
	IPA	62.5	62.5	0.0

4   Conclusion

The objective of this paper was threefold. First, the paper sought to present results from comprehensive evaluations of two known prudence methods: Rated MCRDR and RDMs. Secondly, the paper introduced a novel prudence type known as Integrated Prudence, which is formed by combining the structural and attribute-based prudence methods. Consequently, a new prudence system IPA was developed from a merger of Rated MCRDR and RDMs. The third objective of this paper was to determine if the Integrated Prudence method had any prudence accuracy advantage over its constituent methods: structural and attribute-based prudence.

Results have shown that combining the two methods in some fashion does produce improved Prudence Accuracy results. IPA consistently had the highest Prudence Accuracy in five of seven public datasets used and that on average IPA had the highest Prudence Accuracy across both numeric and categorical datasets. As for RDMs and Rated MCRDR, the two systems' performances were complementary, with the latter recording a better average Prudence Accuracy in categorical data and the former scoring a higher average Prudence Accuracy in numeric data. Further work in this area could expand these evaluations to include domain specific and mixed dataset.

References

1. Maruatona, O., Vamplew, P., Dazeley, R., Watters, P.A.: Evaluating accuracy in prudence analysis for cyber security. In: Proceedings of the 24th International Conference on Neural Information Processing (ICONIP) (2017)
2. Maruatona, O.O., Vamplew, P., Dazeley, R.: RM and RDM: a preliminary evaluation of two prudent RDR techniques. In: The Pacific Rim Knowledge Acquisition Workshop, Kuching, pp. 188–194 (2012)
3. Dazeley, R., Kang, B.: The viability of prudence analysis. In: The Pacific Rim Knowledge Acquisition Workshop, Hanoi, pp. 107–121 (2008)
4. Prayote, A., Compton, P.: Detecting anomalies and intruders. In: International Conference on Artificial Intelligence 2006, Hobart, pp. 1084–1088 (2006)
5. Prayote, A.: Knowledge based anomaly detection. PhD Thesis, University of New South Wales (2007)
6. Dazeley, R., Kang, B.: Rated MCRDR: finding non-linear relationships between classifications in MCRDR. In: 3rd International Conference on Hybrid Intelligent Systems, pp 499–508. IOS Press, Melbourne (2003)
7. Lee, S.J., Watters, P.A.: Cyber budget optimization through security event clustering. In: Proceedings of the 7th IEEE International CYBER Conference, Hawaii, HI (2017)
8. Prichard, J., Watters, P.A., Krone, T., Spiranovic, C., Cockburn, H.: Social media sentiment analysis: a new empirical tool for assessing public opinion on crime? Curr. Issues Crim. Justice 27(2), 217–236 (2015)
9. Suriadi, S., Susnjak, T., Ponder-Sutton, A., Watters, P.A., Schumacher, C.: Using data-driven and process mining techniques for identifying and characterizing problem gamblers. Complex Syst. Inf. Model. Q. 9, 44–66 (2017)
10. UCI: UCI machine learning repository. http://archive.ics.uci.edu/ml/index.html (2012)
11. Hodge, V.J., Austin, J.: A survey of outlier detection methodologies. Artif. Intell. Rev. 22, 85–126 (2004)

A Differentially Private Method for Crowdsourcing Data Submission

Lefeng Zhang[1(✉)], Ping Xiong[1], and Tianqing Zhu[2]

[1] School of Information and Security Engineering, Zhongnan University of Economics and Law, Wuhan, China
lfzhang@tulip.academy, pingxiong@znufe.edu.cn
[2] School of Information Technology, Deakin University, Geelong, Australia
tianqing.e.zhu@gmail.com

Abstract. In recent years, the ubiquity of mobile devices has made spatial crowdsourcing a successful business platform for conducting spatiotemporal projects. However, these platforms present serious threats to people's location privacy, because sensitive information may be leaked from submitted spatiotemporal data. In this paper, we propose a private spatial crowdsourcing data submission algorithm, called PS-Sub. This is a differentially private method that preserves people's location privacy and provides acceptable data utility. Experiments show that our method is able to achieve location privacy preservation efficiently, at an acceptable cost for spatial crowdsourcing applications.

Keywords: Privacy preservation · Spatial crowdsourcing
Differential privacy

1 Introduction

Crowdsourcing has become an effective implementation model since it was first proposed in 2006 [3]. It is a distributed cooperation mode of outsourcing tasks to a crowd of workers to achieve a specific goal, such as information collection and aggregation [2]. In line with the rapid development of smart devices and sensing techniques, *spatial crowdsourcing* (SC), a special form of crowdsourcing, has become increasingly popular. It represents a novel paradigm for performing location-based tasks, also known as spacial tasks. In SC, spatial tasks–such as taking photographs, recording videos or reporting temperatures at specified locations– are assigned to distributed workers, who then physically travel to a locations to complete the work. Accordingly, workers need to submit their exact location to the SC server in advance.

However, exposing of worker locations presents serious risks to personal privacy. Valuable location data may be released for public or research purposes. In fact, studies have demonstrated that sophisticated adversaries can infer an individual's home location, political views, or religious inclinations by exploiting

© Springer Nature Switzerland AG 2018
M. Ganji et al. (Eds.): PAKDD 2018, LNAI 11154, pp. 142–148, 2018.
https://doi.org/10.1007/978-3-030-04503-6_13

location information. Service providers reliability is also a privacy concern. Individual data may be abused or misused by curators for illegal purposes. Further, government agencies may demand that enterprises provide client information for public security purposes, for example, the FBIs attempt to force Apple to unlock a user's iPhone in the US [4].

This paper presents a differentially private SC framework without a trusted third party. In the proposed framework, the SC server publishes a location space in advance that contains all the possible task locations. Further, the location data a worker submits to the server is the position of the potential tasks the worker is willing to accept. Therefore, a worker's location privacy is preserved while task assignment is performed efficiently at an acceptable cost.

The rest of this paper is organized as follows. Section 2 reviews the related literature and background knowledge. Section 3 presents our proposed differentially private framework for SC. Section 4 outlines the performance evaluations, and Sect. 5 offers a conclusion.

2 Background

2.1 Differential Privacy

DP demands that deleting or modifying any record in a dataset result in a negligible effect on the output distribution of a randomized algorithm [1].

Definition 1 (ϵ-DP). *Given any two neighbouring datasets D and D', where their symmetric difference contains at most one record (i.e., $|D \Delta D| \leq 1$), for any possible outcome set Ω, a randomized mechanism $\mathcal{M}$ gives ϵ-DP if*

$$\Pr[\mathcal{M}(D) \in \Omega] \leq \exp(\epsilon) \cdot \Pr[\mathcal{M}(D') \in \Omega] \tag{1}$$

Many mechanisms have been proposed for different privacy preservation scenarios, such as the Laplace and exponential mechanisms.

Definition 2 (*Laplace mechanism*). *Given a function $f : D \to \mathbb{R}$, S is the sensitivity of f, the following mechanism provides ϵ-DP.*

$$\mathcal{M}(D) = f(D) + Laplace(\frac{S}{\epsilon}) \tag{2}$$

where S is the *sensitivity* and represents the maximal change in the query output when deleting or modifying one record from the dataset D.

The exponential mechanism is often accompanied by an application dependent score function $q(D, \psi)$.

Definition 3 (*Exponential mechanism*). *An exponential mechanism $\mathcal{M}$ satisfies ϵ-privacy if*

$$\mathcal{M} = \{return\ \psi\ with\ probability \propto \exp(\frac{\epsilon \cdot q(D, \psi)}{2 \Delta q})\} \tag{3}$$

where Δq represents the sensitivity of q.

3   Method

This section first provides an overview of the proposed private SC method's framework, then presents the private SC data submission algorithm (*PS-Sub*). The details of the algorithm are provided in subsequent subsections.

3.1   Location Randomization

In our proposed SC framework, we introduce an additional component, *task location space* (TLS). The TLS is a distribution of the locations of all possible tasks. Any set of published spatial tasks with locations is a subset of the TLS.

The frequently visited location set of worker u might imply a routine location pattern. Therefore, to prevent such a pattern from being re-built by adversaries, the Laplace mechanism randomizes each frequently visited location set A by inserting or deleting some points, which generates the noisy location set A'. Suppose the number of locations in A is n_a and the privacy budget is ϵ. According to Definition 2, the number of location points in noisy set A', denoted as n'_a, is

$$n'_a = n_a + laplace(\frac{3S}{\epsilon}) \tag{4}$$

where S represents the sensitivity of the function that counts n_a in location set A. In addition, deleting any location in A will create a change of 1, at most, in the count result. Thus, the sensitivity S is 1.

3.2   Location Clustering

Adding noise to a worker's frequently visited location set A generates the noisy location set A'. A' cannot be submitted directly to the SC server, as it still contains the exact worker locations.

The location points in the noisy location set A', $a_1, a_2, \ldots, a_{n_a}$, are considered to be the *centroids* of each potential cluster $C_1, C_2, \ldots, C_{n_a}$. The goal of location clustering process is to assign each TLS point to the nearest cluster. The distance between two locations (x_i, y_i) and (x_j, y_j) is measured in *Euclidean* distance

$$dis(i, j) = \sqrt{(x_i - x_j)^2 + (y_i - y_j)^2}, \tag{5}$$

where x_i, x_j and y_i, y_j represent the location coordinates.

Then the proposed framework assigns a TLS point to a cluster using the exponential mechanism to ensure DP. According to the Definition 3, the distribution of specific assignment probability is calculated as

$$\Pr[P_k \in C_i] = \frac{\exp\left(\frac{-\epsilon \cdot dis(P_k, a_i)}{6 \cdot \Delta q}\right)}{\sum_{j=1}^{n'_a} \exp\left(\frac{-\epsilon \cdot dis(P_k, a_j)}{6 \cdot \Delta q}\right)}, \tag{6}$$

where Δq is the worker-specified sensitive radius that represents the range within which the worker feels secure.

3.3 Location Substitution

The noisy set A', cannot be submitted directly to the SC server because it contains a worker's exact locations. To protect a worker's location from being re-identified by the SC server or by adversaries, each point in A' for every worker is replaced with a TLS location from the relevant cluster. As the TLS is only relative to the task locations, which can be considered public information, it is unrelated to a worker's preferences and routine location pattern.

Therefore, an exponential mechanism is used to perform this substitution process. For each location a_i in A', a substitution point (P_i) is sampled from its cluster, while a point closer to a_i will be sampled with a higher probability. According to Definition 3, the probability of sampling a substitution P_k from related cluster C_k for each a_i in A' is

$$\Pr[P_k] = \frac{\exp\left(\frac{-\epsilon \cdot dis(a_i, P_k)}{6 \cdot \Delta q}\right)}{\sum_{j=1}^{|C_k|} \exp\left(\frac{-\epsilon \cdot dis(a_i, P_j)}{6 \cdot \Delta q}\right)} \tag{7}$$

where $|C_k|$ is the number of TLS locations in cluster C_k, and Δq is the sensitive radius specified by SC workers.

Performing the above process for all points in A', each sampled P_i is added to the set B, which is finally submitted to the SC server.

4 Experiment

4.1 Evaluation Metrics

The real-world datasets Gowalla[1] was used in these experiments. Two evaluation metrics were used to evaluate the performance of the proposed algorithm compared to the non-private method.

Definition 4 (ASR). *The task ASR is the ratio of accepted tasks to the total number of tasks in an SC project.*

$$ASR = \frac{number\ of\ accepted\ tasks}{total\ number\ of\ tasks}. \tag{8}$$

We defined the distance that a high percentage of workers are willing to travel in a non-private SC as the *maximum travel distance* (MTD) [5]. A task with a travel distance larger than MTD will not be accepted by any worker.

ADT is another important metric in SC. It represents the average travel cost for workers when performing a task, i.e., $ADT = \frac{1}{N}\Sigma_{i=1}^{N} d_i, \quad d_i \leq MTD$.

Definition 5 (IRA). *The IRA is the extra ADT introduced by the privacy preservation mechanism divided by the ADT in a non-private assignment strategy*

$$IRA = \frac{ADT_{private} - ADT_{non-private}}{ADT_{non-private}} \tag{9}$$

where $ADT_{non-private}$, $ADT_{private}$ represent the ADT in non-private and private scenarios, respectively.

---

[1] http://snap.stanford.edu/data/loc-gowalla.html.

4.2   Experiment Results and Comparisons

ASR Performance. Variations in the ASR tendencies, along with other parameters, are shown in Fig. 1. Specifically, Fig. 1a shows that the ASR increases as the privacy budget ϵ increases. Figure 1b shows that the ASR decreases when the sensitive radius is increased.

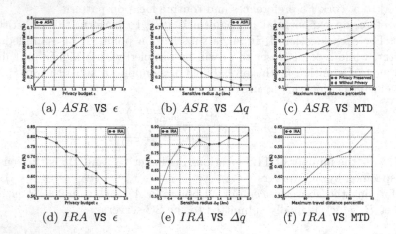

(a) ASR VS ϵ (b) ASR VS Δq (c) ASR VS MTD

(d) IRA VS ϵ (e) IRA VS Δq (f) IRA VS MTD

Fig. 1. ASR and IRA estimation of the Gowalla dataset

Figure 1c illustrates the change in ASR when the MTD varies. Here, the ASR in a privacy-preserving situation is always lower than that in a non-private situation. Figure 2 illustrates the variation tendencies of ASR when MTD is set to 90 (Fig. 2a) and Δq is set to 0.2 (Fig. 2b). Figure 2a reveals that a small sensitive radius usually leads to an improved ASR. Figure 2b indicates a larger MTD usually provides better results when other parameters are fixed.

Performance of IRA. Figure 1d describes the IRA for SC workers with a change in the privacy budget. Here, IRA increases with an increase of ϵ. In Fig. 1e, the IRA increases with increased Δq as achieving location privacy protection in a large area will introduce more noises.

Figure 1f shows the value of IRA with the MTD percentile varying from 75 to 95. Here, IRA increases with an increased MTD. Figures 2c and d compare IRA in scenarios with various sensitive radiuses and MTDs. Figure 2c suggests workers define a small Δq if they do not want to travel too far at a certain privacy level. Figure 2d verifies that large a MTD will generally lead to greater travel distances for workers.

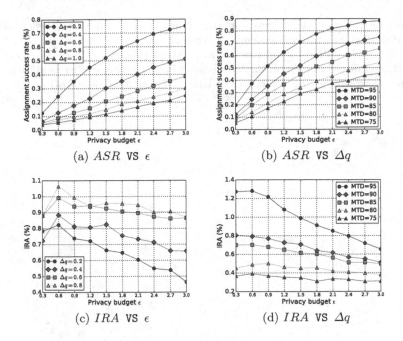

(a) *ASR* vs ϵ

(b) *ASR* vs Δq

(c) *IRA* vs ϵ

(d) *IRA* vs Δq

Fig. 2. Comparisons of ASR and IRA on a Gowalla dataset

5 Conclusion

The popularity of spatial crowdsourcing (SC) has led to increased privacy concerns. In this paper, we propose a novel differentially private method for SC workers to submit their routine points to a server. Our proposed framework differs from existing methods in that no third party is used as a communication agent between the workers and the server. Comparisons between the proposed method and non-privacy task assignment strategies demonstrate that our proposed method's performance equates to that achieved by non-privacy methods, while also protecting the privacy of worker locations.

References

1. Dwork, C.: Differential privacy. In: Bugliesi, M., Preneel, B., Sassone, V., Wegener, I. (eds.) ICALP 2006. LNCS, vol. 4052, pp. 1–12. Springer, Heidelberg (2006). https://doi.org/10.1007/11787006_1
2. Gao, J., Li, Q., Zhao, B., Fan, W., Han, J.: Mining reliable information from passively and actively crowdsourced data. In: Proceedings of the 22nd ACM SIGKDD International Conference on Knowledge Discovery and Data Mining, KDD 2016, pp. 2121–2122. ACM, New York, NY, USA (2016)

3. Howe, J., Robinson, M.: Crowdsourcing: a definition. wired blog network: crowd-sourcing (2006). http://crowdsourcing.typepad.com
4. Lee, D.: Apple v FBI: Us debates a world without privacy (2016)
5. To, H., Ghinita, G., Shahabi, C.: A framework for protecting worker location privacy in spatial crowdsourcing. Proc. VLDB Endow. **7**(10), 919–930 (2014)

An Anomaly Intrusion Detection System Using C5 Decision Tree Classifier

Ansam Khraisat[(✉)], Iqbal Gondal, and Peter Vamplew

Internet Commerce Security Laboratory (ICSL), Federation University Australia,
Ballarat, Australia
{a.khraisa,iqbal.gondal,p.vamplew}@federation.edu.au

Abstract. Due to increase in intrusion activities over internet, many intrusion detection systems are proposed to detect abnormal activities, but most of these detection systems suffer a common problem which is producing a high number of alerts and a huge number of false positives. As a result, normal activities could be classified as intrusion activities. This paper examines different data mining techniques that could minimize both the number of false negatives and false positives. C5 classifier's effectiveness is examined and compared with other classifiers. Results should that false negatives are reduced and intrusion detection has been improved significantly. A consequence of minimizing the false positives has resulted in reduction in the amount of the false alerts as well. In this study, multiple classifiers have been compared with C5 decision tree classifier using NSL_KDD dataset and results have shown that C5 has achieved high accuracy and low false alarms as an intrusion detection system.

Keywords: Malware · Intrusion detection system · NSL_KDD
Anomaly detection

1 Introduction

Anomaly Intrusion Detection Systems (AIDS) [3] have attracted the interest of many researchers due to their potential to detect a zero-day attack. AIDS recognizes abnormal user behavior on a computer system. The assumption for this technique is that attacker activity differs from normal user activity. AIDS [4] creates a behavior profile of normal user's activity by using selected features and machine learning approaches. It then examines the behaviors of new data with the predefined normal behavior profile and tries to identify abnormalities. Those behaviors of users which are unusual are identified as potential attacks.

In this research work, a range of data mining techniques including SVM, Naive Bayes, C4.5 implemented in the WEKA package (developed by the University of Waikato, New Zealand) as well as the C5 algorithm [10] were applied on the NSL-KDD dataset.

© Springer Nature Switzerland AG 2018
M. Ganji et al. (Eds.): PAKDD 2018, LNAI 11154, pp. 149–155, 2018.
https://doi.org/10.1007/978-3-030-04503-6_14

The rest of the paper is organized as follows. Related worked is discussed in Sect. 2. The IDS model with the dataset details is discussed in Sect. 3. Conceptual framework of our IDS model is proposed in Sect. 4. In Sect. 5, the experiment details are given and evaluation results are presented and discussed. Finally, we conclude the paper in Sect. 5.

2   Related Works

Some prior research has examined the use of different techniques to build AIDSs. Chebrolu et al. examined the performance of two feature selection algorithms involving Bayesian networks (BN) and Classification Regression Trees (CRC), and combined methods [2]. Karan et al. proposed a technique for feature selection using a combination of feature selection algorithms such as Information Gain (IG) and Correlation Attribute evaluation then they tested the performance of the selected feature by applying different classification algorithms such as C4.5, Naive Bayes, NB-Tree and Multi-Layer Perceptron [1]. Subramanian et al. propose classifying NSL-KDD dataset using decision tree algorithms to construct a model with respect to their metric data and studying the performance of decision tree algorithms [11].

C5 algorithm's performance is explored very well in a different domain such as modelling landslide susceptibility. Miner et al. used data mining techniques in the topic of landslide susceptibility mapping. They used C5 classifier to handle the complete dataset and address some limitations of WEKA, one of the best results were obtained from C5 applications [9].

3   IDS Model

A prediction model has two main components which are training phase and testing phase. In the training phase the normal profile is created, and in the testing phase the user actions are verified against the corresponding profile. We classify each of the collected data records obtained from the feature phase as normal or an anomaly. In the testing stage, we examine each model.

3.1   Classification

A classification technique is a systematic approach for building classification models from an input data set. Classification is the task of mapping a data item into one of a number of predefined classes [7]. Figure 1 shows a general approach for applying classification techniques.

Decision Trees. are considered one of the most popular classification techniques. Quinlan (1993) has advocated for the decision tree approach and the latest implementation of Quinlan's model is C5 [10]. In this paper we will apply C5 classifier, the algorithm has many advantages like:

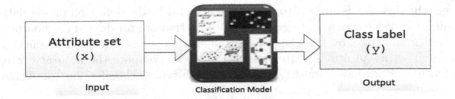

Fig. 1. Classification techniques

- Easy to understand the tree, as the large decision tree can be viewed as a set of rules. C5 can provide the knowledge for the noisy or missing data.
- Addresses over fitting and error pruning issues. Winnowing technique in C5 classifier can predict which attributes are relevant and which are not in the classification. It is useful while dealing with big datasets.

In machine learning, **Naive Bayes** classifiers are a family of least complex probabilistic classifiers based on using Bayes' theorem with robust (naive) independence assumptions between the attributes [8]. It is simple to build, with no complex iterative parameter estimation which makes it suitable for very large datasets. **SVM Model** is a demonstration of the examples as points in space, mapped so that the examples of the separate categories are split by a clear space that is as varied as possible. New examples are then matched into that similar space and predicted to belong to a group based on which side of the gap they belong to [6].

3.2 Framework of Intrusion Detection System

Our purpose is to examine different machine learning techniques that could minimize both the number of false negatives and false positives and to understand which techniques might provide the best accuracy for each category of attack patterns. Different classification algorithms have been applied and evaluated. Figure 2 shows a conceptual framework of our IDS.

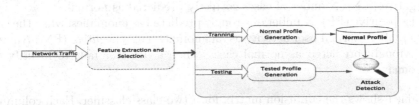

Fig. 2. Overall approach

Collected data is a network traffic, which is used to do feature extraction and selection. In the training phase, a normal profile is developed and in this stage, the classifier is trained to detect the attacks. In the detection phase, data mining techniques are used to generate rule sets that are considered as abnormal activities and used by the classification algorithm already learned to classify the item set as an attack. After testing stage, we compute the accuracy rate, and other performance statistics to distinguish which classifier has predicted successfully.

4   Experimental Analysis

WEKA platform is used [5] to study J48, Naive Bayes and SVM. A commercial system from RuleQuest Research is used for C5 algorithm's [10]. NSL-KDD dataset is used [12]. We compared four different classifiers: C4.5, SVM, Naive Bayes and C5 to evaluate the performance of classification techniques.

4.1   Dataset Description

NSL-KDD data set has been used to overcome KDD cup99 dataset problem. A statistical analysis have been done on KDD cup99 dataset and found issues which have affected the ability to evaluate anomaly detection approaches. It is revealed the main issue is that KDD cup99 dataset has a huge number of redundant records [17]. NSL-KDD is considered as benchmark dataset in evaluating the performance of intrusion detection techniques [12].

The amount of training and testing records in NSL-KDD dataset are significant so the performance of classifiers can be evaluated reliably. The dataset has 125,973 records, where 67,343 are normal cases and 58,630 are anomalies. The dataset contains 22 types of attack, and 41 features.

4.2   Model Evaluation and Results

Our model will be evaluated based on the following standard performance measures:

- True positive (TP): Number of cases correctly predicted as anomaly. True negative (TN): Number of cases correctly predicted as normal.
- False positive (FP): Number of wrongly predicted as anomalies, when the classifier labels normal user activity as an anomaly. False negative (FN): Number of wrongly predicted as normal cases, when a detector fails to identify the anomaly.

Table 1 shows the confusion matrix for a two-class classifier. Each column of the matrix represents the instances in a predicted class, while each row represents the instances in an actual class.

In the paper, we have used k-fold cross validation technique for performance evaluation. In this technique, dataset is randomly divided into k different parts.

In the evaluation, we measured the effectiveness and efficiency of different classification algorithms that wrongly identify the percentage of the False Negative alarm Rate (FN rate) and False Positive (FP rate). Table 2 provide the overall results of our experiments, which indicate that C5 classifiers are best at classifying the intrusions; it has successfully distinguished between normal and anomalous activity with minimum number of false alarm.

Table 1. Confusion matrix for an anomaly detection system

Actual class	Predicted class	
	Normal	Attack
Normal	True Negative(TN)	False Positive(FP)
Attack	False Negative(FN)	True Positive(TP)

Table 2. Confusion matrix for different classification algorithms

Classification algorithm	C5		C4.5		SVM		Naive Bayes	
Classified as	a	b	a	b	a	b	a	b
a = normal	67249	94	67200	143	66370	973	63060	4283
b = anomaly	121	58509	132	58498	2296	56334	7832	50798

Table 3 showed the accuracy for all the classifiers and shows that C5 classifiers have outperformed other classifiers in the study. C5 classifier has the highest accuracy of 99.82% which is followed by C4.5, SVM and Naive Bayes respectively. The number of false alarms, accuracy and time of building IDSs should be considered for IDS evaluation. Although C5 decision tree classifier wasn't faster classifier as shown in Table 4 C5 is the best in term of the accuracy and low false alarm. Naive Bayes is the fastest, but has the lowest accuracy by a substantial margin. The time that takes for generating the ruleset in C5 is 2.06, while the time that takes for generating the ruleset in c4.5 is 29.98, which is slower than C5. The reasons for this, in C5 the rules are generated separately.

Table 3. Accuracy in detection by using different algorithms

Classification algorithm	Accuracy
C5	99.82%
C4.5	99.78%
SVM	97.40%
Naive Bayes	90.38%

Table 4. Time Consuming for each classifier in seconds

Classification algorithm	Time
C5	70.6
C4.5	27.35
SVM	1423.92
Naive Bayes	1.02

5   Discussion and Conclusion

In this paper, an AIDS is proposed with the use of C5 classifier to detect both the normal and anomalous activities accurately. The aim of this approach is to identify attacks with enhanced detection accuracy and decreased false-alarm rates. We have established the robustness of our proposed techniques for intrusion detection by testing them on a NSL-KDD dataset that contains various types of intrusions. Our proposed method is evaluated on NSL-KDD dataset. Our experimental results indicate that our approach can detect malware traffic with a high detection rate of 99.82%. This demonstrates the significance of using C5 classifier in AIDS and makes the detection more effective. C5 are more powerful than C4.5, SVM and Naive Bayes because the memory usage is minimum, good speed and it also has excellent accuracy. In other words, C5 classifier provides high computational efficiency for classifier training and testing.

References

1. Bajaj, K., Arora, A.: Dimension reduction in intrusion detection features using discriminative machine learning approach. IJCSI Int. J. Comput. Sci. Issues **10**(4), 324–328 (2013)
2. Chebrolu, S., Abraham, A., Thomas, J.P.: Feature deduction and ensemble design of intrusion detection systems. Comput. Secur. **24**(4), 295–307 (2005)
3. Denning, D.E.: An intrusion-detection model. IEEE Trans. Softw. Eng. **2**, 222–232 (1987)
4. Garcia-Teodoro, P., Diaz-Verdejo, J., Maciá-Fernández, G., Vázquez, E.: Anomaly-based network intrusion detection: techniques, systems and challenges. Comput. Secur. **28**(1–2), 18–28 (2009)
5. Hall, M., Frank, E., Holmes, G., Pfahringer, B., Reutemann, P., Witten, I.H.: The weka data mining software: an update. ACM SIGKDD Explor. Newsl. **11**(1), 10–18 (2009)
6. Hearst, M.A., Dumais, S.T., Osuna, E., Platt, J., Scholkopf, B.: Support vector machines. IEEE Intell. Syst. Appl. **13**(4), 18–28 (1998)
7. Lee, W., Stolfo, S.J., Mok, K.W.: A data mining framework for building intrusion detection models. In: Proceedings of the 1999 IEEE Symposium on Security and Privacy, pp. 120–132. IEEE (1999)
8. McCallum, A., Nigam, K., et al.: A comparison of event models for Naive Bayes text classification. In: AAAI-98 Workshop on Learning for Text Categorization, vol. 752, pp. 41–48. Citeseer (1998)

9. Miner, A., Vamplew, P., Windle, D., Flentje, P., Warner, P.: A comparative study of various data mining techniques as applied to the modeling of landslide susceptibility on the Bellarine Peninsula, Victoria, Australia (2010)

10. Quinlan, R.: Data mining tools See5 and C5. 0 (2004)

11. Subramanian, S., Srinivasan, V.B., Ramasa, C.: Study on classification algorithms for network intrusion systems. J. Commun. Comput. **9**(11), 1242–1246 (2012)

12. Tavallaee, M., Bagheri, E., Lu, W., Ghorbani, A.A.: A detailed analysis of the KDD CUP 99 data set. In: IEEE Symposium on Computational Intelligence for Security and Defense Applications, 2009. CISDA 2009, pp. 1–6. IEEE (2009)

Deep Learning for Classifying Malicious Network Traffic

K. Millar[(✉)], A. Cheng, H. G. Chew, and C.-C. Lim

School of Electrical and Electronic Engineering, University of Adelaide,
Adelaide, Australia
{kyle.millar,adriel.cheng,honggunn.chew,cheng.lim}@adelaide.edu.au

Abstract. As the sophistication of cyber malicious attacks increase, so too must the techniques used to detect and classify such malicious traffic in these networks. Deep learning has been deployed in many application domains as it is able to learn patterns from large feature sets. Given that the implementation of deep learning for network traffic classification is only just starting to emerge, the question of how best to utilise and represent network data to such a classifier still remains. This paper addresses this question by devising and evaluating three different ways of representing data to a deep neural network in the context of malicious traffic classification. We show that although deep learning does not show significant improvement over other machine learning techniques using metadata features, its use on payload data highlights the potential for deep learning to be incorporated into novel deep packet inspection techniques. Furthermore, we show that useful predictions of malicious classes can still be made when the input is limited to just the first 50 bytes of a packet's payload.

Keywords: Deep learning · Convolutional neural networks
Internet traffic classification · Malicious traffic detection

1 Introduction

As the number of users who rely on the Internet in their professional and personal lives increases, so too does the profit in exploiting the vulnerabilities of these networking systems. Recent years have shown an unprecedented number of malicious attacks worldwide. By design, malicious attacks are not easily identifiable, with each passing year producing new ways to foil existing systems of detection. Therefore, in today's constantly evolving networks, there arises the need for a classification system capable of adapting to these changes.

This research is supported by the Commonwealth of Australia as represented by the Defence Science and Technology Group of the Department of Defence. The authors acknowledge the following for their contributions in gathering the results discussed in this paper: Clinton Page, Daniel Smit, and Fengyi Yang.

© Springer Nature Switzerland AG 2018
M. Ganji et al. (Eds.): PAKDD 2018, LNAI 11154, pp. 156–161, 2018.
https://doi.org/10.1007/978-3-030-04503-6_15

Deep learning has become highly prominent in the machine learning community due to the availability of big data and the specialised hardware required to utilise it. Deep learning's advantage lies in its ability to learn and adapt to complex patterns from a given data set without the need to first define the important features by hand. This allows for large features sets to be analysed with the possibility of learning unprecedented ways to represent the underlying patterns within the data.

Although deep learning is capable of a generalised form of pattern detection, the performance of any learning technique is still only as good as the data it is trained on. As the implementation of deep learning on network traffic classification is only just starting to emerge, the question of how to best represent network data to such a classifier for effective training still remains. This paper aims to address this question by exploring three different ways of representing data to a deep neural network. The contributions of this paper are as follows:

1. We devise, evaluate and discuss three different representations of network data for use in a deep learning classifier.
2. We highlight the potential for novel deep packet inspection techniques based on deep learning and show that useful predictions of malicious classes can still be made within the first 50 bytes of a packet's payload.
3. We show that deep learning achieves comparable results to other machine learning methods when using only metadata features.

2   Related Work

Network traffic classification techniques typically fall into two categories: payload-based classification, where the traffic is classified based on the distinct signatures found in the message content of the packet; and statistical-based classification, where the traffic is classified based on a collection of metadata features, such as the number of packets sent. The latter of these two options has been typically favoured in machine learning research in recent years due to the ease of generating well defined features sets [2]. While this technique displays desirable results on detecting malicious network traffic, they rely on existing knowledge of what features certain attacks are likely to exhibit and thus are unlikely to lead to any additional insights into the data.

Payload-based classification has been typically left for non-learning algorithms such as deep packet inspection (DPI), which searches the packets for a list of predefined signatures. This method benefits from high classification accuracies but is only as accurate as its knowledge base of such attacks. This static approach remains popular for commercial systems that have the resources to manage such a large repository of known attacks. However, with the introduction of deep learning, payload-based techniques are beginning to make use of learning algorithms that facilitate a more adaptive signature-based detection.

Wang [7] showcased deep learning's potential to achieve an automation of DPI methods, acquiring high classification accuracies for application traffic using the first 1000 bytes of a network flow's payload as the input to a deep (multi-layered) neural network. Likewise, Wang *et al.* [6] used a convolutional neural network (CNN), a deep learning approach typically utilised for image classification, to detect patterns within the payload data. This approach was reported to achieve high classification accuracies for both application and malicious traffic but was unable to be formally validated in comparison to the standard techniques of classification.

In this paper we explore both statistical and payload-based approaches for deep learning in the context of cyber malicious classification and detection. Through our analysis, we show that both techniques have different strengths and weakness, and conclude that a union of the two would result in a more robust classifier.

3   Network Data Representation for Deep Learners

The performance of any machine learning technique is only as good as the input data it receives. Although deep learning is capable of extracting useful features from a larger feature set, an effective representation of the input data will further aid the classifier to perform better on the given task. This section defines three representations of network data for use in a deep learning classifier.

3.1   Payload Data

The main disadvantage of implementing deep packet inspection in a real-world scenario is that the signatures used for classification can be subject to regular change. This is especially evident in malicious attacks as their implementation is constantly evolving in order to bypass existing systems of detection. However, a way of automating this feature selection process is proposed using deep learning. By allowing the deep learning classifier to train from the same data commonly used in DPI, the classifier can attempt to learn the underlying patterns within the payload content rather than just searching for pre-defined signatures. Like DPI, a training process will still need to be re-administered to keep up with the constant evolution of network traffic but deep learning can remove much of the human effort involved in this process.

As the inputs to a deep learning classifier must remain fixed, a selection of the payload data must first be made. In this investigation the first 50 bytes of each traffic flow were mapped to the inputs of the deep learner; each byte value mapping to a corresponding input node. To evaluate this input strategy, a dense neural network was created with two hidden layers of 1000 nodes each. This input method expands upon research outlined in [5].

3.2   Flow Image

The largest defining factor between deep learning and other machine learning techniques is its ability to generate insight from a large source of data. To this effect, the method of data representation described herein aims to maximise the amount of input data seen by the classifier in order for a more complex evaluation of the traffic to be made. However, due to the large size of this input method, a standard dense neural network would not be able to efficiently evaluate this input and therefore another deep learning method must be selected.

Convolutional neural networks (CNNs) have been at the forefront of deep learning due to their performance in image processing tasks. Given that image classification requires the processing of large input sizes, CNNs have developed many ways of increasing their efficiency in these tasks such as utilising dimension reduction layers known as pooling.

In order to explore network traffic with a CNN, each flow from the captured network traffic was first converted to an image with each pixel representing a byte of data in the network. Each row of the image is a new packet in the flow, with the bytes contained in the respective packet filling out the columns.

3.3 Flow Statistics

In recent years, flow statistics has been the conventional subject of analysis when applying machine learning techniques to the field of network traffic classification. This method of classification is based on the assumption that the traffic's metadata can be used to identify the distinguishing behaviours of certain traffic. For example, frequent short messages could be an indicator of a denial of service (DoS) attack. As metadata collected from the traffic is seen as an intrinsic property, this method of analysis was aimed at creating a method of detection that was resilient to obfuscation.

To investigate this method of data representation, 24 features were collected from the traffic. These features consisted of data relating to four main categories, temporal, data transferal, TCP window and IP flags. To analyse these features, a dense neural network was created consisting of two hidden layers of five hundred nodes each. In deep learning, it is common practise to use low level features as inputs to the neural network such that higher order features can be learnt during the training process [1]. However, using these high order features directly was explored to compare deep learning to the more traditional practises.

4 Data Set

Deep learning benefits from a large and extensive data set. For this reason and the need for labelled data, the UNSW-NB15 data set [3, 4] was chosen as the subject of this experiment. Contained in this data set was a set of nine synthetically generated malicious classes. However, as deep learning requires a large number of samples to generate an accurate prediction, four of the smallest malicious classes were merged into a single class entitled OTHER-MALICIOUS. The remaining malicious classes and their distributions within the data set are presented in Table 1. A non-malicious class was also included such that the classifier's ability to distinguish between malicious attacks and normal traffic behaviour could be investigated. This non-malicious class was made up of traffic likely to be in abundance on a typical network such as HTTP, P2P, and FTP.

Table 1. UNSW-NB15 malicious classes.

Malicious classes	Number of samples	Percentage of total
DoS	3,603	3.1%
Exploits	25,274	21.8%
Fuzzers	19,240	16.6%
Generic	3,798	3.3%
Reconnaissance	11,671	10.1%
Other-malicious	2,249	1.9%
Non-malicious	50,000	43.2%

5   Results

In this section, the inputs methods (Sects. 3.1, 3.2 and 3.3) are evaluated based on their ability to detect and classify malicious traffic. In order to provide a baseline comparison, three machine learning classifiers, support-vector-machine (SVM), decision tree-based J48, and random forests, were also trained using the Flow Statistics input method.

The F1 scores for malicious traffic classifications are shown in Table 2. By examining the weighted-average row in this table, it can be seen that the two highest performing methods are payload-based. As payload-based methods typically rely on security experts to predefine and manage a set of known signatures, deep learning's performance on this input method shows the potential of these techniques to augment or automate this laboursome process. Furthermore, the Payload Data method showcases that a useful prediction can still be achieved with just the first 50 bytes of a packet's payload, allowing for faster predictions to be made.

Using the flow statistics method to compare deep learning to other machine learning techniques, it was shown that given a well-defined task and a strong representative feature set, other forms of tree-based machine learning may outperform certain deep learning techniques. While deep learning can achieve comparable results on this same feature set, due to the complexity in how it correlates input data, we suspect overly deep neural networks may unintentionally obscure some of the underlying features critical for distinguishing malicious traffic.

Although low accuracies are shown for DoS attacks for all methods, its low detection rate in classifiers which showed an overall high performance (i.e. Payload Data and Flow Image) highlights the issue of choosing one technique over the other. As a DoS attack will typically flood the network with benign packets, a detection system which only analyses payload data will not be able to recognise such an attack.

Table 2. Malicious classification comparison with F1 scores.

Malicious classes	Deep learning methods			Other machine learning methods		
	Payload data	Flow image	Flow statistics	SVM	J48	Random forest
DoS	36.4	36.5	39.3	0.5	**52.2**	**52.2**
Exploits	88.8	**92.2**	85.3	72.2	87.6	88.0
Fuzzers	94.6	**95.1**	86.3	61.3	90.9	89.7
Generic	**88.7**	80.6	69.6	0.0	83.9	87.2
Reconnaissance	96.9	**98.2**	83.6	47.0	84.5	83.7
Other-malicious	86.6	**92.8**	37.4	0.0	54.0	54.1
Non-malicious	97.7	**99.1**	98.9	98.9	98.9	98.9
Weighted-average	92.7	**94.2**	88.3	73.4	90.8	90.8

6 Conclusion and Future Work

In this paper three different ways of representing network data to a deep learning classifier were explored in the context of malicious traffic classification. As to be expected, there is no 'one size fits all' solution, with different malicious attacks exhibiting different defining characterises. While deep learning does not show a significant improvement to other, more conventional machine learning approaches for statistical-based malicious traffic detections, its introduction has paved new grounds for the automation of payload-based detection. In future works, the combination of statistical and payload-based inputs will be explored.

References

1. Bromley, J., Guyon, I., LeCun, Y., Sckinger, E., Shah, R.: Signature verification using a "siamese" time delay neural network. In: Advances in Neural Information Processing Systems, pp. 737–744 (1994)
2. Divyatmika, Sreekesh, M.: A two-tier network based intrusion detection system architecture using machine learning approach. In: 2016 International Conference on Electrical, Electronics, and Optimization Techniques (ICEEOT), pp. 42–47. https://doi.org/10.1109/ICEEOT.2016.7755404
3. Nour, M., Slay, J.: UNSW-NB15: a comprehensive data set for network intrusion detection systems (UNSW-NB15 network data set). In: Military Communications and Information Systems Conference (MilCIS) (2015)
4. Nour, M., Slay, J.: The evaluation of network anomaly detection systems: statistical analysis of the UNSW-NB15 data set and the comparison with the KDD99 data set. Inf. Secur. J.: Glob. Perspect. **25**, 1–14 (2016)
5. Smit, D., Millar, K., Page, C., Cheng, A., Chew, H.G., Lim, C.C.: Looking deeper - using deep learning to identify internet communications traffic. In: 2017 Australasian Conference of Undergraduate Research (ACUR) (2017)
6. Wang, W., Zhu, M., Zeng, X., Ye, X., Sheng, Y.: Malware traffic classification using convolutional neural network for representation learning. In: 2017 International Conference on Information Networking (ICOIN), pp. 712–717. https://doi.org/10.1109/ICOIN.2017.7899588
7. Wang, Z.: The applications of deep learning on traffic identification (2015). https://goo.gl/WouIM6

A Server Side Solution for Detecting WebInject: A Machine Learning Approach

Md. Moniruzzaman[1(✉)], Adil Bagirov[1], Iqbal Gondal[1], and Simon Brown[2]

[1] Internet Commerce Security Laboratory (ICSL), Federation University Australia, Ballarat, Australia
m.moniruzzaman@federation.edu.au
[2] Westpac Banking Corporation, Sydney, Australia

Abstract. With the advancement of client-side on the fly web content generation techniques, it becomes easier for attackers to modify the content of a website dynamically and gain access to valuable information. A majority portion of online attacks is now done by WebInject. The end users are not always skilled enough to differentiate between injected content and actual contents of a webpage. Some of the existing solutions are designed for client side and all the users have to install it in their system, which is a challenging task. In addition, various platforms and tools are used by individuals, so different solutions needed to be designed. Existing server side solution often focuses on sanitizing and filtering the inputs. It will fail to detect obfuscated and hidden scripts. In this paper, we propose a server side solution using a machine learning approach to detect WebInject in banking websites. Unlike other techniques, our method collects features of a Document Object Model (DOM) and classifies it with the help of a pre-trained model.

Keywords: WebInject · Machine learning · Server side detection

1 Introduction

JavaScript has become one of the most effective and common tools for the attackers due to its flexibility and dynamic characteristics [5]. Usual a JavaScript or HTML code is injected by a malware configuration file which is saved in infected computer [7]. The malware evolves over the time to bypass new defense mechanisms. The recent malware has the capabilities of bypassing Transaction Authentication Number (TAN) and some of the malware from Banatrix family can temper computer's memory to alter the destination bank account [6]. Cross-Site Scripting (XSS) is also a popular method to attack a client-side page. Similar to SQL injection, XSS also exploits the hole in the web page that was introduced due to improper form validation and lack of sanitizing data before executing on the server side [12]. JavaScript is not limited to client-side development only.

© Springer Nature Switzerland AG 2018
M. Ganji et al. (Eds.): PAKDD 2018, LNAI 11154, pp. 162–167, 2018.
https://doi.org/10.1007/978-3-030-04503-6_16

NoSQL and Node.js are now being used to design JavaScript based web servers. The various kind of server-side JavaScript injection includes denial of service, file system access, execution of binary files, NoSQL injection and much more [14].

In broad category there are three types of XSS attacks: Non-persistent or reflective XSS, persistent orstored XSS and DOM-based XSS [3,10]. Our research focus is to analyze Document Object Model (DOM) and identify the content that is injected by malware.

Ensuring a safe browsing experience for everyone is a challenging task. Not every individual are always very secured and often get infected in various ways, i.e malicious email, phishing websites, contagious web link and sometimes via physical devices. Another challenge is to deal with various implementation methods of malware and also to identify new and unknown malwares.

This paper is organized in the following sections. Section 2 discusses some earlier works in this area and highlights their strengths and weaknesses. The details of our proposed method is discussed in Sect. 3. In Sect. 4 we discussed experiment results of our proposed method. Finally, we conclude the paper with the summary of our work and highlighting potential research scope in this area.

2 Related Works

Malware codes are very often obfuscated and their implementation strategy changes frequently over time but the behavior of malware remains similar regardless of their development environment. That is why behavior based malware analysis brought the attention of recent researchers. In DOM based XSS, malware injects additional contents on the fly inside the DOM of a web page. Criscione et al. proposed a method for extracting WebInject signatures by inspecting DOM [2].

Continella et al. also used the DOM comparison to find the changes done by malware [1]. Another client-side solution is to integrate the solution to the browser [9]. Lekies et al. modified the source code for Chromium browser to use it for identifying DOMbased XSS issues. Stock et al. worked on top Alexa websites and based on their defined matrices, they find the website that contains at least one XSS vulnerabilities [13]. Fattori et al. used the general interactions between benign programs with the operating system to make a behavioral model [4].

Sandboxing mechanism is used to ensure secure browsing experience by limiting permission of JavaScript and other processes. Same-origin policy also restricts instances of one process to use resources of its own origin only [8]. Even with the presence of these technique, XSS can still attack user and steal valuable information. This can be done by tricking user to click malicious link and downloads malicious JavaScript code into trusted site and execute it inside it. Which allows that malicious code to use the resources of the trusted site and collect the valuable information.

A custom firewall system is proposed by Kirda et al. for the client side which will decide whether a web page request from browser is secured or not based on some rules [8]. User can make their own custom rule depending on their needs.

Dalai et al. proposed a server side solution which will filter the data before executing based on some criteria [3]. They tested using 230 attack vectors. Some of them is not functioning due to the change of browser policy and functions. They tested using five different browsers. Their solution introduces a little overhead in terms of execution time. The papers [10,11] used genetic algorithm to generate suitable XSS from an initial random XSS by applying crossover and mutation and tries to find whether any path in the webpage executes that code.

3   Proposed Approach

We propose a server-side solution to detect WebInject caused by malware at client side. We experimented with both real and simulated banking environment. After analyzing the DOM in the client's browser of a website to find the features and send it back to the server. A pre-trained machine learning model will help to classify the page based on the collected features.

Figure 1 shows basic architecture of our proposed system. A feature collector using JavaScript is designed and it will be sent along with the actual webpage to collect features. We designed feature extractor as a browser extension for google chrome which analyzes the DOM to collect features.

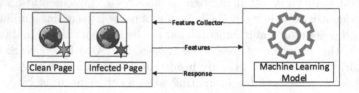

Fig. 1. Basic architecture of our proposed system

After receiving the feature response, server classifies the current page. If server detects unusual content in the web page then it will send a negative response to the client, otherwise it will give a positive response indicating that the page is safe to browse.

Our proposed approach is designed for server side and it will be easily reachable to all the users. Irrespective of which platform is used, our approach will still be able to peform well as it doesn't need to be deployed in client's device. Our proposed method will be able to deal with different implementaion of malwares also, as it does not analyze the source code, rather than it looks for behaviour or the features that have been introduced by an injection. The activities of an WebInject remains almost similar no matter how they are implemented. Selecting the features carefully and designing a good behavioral model will ensure higher detection rate.

4   Evaluation

We experimented the proposed approach by simulating a local client server setup. A simulated banking environment is created resembling a real life banking website. We also used three different banks' live webpage to collect features.

A customized browser extension is used to simulate web injection. We collected features from 793 web-pages and among them 693 pages were clean and rest 100 pages contains injection. Like real world, we tested our system with imbalanced dataset where number of infected instances are lower than number of clean instances. We also evaluated our system with a reduced dataset using 307 clean and 100 infected instances. Classifiers based on different approaches are used to identify the one suitable for our dataset. Classification accuracy of classifiers are computed using 10-fold cross validation technique. We apply classifiers implemented in WEKA. Results are summarized in Tables 1 to 3.

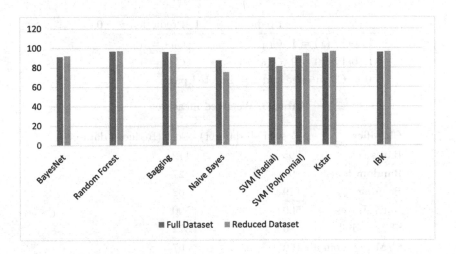

Fig. 2. Accuracy of various classifiers

Figure 2 shows the accuracy of of classifiers for both full and reduced dataset. We observed that Random Forest provided highest accuracy for the full dataset and Bayes Net provided highest accuracy for the reduced dataset. There is scope to work on the feature selection and that will impact on the classification process.

Table 1 shows the precision, recall and F-measure for the clean and infected webpages.

We calculated cost of our classification using cost matrix shown in Table 2. As the dataset is imbalanced and one class have more significance over another one so we assigned different costs for various prediction. We assumed right prediction for both clean and infected data doesn't add any cost and false prediction of an infected page carrys five times more cost than false prediction of a clean page. Table 3 shows the cost of our prediction for both datasets. As we can see even

Table 1. Precision, recall and F-Measure for clean and infected pages

Classifier	Clean page			Infected page		
	Precision	Recall	F-Measure	Precision	Recall	F-Measure
BayesNet	0.924	0.978	0.950	0.746	0.440	0.553
Random forest	0.976	0.987	0.981	0.902	0.830	0.865
Bagging	0.969	0.987	0.978	0.897	0.780	0.834
Naive Bayes	0.874	1.000	0.933	0.000	0.000	0.000
SVM radial	0.900	0.999	0.947	0.958	0.230	0.371
SVM polynomial	0.965	0.942	0.953	0.655	0.760	0.704
Kstar	0.970	0.970	0.970	0.790	0.790	0.790
IBK	0.845	0.820	0.832	0.974	0.978	0.976

Table 2. Cost matrix

	Prediction $y = 1$	Prediction $y = 0$
Label $h(x) = 1$	$C_{1,1} = 0$	$C_{0,1} = 5$
Label $h(x) = 0$	$C_{1,0} = 1$	$C_{0,0} = 0$

$^{*}0$ = Clean page and 1 = Infected page

Table 3. Weighted measure

Classifier	Cost (Full dataset)	Cost (Reduced dataset)
BayesNet	295	112
Random forest	94	27
Bagging	119	68
Naive Bayes	500	500
SVM radial	386	372
SVM polynomial	160	67
Kstar	126	37
IBK	105	51

though some of the classifiers shows better accuracy than others, but their cost is higher. So depending on the need we have to select the most suitable one.

5   Conclusion and Future Work

A server-side detection needs the solution to be implemented in server only, which allows the solution to be available for every user. This will add some overhead in loading time of the page due to the feature extraction and classification process. It is also a challenging task to secure the solution from being exposed. Further works can be done to improve the user experience by reducing the time and also there is a scope for improving the way of transferring the solution from server to client more securely.

References

1. Continella, A., Carminati, M., Polino, M., Lanzi, A., Zanero, S., Maggi, F.: Prometheus: analyzing webinject-based information stealers. J. Comput. Secur. **25**(2), 117–137 (2017)
2. Criscione, C., Bosatelli, F., Zanero, S., Maggi, F.: ZARATHUSTRA: extracting webinject signatures from banking trojans. In: 2014 Twelfth Annual International Conference on Privacy, Security and Trust (PST), pp. 139–148. IEEE (2014)
3. Dalai, A.K., Ankush, S.D., Jena, S.K.: XSS attack prevention using DOM-based filter. In: Sa, P.K., Sahoo, M.N., Murugappan, M., Wu, Y., Majhi, B. (eds.) Progress in Intelligent Computing Techniques: Theory, Practice, and Applications. AISC, vol. 519, pp. 227–234. Springer, Singapore (2018). https://doi.org/10.1007/978-981-10-3376-6_25
4. Fattori, A., Lanzi, A., Balzarotti, D., Kirda, E.: Hypervisor-based malware protection with accessminer. Comput. Secur. **52**, 33–50 (2015)
5. Heiderich, M., Frosch, T., Holz, T.: IceShield: detection and mitigation of malicious websites with a frozen DOM. In: Sommer, R., Balzarotti, D., Maier, G. (eds.) RAID 2011. LNCS, vol. 6961, pp. 281–300. Springer, Heidelberg (2011). https://doi.org/10.1007/978-3-642-23644-0_15
6. Kałużny, J., Olejarka, M.: Script-based malware detection in online banking security overview. Black Hat Asia (2015)
7. Kharouni, L.: Automating online banking fraud, automatic transfer system: the latest cybercrime toolkit feature. Technical report, Trend Micro Incorporated (2012)
8. Kirda, E., Kruegel, C., Vigna, G., Jovanovic, N.: Noxes: A client-side solution for mitigating cross-site scripting attacks. In: Proceedings of the 2006 ACM Symposium on Applied Computing SAC 2006, pp. 330–337. ACM, New York (2006)
9. Lekies, S., Stock, B., Johns, M.: 25 million flows later: large-scale detection of DOM-based XSS. In: Proceedings of the 2013 ACM SIGSAC Conference on Computer & Communications Security, pp. 1193–1204. ACM (2013)
10. Marashdih, A.W., Zaaba, Z.F.: Detection and removing cross site scripting vulnerability in PHP web application. In: 2017 International Conference on Promising Electronic Technologies (ICPET), pp. 26–31, October 2017
11. Marashdih, A.W., Zaaba, Z.F., Omer, H.K.: Web security: detection of cross site scripting in PHP web application using genetic algorithm. Int. J. Adv. Comput. Sci. Appl. (ijacsa) **8**(5) (2017)
12. Saha, S., Jin, S., Doh, K.G.: Detection of DOM-based cross-site scripting by analyzing dynamically extracted scripts. In: The 6th International Conference on Information Security and Assurance (2012)
13. Stock, B., Pfistner, S., Kaiser, B., Lekies, S., Johns, M.: From facepalm to brain bender: exploring client-side cross-site scripting. In: Proceedings of the 22nd ACM SIGSAC Conference on Computer and Communications Security, pp. 1419–1430. ACM (2015)
14. Sullivan, B.: Server-side Javascript injection. Black Hat USA (2011)

A Novel LSTM-Based Daily Airline Demand Forecasting Method Using Vertical and Horizontal Time Series

Boxiao Pan[2], Dongfeng Yuan[1(✉)], Weiwei Sun[1], Cong Liang[1],
and Dongyang Li[1]

[1] Shandong University, Jinan, China
dfyuan@sdu.edu.cn, sunweiwei931229@gmail.com, liangyacongyi@163.com,
15665863197@163.com
[2] South China University of Technology, Guangzhou, China
pankobe24@gmail.com

Abstract. In this paper, we propose a LSTM-based model to cope with airlines' needs for daily demand forecasting. For short-term (e.g. one day in advance) forecasting, we followed the traditional horizontal time series. But for long-term (e.g. half a month in advance) forecasting, the horizontal time series is no longer capable of doing this due to the lack of input data. So we came up with a novel vertical time series, which is also our main contribution in this paper. The vertical time series we propose possesses great application value and has big potential for future research. Empirical analysis showed that our LSTM-based model achieved the state-of-the-art prediction accuracy among all the tested models in both time series. Developed on a dataset from airline industry though, our approach can be applied to all sales scenarios where sale data is recorded continuously for a fixed period before the sale closes. So our research has big value for the industry.

1 Introduction

Demand forecasting is the foundation of revenue management for airlines [1]. Because empty seats quickly lose value after a flight takes off [2], it is critical for airlines to cut down the number of unsold tickets, but at a reasonable price. To achieve this goal, airlines need to first predict demands so as to adjust their supply of services accordingly [3]. Although such need is urgent, there were surprisingly little efforts put on solving it. Most research works about micro-level forecasting focused on predicting the total number of tickets sold [4,5] rather than daily demand, thus providing little help with the need presented above. To the best of our knowledge, our research is the first attempt to do daily airline demand forecasting.

Based on how early the forecasting is made, it can be further classified into long-term forecasting and short-term forecasting. In practice, airlines first need to have a rough idea of how many tickets can be sold by each day in a presell

© Springer Nature Switzerland AG 2018
M. Ganji et al. (Eds.): PAKDD 2018, LNAI 11154, pp. 168–173, 2018.
https://doi.org/10.1007/978-3-030-04503-6_17

period before this flight begins selling tickets, thus being able to make a general plan of pricing and plane arranging. Then within a presell period, airlines want to forecast the demand by the next day more accurately so as to make finer tuning. In order to meet both needs, we proposed a vertical time series for long-term forecasting and a horizontal time series for short-term forecasting. Based on these two time series, we developed a LSTM-based [6] model. Experimental results showed that this method achieved the best prediction accuracy among many powerful methods including Support Vector Regression (SVR) [7], Random Forest Regression (RFR) [8] and Decision Tree Regression (DTR) [9].

The remainder of this paper is organized as follows. Section 2 provides a detailed description of raw data and the two time series. In Sect. 3, we give a dissection of the model we developed. Section 4 presents the experimental results and corresponding analysis. Finally, Sect. 5 draws the conclusion.

2 Data Description

2.1 Raw Data Description

The dataset we use is constrained to one airline and to the route from Dalian to Jinan. The raw data contains tens of indices such as the flight number, plane type, group numbers, and so on. After careful selection, we picked out 7 useful indices and discarded the rest. We provide a detailed explanation of these 7 indices in Table 1.

Table 1. An explanation of the selected indices

Index	Type	Value range	Note
FLIGHT_NO	string	19 unique values	The flight number of a physical plane
DEP_DATE	datetime	2003-12-31–2017-07-23 (inclusive)	The departure date of a flight
EX_DATE	datetime	2004-01-01–2007-07-07 (inclusive)	The date when a record was extracted
EX_DIF	int	16––1 (inclusive)	The date difference between DEP_DATE and EX_DATE. Since all data was extracted at 0 am on the next day, so a record whose EX_DIF equals to −1 actually reflects the situation on the departure day
CAP	int	44 unique values	The total number of available seats on a plane. Some examples include 50, 130 and 174
BKD	int	0–CAP (inclusive)	The accumulative number of tickets sold by EX_DATE of a flight
GRS	int	0–BKD (inclusive)	The accumulative number of tickets sold to groups by EX_DATE of a flight

2.2   The Two Hidden Time Series

Horizontal time series: In this paper, we name the period that EX_DIF taking a range of values from 16––1 as the presell period of a given flight. For each flight, its presell period constitutes a horizontal time series.

Vertical time series: We split all data into 18 groups, within each group the data shares a common EX_DIF. And we name the time series along each group as the vertical time series.

We provide a more intuitive explanation combining both time series in Fig. 1.

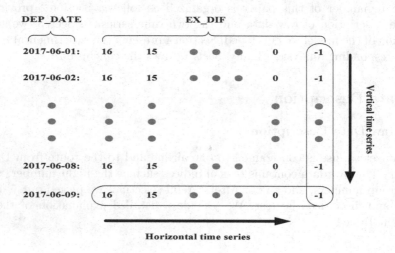

Fig. 1. The two time series

3   Model Dissection

Combining "seq2seq" [10,11], the LSTM-based model we used consists of 2 LSTM layers and 2 Dense layers. The network structure is depicted in Fig. 2.

4   Experimental Results and Analysis

First, we need to clarify our choice of the hyper-parameter seq_len. We tested the performance of our model under the conditions of seq_len taking an integer value from 1 to 10. The best prediction accuracy appeared when seq_len equalled to 7. We think this might be because for most people, a week is their period for work and life, and this can be reflected in their ways of booking tickets.

As stated earlier, our research is the first attempt to do daily airline demand forecasting, so there is no baseline method for us to compare with. But we have implemented several other models for comparison, from which our LSTM-based method obviously stands out. We will present the results obtained using the two time series separately.

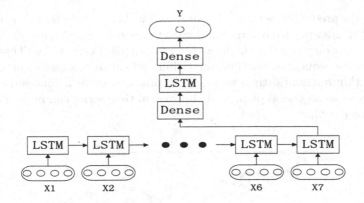

Fig. 2. The structure of our network

4.1 Horizontal Time Series

For horizontal time series, we ran tests on the test set using all the compared models for ten times, and recorded their best results in the ten tests. For comparison purpose, we used the same datasets (training, validation and test) for all models. The results are listed in Table 2. Within the table, Y_TEST_MEAN stands for the arithmetic average of corresponding real values.

Table 2. Comparisons for horizontal time series

Model	MAE	Y_TEST_MEAN	Accuracy
Our model	**4.60**	**42.25**	**89.11%**
SVR	4.77	41.35	88.46%
RFR	5.11	41.35	87.64%
DTR	5.92	41.35	85.68%

As can be seen from Table 2, our LSTM-based model outperforms all its peers.

4.2 Vertical Time Series

We varied the time gap between the last day in the input sequence and the output day from 1 day to 30 days, and took an average of the results of all EX_DIF. In Table 3, we only picked 3 results for each method to save space. While in Fig. 3, we drew the curves using all the 30 results. Here we defined a new variable PRE_DAY, which denotes how many days the prediction is made in advance.

As you can see, when using vertical time series, our LSTM-based model also performs the best among the tested models, and both the table and figure clearly show that the prediction accuracy of our model will not decrease drastically while PRE_DAY increasing from 1 to 30. However, when PRE_DAY equals to 1, the prediction accuracy is not as high as that of horizontal time series. That should be because the sequential relationship within vertical time series is not as strong as that within horizontal time series. So in practice, vertical time series should be used to make a general plan, while horizontal time series can be harnessed to conduct finer tuning.

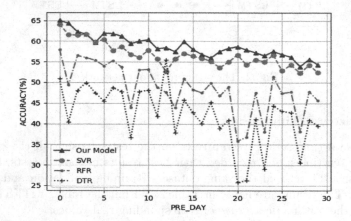

Fig. 3. Comparisons for vertical time series

Table 3. Comparisons for vertical time series

Model	PRE_DAY	MAE	Y_TEST_MEAN	Accuracy
Our model	1	14.40	41.18	**65.03%**
	15	16.24	40.50	**59.90%**
	30	18.51	40.52	**54.32%**
SVR	1	14.89	41.43	64.06%
	15	18.24	42.49	57.07%
	30	19.99	42.02	52.43%
RFR	1	17.42	41.43	57.95%
	15	20.85	42.49	50.93%
	30	22.83	42.02	45.67%
DTR	1	20.34	41.43	50.91%
	15	23.05	42.49	45.75%
	30	25.41	42.02	39.53%

5   Conclusion

In this paper, we propose a novel LSTM-based model using two time series for airline demand forecasting. While the traditional horizontal time series can make near forecasts very well, we also designed a novel vertical time series to help airlines make rougher but more well in advance plans. To the best of our knowledge, this is the first attempt to focus on daily demand forecasting in airlines, and experimental results clearly show that our proposed method can accomplish this goal very well.

References

1. Xie, G., Wang, S., Lai, K.K.: Short-term forecasting of air passenger by using hybrid seasonal decomposition and least squares support vector regression approaches. J. Air Transp. Manag. **37**, 20–26 (2014)
2. Wickham, R.R.: Evaluation of forecasting techniques for short-term demand of air transportation. Ph.D. thesis, Massachusetts Institute of Technology (1995)
3. Doganis, R.: Flying Off Course IV: Airline Economics and Marketing. Routledge, Abingdon (2009)
4. Van Ostaijen, T.: Dynamic booking forecasting for airline revenue management: a Kenya airways case study (2017)
5. Queenan, C.C., Ferguson, M., Higbie, J., Kapoor, R.: A comparison of unconstraining methods to improve revenue management systems. Prod. Oper. Manag. **16**(6), 729–746 (2007)
6. Hochreiter, S., Schmidhuber, J.: Long short-term memory. Neural Comput. **9**(8), 1735–1780 (1997)
7. Cortes, C., Vapnik, V.: Support-vector networks. Mach. Learn. **20**(3), 273–297 (1995)
8. Ho, T.K.: Random decision forests. In: Proceedings of the Third International Conference on Document Analysis and Recognition, vol. 1, pp. 278–282. IEEE (1995)
9. Quinlan, J.R.: Simplifying decision trees. Int. J. Man-Mach. Stud. **27**(3), 221–234 (1987)
10. Sutskever, I., Vinyals, O., Le, Q.V.: Sequence to sequence learning with neural networks. In: Advances in Neural Information Processing Systems, pp. 3104–3112 (2014)
11. Cho, K., et al.: Learning phrase representations using RNN encoder-decoder for statistical machine translation (2014). arXiv preprint arXiv:1406.1078

A Transfer Metric Learning Method for Spammer Detection

Hao Chen[1,2(✉)], Jun Liu[2,3], and Yanzhang Lv[1,3]

[1] National Engineering Lab for Big Data Analytics, Xi'an Jiaotong University,
Xi'an 710049, Shaanxi, China
chenhaogs@stu.xjtu.edu.cn
[2] School of Electronic and Information Engineering, Xi'an Jiaotong University,
Xi'an 710049, Shaanxi, China
[3] Shaanxi Province Key Laboratory of Satellite and Terrestrial Network Tech. R&D,
Xi'an, China

Abstract. Microblogs open opportunities for social spammers, who are threatening for microblog services and normal users. Therefore, detecting spammers is an essential task in social network mining. However, existing methods are difficult to achieve desired performance in real applications. The underlying causes are the insufficiency of knowledge learned from limited training examples and the differences between data distributions on training and test examples. To address these, in this paper, we present a transfer metric learning method to extract more informative knowledge underlying training instances by similarity learning and transfer this knowledge to test instances using importance sampling in a unified framework. We evaluate the proposed method on real-world data. Results show that our method outperforms many baselines.

Keywords: Metric learning · Transfer learning · Spammer detection

1 Introduction

Social network services, such as Microblog and Twitter, give us a wonderful platform to share and enjoy life. Unfortunately, it also opens opportunities for spammers. In detail, spammers may spread mass advertisement posts in a discussing topic; mislead readers with fake content; or lurk as predators for some personal gains. Due to its seriousness, spammer detection has attracted a lot

Supported by "The Fundamental Theory and Applications of Big Data with Knowledge Engineering" under the National Key Research and Development Program of China with grant number 2016YFB1000903; National Science Foundation of China under Grant Nos. 61532004, 61532015, 61572399, 61672419 and 61672418; Innovative Research Group of the National Natural Science Foundation of China (61721002); Project of China Knowledge Centre for Engineering Science and Technology; Ministry of Education Innovation Research Team No. IRT17R86.

M. Ganji et al. (Eds.): PAKDD 2018, LNAI 11154, pp. 174–180, 2018.
https://doi.org/10.1007/978-3-030-04503-6_18

of attention. Wang et al. [11] proposed a graph based method to capture intricate relationships among different entities for detecting spammers in a product review site. Hu et al. [5] introduced an optimization method to detect spammers by collectively using network and content information. Alvarez et al. [7] proposed to detect spammers by comparing the neighborhoods in different views.

However, evolutive spammers are easy to escape the detection by changing their spamming strategies, that is to say, the data distributions of training and test sets are not always the same as well as the training set is changing itself. This makes existing fixed machine learning methods difficult to achieve an ideal performance. In short, the evaluation of spammers brings the following three practical challenges and difficulties: (1) the information used as clues to identify a spammer should be completed; (2) it is hard to label enough instances for learning in time; (3) the last and the most significant is that spammers' diversity spamming strategies make insightful knowledge for spammers detection is changing and hard to capture thoroughly.

To resolve the above dilemmas, we propose to apply a transfer metric spammer detection (TMSD) method which is designed with two considerations in mind: (1) good similarity measures may also provide insight into the underlying knowledge, this has been proven by many metric learning based approaches [2]. (2) transfer learning method can release the problem of difference between training and test data [8].

2   Method

2.1   Metric Learning for Evolutive Spammer Detection

Techniques used in our method is inspired by the previous work [12]. We first assume that the training and test instances are drawn from the same distribution just like most work in this area and only consider the problem of learning a similarity matrix on training set for using the information brought by training instances [9]. Assuming a set of data samples $X = \{x_1, x_2, \cdots, x_n\} \in \mathbb{R}^d$ is given, we aim to learn a similarity function of the following form:

$$K_M(x_i, x_j) = x_i^T M x_j, \tag{1}$$

where $M \in \mathbb{R}^{d \times d}$ is a symmetric matrix. $K_M(x_i, x_j)$ measures the notion of the similarity between a pair of training instances x_i and x_j. Among the training instances, some of them share similar labels while some of them have different labels. Hence, we explicitly assume two sets of pairwise similarity constraints as the follow:

$$
\begin{aligned}
S &= \{(x_i, x_j) : x_i \text{ and } x_j \text{ should be similar}\}, \\
D &= \{(x_i, x_j) : x_i \text{ and } x_j \text{ should be dissimilar}\},
\end{aligned}
\tag{2}
$$

where users with the same label, normal or anomalous, are considered to be similar while users with different labels are dissimilar. And we define a binary label for each pair of $(\boldsymbol{x}_i, \boldsymbol{x}_j) \in \mathcal{S} \bigcup \mathcal{D}$ as:

$$s_{ij} = \begin{cases} 1, & (\boldsymbol{x}_i, \boldsymbol{x}_j) \in \mathcal{S}, \\ -1, & (\boldsymbol{x}_i, \boldsymbol{x}_j) \in \mathcal{D}. \end{cases} \tag{3}$$

Our goal is to learn the optimal similarity function parameterized by M such that it best agrees with the above constraints in an effort to approximate the underlying semantic similarity. We propose to formulate the optimization problem by solving the following objective function:

$$\arg \min_{M \in \mathbb{R}^{d \times d}} \sum_{(\boldsymbol{x}_i, \boldsymbol{x}_j) \in \mathcal{S} \bigcup \mathcal{D}} \mathcal{L}(\boldsymbol{x}_i, \boldsymbol{x}_j; s_{ij}, K_M) + \lambda ||M||_{\mathcal{F}}^2, \tag{4}$$

where $||\cdot||_{\mathcal{F}}$ denotes the Frobenius norm and $||M||_{\mathcal{F}}^2$ is a regularizer on the parameters of the learned similarity function. λ is the trade-off parameter to balances the loss function and regularization. $\mathcal{L}(\boldsymbol{x}_i, \boldsymbol{x}_j; s_{ij}, M)$ is the loss function that incurs a plenty when training constraints are violated. Here, we empirically choose log loss used in our problem, and the loss function is formulated as:

$$\mathcal{L}(\boldsymbol{x}_i, \boldsymbol{x}_j; s_{ij}, K_M) = log(1 + exp(-s_{ij}(K_M(\boldsymbol{x}_i, \boldsymbol{x}_j) - 1))). \tag{5}$$

2.2   Transfer Metric Learning for Evolutive Spammer Detection

Unfortunately, in practice, the training instance and test instance are usually drawn from different distributions. Thus we have

$$x_{tr} \sim P_{tr}(X), \quad x_{te} \sim P_{te}(X). \tag{6}$$

Therefore, the similarity function learned from training set may not be efficient in test data due to some changes of data. This phoneme is named *covariate shift*. The covariate shift problem can be solved by importance sampling methods [10]. Here we adapt this kind of methods to the similarity learning problem based on Theorem 1.

Theorem 1. Suppose that x_i and x_j are drawn independently from $P_{tr}(X)$ and $P_{te}(X)$. $\ell_{te}(\boldsymbol{X}; k_M)$ is the loss on test set. If a similarity learning method is without covariate shift, then the Eq. (7) will set up.

$$\ell_{te}(\boldsymbol{X}; k_M) = \sum_{i,j} \beta_{ij} \mathcal{L}(\boldsymbol{x}_i, \boldsymbol{x}_j; s_{ij}, K_M). \tag{7}$$

where $\beta_{ij} = \dfrac{P_{te}(\boldsymbol{x}_i) P_{te}(\boldsymbol{x}_j)}{P_{tr}(\boldsymbol{x}_i) P_{tr}(\boldsymbol{x}_j)}$ is the importance weight.

Proof. We use the constructing proof method relative to method in [3]. Thus,

$$
\begin{aligned}
\ell_{te}(X; k_M) &= \mathbb{E}_{x_i, x_j \sim \mathrm{P}_{te}(X)}[\mathcal{L}(x_i, x_j; s_{ij}, K_M)] \\
&= \int \mathcal{L}(x_i, x_j; s_{ij}, K_M)\mathrm{P}_{te}(x_i)\mathrm{P}_{te}(x_j)\mathrm{d}x_i\mathrm{d}x_j \\
&= \int \beta_{ij}\mathcal{L}(x_i, x_j; s_{ij}, K_M)\mathrm{P}_{tr}(x_i)\mathrm{P}_{tr}(x_j)\mathrm{d}x_i\mathrm{d}x_j \\
&= \mathbb{E}_{x_i, x_j \sim \mathrm{P}_{tr}(X)}[\beta_{ij}\mathcal{L}(x_i, x_j; s_{ij}, K_M)] \ .
\end{aligned}
$$

This shows the conclusion holds.

As we analyzed above, the corresponding similarity learning with covariate shift problem can be formulated as:

$$
\arg \min_{M \in \mathbb{R}^{d \times d}} \sum_{i,j} \beta_{ij}\mathcal{L}(x_i, x_j; s_{ij}, K_M) + \lambda \|M\|_{\mathcal{F}}^2, \tag{8}
$$

where $x_i, x_j \sim P_{tr}(X)$. In the following, we give a possible way to estimate the importance weights $\beta(x_i, x_j)$ directly and solve the optimization problem (8).

In this paper, we use an efficient algorithm kernel-mean matching (KMM) proposed by Huang et al. in [6] to estimate the importance weights. Readers can refer to [6] for more details. Then, the optimization problem (8) can be carried out efficiently using recent advances in optimization over matrix manifolds [1]. At each iteration, the procedure finds a descent direction in the tangent space of the current solution.

3   Experiments

3.1   Data for the Experiments

Data Crawling. A developed asynchronous parallel crawler for *Sina Weibo* was used from February 2014 to November 2016 to collect data from user profile, user generated content and interaction record. The crawler firstly obtains raw HTML data from web and then extracts metadata based on HTML tags.

Real-World Ground-Truth Data. The real-world data are divided into different subsets according to time periods and topics, including ten subsets. The statistic of the dataset is presented in Table 1.

Feature Extraction. For features used in the classification method, we follow the previous work [4] to extract and organize them. These discriminatory features are applicable for our method and the baseline methods.

3.2   Evaluation Methodology and Experimental Setup

We use the area under the curve (AUC) to evaluate TMSD and the baseline methods. Each experiment is repeated ten times independently and the reported numbers represent *mean* and *standard deviation*. To verify the performance of our proposed method TMSD, we consider the following different methods for comparison:

MVSD[9]: a generalized spammer detection framework by jointly modeling multiple view information and a social regularization term into a classification model.

SMCD [13]: a framework for spammer and spam message co-detection.

SSDM [5]: a matrix factorization-based model that integrates social- aspect features and content-aspect features as a long vector for social spammer detection. In the experiments, we empirically set parameters $\lambda_1 = 0.1$, $\lambda_2 = 0.1$ and $\lambda_s = 0.1$.

kNN+TMSD: our method. The output of our proposed method TMSD is a similarity matrix. We use k-Nearest-Neighbor (kNN) as the basic classifier in our method. The result of kNN is the values for k ranging from 1 to 5.

3.3   Experimental Results and Analysis

We simultaneously conduct experiments on the data which crawled from different time periods and topics. We conduct the experiments on a quad-core 3.20 GHz PC operating on Windows 7 Professional (64 bit) with 8 GB RAM and 1 TB hard disk.

Performance for Different Time Periods. We use data set Period-V as the training set to learn a classifier model and then test it on data sets of Period-I, Period-II, Period-III, Period-IV and Period-V, respectively. It is evident from the Table 2 that our method outperforms other methods, especially when the training data and testing data from different time periods. The proposed method is able to achieve of AUC stably while baseline methods suffer a performance fluctuation with the best AUC < 0.850.

Performance for Different Topics. We use data set of Music topic as the training set to learn a classifier and apply it on data sets of Sports, Science, Politics and Business topics. The reported performance of all methods is shown in Fig. 1. The first three bars represent the performance of baselines without considering learning a proper instance-based distance under covariate shift. The last bar is our proposed method. From the figure, we observe that, with transfer metric learning in a unified model, the proposed method achieves better performance than baseline ones.

Table 1. Description of data sets

	Period-I	Period-II	Period-III	Period-IV	Period-V
Spammer	303	287	366	299	205
Normal user	1226	1599	1738	2002	1365
	Sports	Science	Politics	Business	Music
Spammer	857	776	415	403	628
Normal user	2086	2553	1978	2376	3260

Table 2. AUC values for different time periods

Methods	AUC values for different time periods				
	Period-I	Period-II	Period-III	Period-IV	*Period-V*
MVSD	0.811 ± 0.010	0.799 ± 0.012	0.827 ± 0.012	0.830 ± 0.008	*0.857 ± 0.004*
SMCD	0.802 ± 0.022	0.792 ± 0.012	0.815 ± 0.009	0.816 ± 0.011	*0.846 ± 0.012*
SSDM	0.761 ± 0.007	0.746 ± 0.021	0.758 ± 0.013	0.793 ± 0.008	*0.803 ± 0.007*
KNN+TMSD	0.871 ± 0.021	0.824 ± 0.013	0.852 ± 0.011	0.899 ± 0.006	*0.903 ± 0.013*

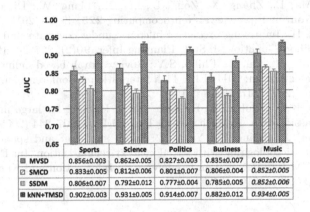

	Sports	Science	Politics	Business	Music
MVSD	0.856±0.003	0.862±0.005	0.827±0.003	0.835±0.007	0.902±0.005
SMCD	0.833±0.005	0.812±0.006	0.801±0.007	0.806±0.004	0.852±0.005
SSDM	0.806±0.007	0.792±0.012	0.777±0.004	0.785±0.005	0.852±0.006
kNN+TMSD	0.902±0.003	0.931±0.005	0.914±0.007	0.882±0.012	0.934±0.005

Fig. 1. AUC values for different topics

4   Conclusion

In this paper, the transfer metric learning method is designed to address the problem that knowledge may be not sufficiently extracted from training examples. Our proposed method TMSD combines similarity learning and instance-based transfer learning in a unified framework. Evaluations on real-world data show that TMSD is capable to improve the spammer detection performance.

References

1. Absil, P.A., Mahony, R., Sepulchre, R.: Optimization Algorithms on Matrix Manifolds. Princeton University Press, Princeton (2009)
2. Bellet, A., Habrard, A., Sebban, M.: A survey on metric learning for feature vectors and structured data. arXiv preprint arXiv:1306.6709 (2013)
3. Cao, B., Ni, X., Sun, J.T., Wang, G., Yang, Q.: Distance metric learning under covariate shift. In: Proceedings of the Twenty-Second International Joint Conference on Artificial Intelligence, Barcelona, Catalonia, Spain, p. 1204 (2011)
4. Chen, H., Liu, J., Lv, Y., Li, M.H., Liu, M., Zheng, Q.: Semi-supervised clue fusion for spammer detection in Sina Weibo. Inf. Fusion **44**, 22–32 (2018)
5. Hu, X., Tang, J., Zhang, Y., Liu, H.: Social spammer detection in microblogging. IJCAI **13**, 2633–2639 (2013)

6. Huang, J., Gretton, A., Borgwardt, K.M., Schölkopf, B., Smola, A.J.: Correcting sample selection bias by unlabeled data. In: Advances in Neural Information Processing Systems, pp. 601–608 (2007)
7. Marcos Alvarez, A., Yamada, M., Kimura, A., Iwata, T.: Clustering-based anomaly detection in multi-view data. In: Proceedings of the 22nd ACM International Conference on Information & Knowledge Management, pp. 1545–1548. ACM (2013)
8. Pan, S.J., Yang, Q.: A survey on transfer learning. IEEE Trans. Knowl. Data Eng. 22(10), 1345–1359 (2010)
9. Shen, H., Ma, F., Zhang, X., Zong, L., Liu, X., Liang, W.: Discovering social spammers from multiple views. Neurocomputing 225, 49–57 (2017)
10. Shimodaira, H.: Improving predictive inference under covariate shift by weighting the log-likelihood function. J. Stat. Planning Infer. 90(2), 227–244 (2000)
11. Wang, G., Xie, S., Liu, B., Philip, S.Y.: Review graph based online store review spammer detection. In: 2011 IEEE 11th International Conference on Data Mining (ICDM), pp. 1242–1247. IEEE (2011)
12. Weinberger, K.Q., Saul, L.K.: Distance metric learning for large margin nearest neighbor classification. J. Mach. Learn. Res. 10(Feb), 207–244 (2009)
13. Wu, F., Shu, J., Huang, Y., Yuan, Z.: Social spammer and spam message co-detection in microblogging with social context regularization. In: Proceedings of the 24th ACM International on Conference on Information and Knowledge Management, pp. 1601–1610. ACM (2015)

Dynamic Whitelisting Using Locality Sensitive Hashing

Jayson Pryde[✉], Nestle Angeles, and Sheryl Kareen Carinan

Trend Micro Inc., PH, 7th Floor, Tower 2, Rockwell Business Center, Ortigas Ave.,
Pasig, Philippines
{jayson_pryde,nestle_angeles,sheryl_carinan}@trendmicro.com

Abstract. Computer systems may employ some form of whitelisting
for execution control, verification, minimizing false positives from other
detection methods or other purposes. A legitimate file in a whitelist may
be represented by its cryptographic hash, such as a hash generated using
an SHA1 or MD5 hash function. Due to the fact that any small change to
a file in a cryptographic hash results in a completely different hash, a file
with a cryptographic hash in a whitelist may no longer be identifiable in
the whitelist if the file is modified even by a small amount. This prevents
a target file from being identified as legitimate even if the target file is
simply a new version of a whitelisted legitimate file.

Locality Sensitive Hashing is a state of the art method in big data
and machine learning for the scalable application of approximate nearest neighbor search in high dimensional spaces [9]. The identification of
executable files which are very similar to known legitimate executable
files fits very well within this paradigm.

In this paper, we show the effectiveness of applying TLSH [1,2]; Trend
Micro's implementation of locality sensitive hashing, to identify files similar to legitimate executable files. We start with a brief explanation of
locality sensitive hashing and TLSH. We then proceed with the concept
of whitelisting, and describe typical modifications made to legitimate
executable files such as security updates, patches, functionality enhance-
ments, and corrupted files. We will also describe the scalability problems
posed by all the legitimate executable files available on the Windows
OS. We will also show results of similarity testing against malicious files
(malwares). Data will be provided on the efficacy and scalability of this
approach. We will conclude with a discussion of how this new methodology may be employed in a variety of computer security applications to
improve the functionality and operation of a computer system. Examples may include whitelisting, overriding malware detection performed
by a machine learning system, identifying corrupted legitimate files, and
identifying new versions of legitimate files.

Keywords: Whitelisting · Locality sensitive hashing · TLSH

© Springer Nature Switzerland AG 2018
M. Ganji et al. (Eds.): PAKDD 2018, LNAI 11154, pp. 181–185, 2018.
https://doi.org/10.1007/978-3-030-04503-6_19

1   Whitelisting

Due to the very fast evolving landscape of malwares, whitelisting has garnered significant attention, to be placed in the main stream of computer security protection. With advancements in packing technologies, malicious writers can easily mass replicate and reproduce their malicious codes, allowing them to both spread their malwares significantly, and evade the common signature based approach used by security industries. Hence, checking if a particular file is present in a whitelist database or repository maybe more suitable.

There are a number of approaches to Application Whitelists [5]. The two main approaches are a list of cryptographic hashes of allowed files, and a list of the allowed certificate signers (including their full certificate chain), and combinations that include file paths and other metadata [5]. The first of these approaches is often considered too restrictive as it can fail to keep up with the constantly changing legitimate files. Relying purely on certificates for whitelisting has had problems in the past with a number of incidents including signed malware [6], security companies signing malware [7] and Microsoft's certificates being able to be 'borrowed' [8]. This paper adopts the position that certificates combined with meta data about the file (in the form of a locality sensitive hash) are significantly safer than certificates alone [4].

However, legitimate software went rapid too, setting their releases, patches, updates, and daily builds in a fast pace, especially those that are open sourced and free. Two key properties of locality sensitive hashes (and in particular TLSH [1]) that we use are that (1)The hash value of similar files will remain relatively invariant and (2)The distance between files can be measured by the hamming distance.

With this challenge, a way to cross-check or compute the similarity of a file against a repository of good files is proposed to improve and optimize the whitelisting approach.

2   Locality Sensitive Hashing

Locality Sensitive Hashing (LSH) is an algorithm known for providing scalable, approximate nearest neighbor search of objects [1]. LSH enables a precomputation of a hash that can be quickly compared with another hash to ascertain their similarity. LSH can be applied to optimize data processing and analysis. The transportation company Uber, for instance, implements LSH in the infrastructure that handles much of its data to identify trips with overlapping routes and reduce inconsistencies in GPS data [10].

2.1   Trend Micro Locality Sensitive Hashing (TLSH)

TLSH is a kind of fuzzy hashing that can be employed in machine learning extensions of whitelisting; TLSH can generate hash values which can then be analyzed for similarities. TLSH helps determine if the file is safe to be run on the system based on its similarity to known, legitimate files. Thousands of hashes of different versions of a single application, for instance, can be sorted through and streamlined for comparison and further analysis. Metadata, such as certificates, can then be utilized to confirm if the file is legitimate.

TLSH helps detect and inspect files that are on, or introduced to, computers. It can be utilized to ensure that only legitimate applications or documents can be run, opened, or saved on the system. TLSH provides a mechanism for promptly comparing the similarity digests of unknown files with a searchable repository of similarity digest of known, legitimate, and allowable files. The idea behind similarity digests is to enable a reliable measure of correlation in terms of identifying similar and unique features on each version of a particular file.

Applications can be implemented using TLSH as a local service, or as a web service. For this paper, TLSH running as a web service (Fig. 1) was tested.

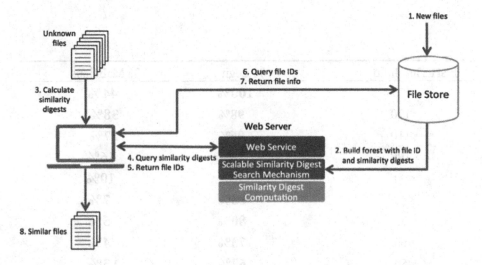

Fig. 1. TLSH as webservice operation.

3   Testing and Results

As mentioned, the webservice implementation was used for the testing. First, the TLSH index (e.g. can be also called as a TLSH pattern) was built and created using selected legitimate files. Below table (Table 1) shows the breakdown of the legitimate files used for the index generation. After building the TLSH pattern, test samples, which are going to be queried against the built TLSH pattern, are to be retrieved. To make sure that the results of the test performed are not biased, it was made sure that each entry of the test set is not included in the TLSH pattern. The legitimate test set was comprised of 1M samples of Microsoft OS files, again, not included in the TLSH pattern. In addition, a malware set, comprised of 25k malicious DLL and 25k malicious EXE samples, were tested against the TLSH pattern. The files were verified to be malicious by means of double-checking in Virus Total. The expectation on this test scenario is to have a very high match in the legitimate test set, and a very low (or 0) match for the malicious test set. For this test scenario, 90 was the TLSH threshold similarity score used. Below figure (Fig. 2) shows the results.

Table 1. Summary of the legitimate files used in building the TLSH index

Company/Vendor	Number of Entries	Description
Microsoft OS files	495975	System/OS files
Microsoft applications	256429	Popular Microsoft applications
False-Alarmed (FA) legitimate files	3140	Incorrectly detected legitimate files based from AVTest and AVComparatives

TLSH Threshold	% Legit	% Malicious
<127	100%	44%
<120	98%	38%
<110	96%	27%
<100	93%	16%
<90	90%	10%
<80	85%	7%
<70	80%	5%
<60	73%	4%
<50	67%	3%
<40	60%	2%
<30	51%	2%
<20	40%	2%
<10	33%	1%

Fig. 2. Results of testing/querying legitimate and malicious test samples set against the TLSH pattern. As can be seen, TLSH similarity score threshold 90 gives 90% detection rate for legitimate files, and 10% detection malicious files.

4   Conclusion

With the results discussed above, TLSH indeed enables smart whitelisting. However, important factors, whether ran in local or as a webservice, are to be considered carefully:

1. Whether ran in local or as a web service, baseline entries of the TLSH index or pattern is very important. That is, placing, as much as possible, the best representation of the entire population in the pattern.
2. In line with consideration 1, TLSH pattern upkeep and maintenance is also of high consideration. Making sure that all unseen, unknown, or unhandled real-world cases will be handled immediately and seamlessly.
3. A TLSH similarity score threshold should be assessed and pinned down thoroughly.
4. Combining TLSH similarity match with other information, can be experimented to improve both detection and misdetection rates.
5. For the case of running as a web service, scalability (auto-scaling) should be considered based on the number of users integrated to the system.

References

1. Open Sources in GitHub. https://github.com/trendmicro/tlsh/
2. Oliver, J., Cheng, C., Chen, Y.: TLSH - A Locality Sensitive Hash. In: 4th Cybercrime and Trustworthy Computing Workshop, Sydney, November 2013. https://github.com/trendmicro/tlsh/blob/master/TLSH_CTC_final.pdf
3. Oliver, J., Forman, S., Cheng, C.: Using randomization to attack similarity digests. In: Batten, L., Li, G., Niu, W., Warren, M. (eds.) ATIS 2014. CCIS, vol. 490, pp. 199–210. Springer, Heidelberg (2014). https://doi.org/10.1007/978-3-662-45670-5_19
4. Oliver, J., Pryde, J.: Smart Whitelisting Using Locality Sensitive Hashing. https://blog.trendmicro.com/trendlabs-security-intelligence/smart-whitelisting-using-locality-sensitive-hashing/
5. Implementing Application Whitelisting. Australian Signals Directorate. https://www.asd.gov.au/publications/protect/application_whitelisting.htm
6. How CyberCrime Exploits Digital Certificates. InfoSec Institute. http://resources.infosecinstitute.com/cybercrime-exploits-digital-certificates/#gref
7. Krebs, B.: Security Firm Bit9 Hacked, Used to Spread Malware. https://krebsonsecurity.com/2013/02/security-firm-bit9-hacked-used-to-spread-malware/
8. IOPI: Borrowing Microsoft Code Signing Certificates. https://blog.consciou shacker.io/index.php/2017/09/27/borrowing-microsoft-code-signing-certificates/
9. Rajaraman, A., Ullman, J.: Mining of Massive Datasets (2010). (Chapter 3)
10. Ni, Y., Chu, K., Bradley, J.: Detecting Abuse at Scale: Locality Sensitive Hashing at Uber Engineering (2017). https://eng.uber.com/lsh/

Payload-Based Statistical Intrusion Detection for In-Vehicle Networks

Takuya Kuwahara[1]([✉]), Yukino Baba[1], Hisashi Kashima[1], Takeshi Kishikawa[2],
Junichi Tsurumi[2], Tomoyuki Haga[2], Yoshihiro Ujiie[2], Takamitsu Sasaki[2],
and Hideki Matsushima[2]

[1] Kyoto University, Kyoto, Japan
taakuu19@gmail.com
[2] Panasonic Corporation, Kadoma, Japan

Abstract. Modern vehicles are equipped with Electronic Control Units
(ECUs), and they communicate with each other over in-vehicle networks.
However, since the Controller Area Network (CAN), a common communication protocol for ECUs, does not have a security mechanism, malicious attackers might take advantage of its vulnerability to inject a malicious message to cause unintended controls of the vehicle. In this paper,
we study the applicability of statistical anomaly detection methods for
identifying malicious CAN messages in in-vehicle networks. To incorporate various types of information included in a CAN message, we apply a
rule-based field classification algorithm for extracting message features,
and then obtain low dimensional embeddings of message features, and
use the reconstruction error as a maliciousness score of a message. We
collected CAN message data from a real vehicle, and confirmed the effectiveness of the methods in practical situations.

1 Introduction

Modern vehicles are equipped with multiple Electronic Control Units (ECUs)
that control the main operations of vehicular components, such as engines or
brakes. There are about 100 ECUs in a present-day vehicle, each of which com-
municates with the others over a bus-topology in-vehicle network. The vehicles are also equipped with external communication devices that allow them to
update the firmware and interact with mobile devices such as the drivers' smart-
phones. These communication devices act as an interface between the in-vehicle
network and the out-vehicle network.

Although the external network connectivity offers various benefits, it also
increases the security risk within the in-vehicle network. The Controller Area
Network (CAN), a widely used communication protocol for ECUs, does not
have a security mechanism to detect improper packets. If attackers exploit the
vulnerability of an ECU and inject a malicious message, they will be able to
control other ECUs to cause incorrect operations within the vehicle. Actually,
in 2010, Koscher *et al.* reported the ability of malicious CAN messages to adver-
sarially control several operations, such as unlocking the doors and stopping

M. Ganji et al. (Eds.): PAKDD 2018, LNAI 11154, pp. 186–192, 2018.
https://doi.org/10.1007/978-3-030-04503-6_20

the engine [3]. With the emergence of telematics services and intelligent navigation systems, and the increasing popularity of connected cars, it has become an
urgent matter to protect in-vehicle networks against security threats.

In this paper, we address the problem of anomaly detection which aims to
identify malicious CAN messages in in-vehicle networks. We focus on intrusion attacks, which is one of several standard attacks wherein an attacker need
not investigate the message patterns on the CAN bus network nor conduct a
reverse-engineering investigation of vehicle functions. Particularly, we aim to
detect whether a *single message* is malicious or not. We define a malicious message as a message which is sent by an attacker to cause incorrect operations
within the vehicle. Despite the practical benefit of malicious message detection,
existing studies for statistical intrusion detection in in-vehicle networks focus on
detecting whether a *message sequence* is malicious. Some of them use only the
timestamps of the messages [1,2,8], while the others use the payload information
of CAN messages [4,7,9].

We propose a statistical method for detecting a malicious message in in-
vehicle networks. Our payload-based method extracts features from a CAN message, which is originally represented as a n-bit payload. A CAN message includes
various information and its semantics would be helpful for the detection. We
apply a rule-based field classification algorithm [5] to infer the semantics and
design three features based on the classification results. We then use principal
component analysis (PCA) to obtain low dimensional embeddings of messages.
The reconstruction error is used as a maliciousness score of each CAN message.

We actually perform attacks on a in-vehicle network by injecting malicious
CAN messages that cause incorrect operations within the vehicle. We evaluate
our method in three attack scenarios: steering for an unintended direction, controlling a speedometer as if the vehicle is idle, and changing a gear for driving
backwards. By applying our method to CAN data collected in each scenario, we
observed that our simple method accurately detected malicious messages and
confirmed that the proposed method performs better than the more complex
LSTM-based method in many cases.

2 Proposed Methods for Intrusion Detection in In-Vehicle Networks

Let us formalize the anomaly detection problem we address in this paper.
Sequences of normal CAN messages N are available for training an anomaly
detection model. Given a new sequence of CAN messages D, our goal is to
determine whether each message in D is a malicious one or not. We especially
focus on intrusion attacks on CAN networks through out-vehicle networks. The
payload of a CAN message is of n-bit length (commonly, $n = 64$); therefore a
CAN message is denoted by $b \in \{0, 1\}^n$.

CAN messages include a variety of information about the status of a car.
Instead of directly using the binary values, considering the semantics of payloads
would be helpful for detecting malicious messages. However, the semantics differs

among types of cars [6], such as Toyota Prius or Ford Escape, and the car companies keep it private. Thus, we resort to infer the semantics from CAN messages themselves. Markovitz and Wool [5] proposed a rule-based algorithm for splitting a message into fields and identifying the field types. We apply this algorithm for extracting a feature of each message.

The algorithm assumes there are three field types: constant, multi-value, and counter/sensor. Formally, the algorithm takes a sequence of messages, $(b_1, \ldots, b_T)$, and converts each message, b_t, to $(c_{t,1}, c_{t,2}, \ldots, m_{t,1}, m_{t,2}, \ldots, s_{t,1}, s_{t,2}, \ldots)$, where $c_{t,i}$ is the value of the i-th constant field, $m_{t,j}$ is the value of the j-th Multi-value field, and $s_{t,k}$ is the value of the k-th counter/sensor value. These values are decoded from the corresponding bits.

By using the identified field values, we represent each message by a feature vector. Specifically, we design three features for the t-th message:

- *All-fields*: this feature incorporates the values of all the field types, i.e., $x_t = (c_{t,1}, c_{t,2}, \ldots, m_{t,1}, m_{t,2}, \ldots, s_{t,1}, s_{t,2}, \ldots)$
- *One-field-history*: this feature focuses on one filed type and incorporates the values of the previous k messages. For example, if we focus on the multi-value field, the feature for the t-th message is represented as follows: $x_t = (m_{t,1}, m_{t,2} \ldots, m_{t-1,1}, m_{t-1,2}, \ldots, m_{t-k,1}, m_{t-k,2}, \ldots)$
- *All-field-history*: this feature incorporates the values of all the field types of the previous k messages, i.e.,

$$x_t = \Big(c_{t,1}, c_{t,2}, \ldots, c_{t-1,1}, c_{t-1,2}, \ldots, c_{t-k,1}, c_{t-k,2}, \ldots,$$
$$m_{t,1}, m_{t,2}, \ldots, m_{t-1,1}, m_{t-1,2}, \ldots, m_{t-k,1}, m_{t-k,2}, \ldots,$$
$$s_{t,1}, s_{t,2}, \ldots, s_{t-1,1}, s_{t-1,2}, \ldots, s_{t-k,1}, s_{t-k,2}, \ldots, \Big)$$

We use the reconstruction error as a maliciousness score of each message. By applying PCA for feature vectors of normal messages, we first build a dimension reduction function, $q(x)$, which returns a low dimensional embedding of x.

In the detection phase, given the target message x_t, we obtain its embedding $z_t = q(x_t)$. We then calculate the reconstruction error of the target message as $e_t = \| x_t - p(z_t) \|^2$, where $p(\cdot)$ is the reconstruction function, which is a by-product of $q(\cdot)$. If the reconstruction error is higher than a threshold, the target message is detected as a malicious one.

3   Experiments

3.1   Datasets

We collected CAN messages sent into the in-vehicle network of an actual automobile. We first prepared a normal dataset, which contained messages recorded while a vehicle was moving. This normal dataset was used for capturing linear projection with principal component analysis (PCA). We prepared nine datasets which differ in the amount of malicious messages.

These scenarios are simulated in the following ways:

- *Small1, Small2, Small3* and *Small4*: A small number of malicious messages are injected right after a normal message having a specific ID. We assume that attackers know the cycle of normal messages and they manage to inject malicious messages in accordance with this cycle.
- *Large1, Large2, Large3, Large4* and *Large5*: A large number of malicious messages are injected independent of the ways that normal messages are present in the network.

We have confirmed that inserting our attack messages actually affect the behavior of the actual vehicle in our experiments; it is worth noticing that we demonstrate the feasibility of the statistical approaches in practical attack scenarios.

3.2 Methods

We prepared three types of features for our method: *All-fields, Checksum-history*, and *All-field-history*. All-fields and All-field-history are explained in Sect. 2. Checksum-history is a variation of One-field-history which targeting the checksum field, that is, a specific byte of a CAN message. Our method takes the number of dimensions for PCA as an input parameter. For each dataset, we choose the number from 2 to $\min(30, \dim(x_t))$ by using the cross-validation, where $\dim(\cdot)$ returns the size of the vector. In addition, for Checksum-history and All-field-history, we chose the number of previous messages from 0 to 30.

We compare our proposed method with a LSTM method [9]. This method was proposed for detecting a malicious message sequence, although we target the detection of a malicious message. Thus, we modify the LSTM method to apply for malicious message detection. Specifically, we prepare three variations for the LSTM method: (1) *LSTM-max*: this method uses the maximum of the n-bit logarithmic loss values as a maliciousness score; (2) *LSTM-min*: this method uses the minimum of them; (3) *LSTM-mean*: this method uses the average of them.

3.3 Results

We first conducted an experiment for ID-1, which is used for in-vehicle communication about steering. Table 1 shows AUC values. We see that LSTM-max, LSTM-mean, and All-fields achieved an AUC of 1.0 in almost all the datasets. Especially, LSTM-mean showed the perfect detection accuracy in all the datasets. Checksum-history and All-field-history were inferior to All-field in most cases; however, they demonstrated better performance than All-fields in large1 and large2. The detection accuracy of LSTM-max and that of All-fields were low in large1 and large2. This is because payloads of malicious messages and those of normal messages are very similar in these datasets.

Table 1. AUC values in steering attack case

	Small1	Small2	Small3	Small4	Large1	Large2	Large3	Large4	Large5
LSTM-max	1.000	1.000	1.000	1.000	0.191	0.341	1.000	1.000	1.000
LSTM-min	0.530	0.449	0.396	0.304	0.873	0.774	0.887	0.796	0.774
LSTM-mean	1.000	1.000	1.000	1.000	1.000	1.000	1.000	1.000	1.000
All-fields	1.000	1.000	1.000	1.000	0.675	0.603	1.000	1.000	1.000
Checksum-history	0.953	0.847	0.852	0.966	0.971	0.925	0.991	0.945	0.942
All-field-history	0.934	0.875	0.764	0.906	0.935	0.890	0.954	0.899	0.895

We next performed an experiment for ID-2, which is used for speedometer. Table 2 shows AUC values. The proposed method demonstrated better detection performance than the LSTM method in most cases. The reason of lower performance of the LSTM method in this experiment is explained as follows. In the large datasets, malicious messages are inserted consecutively and the average size of malicious message sequences is 24. When the LSTM method is given a malicious message as an input, it predicts that the next message would be almost identical to the input; if the actual next message is a malicious one, the logarithmic loss is small and thus the message is wrongly detected as a normal one. Additionally, a large population of malicious messages (for example, one dataset has 39,384 malicious messages out of 43,308) results inferior performance of LSTM method as well. In the small datasets, inserted malicious messages are identical to normal messages except only the last two bytes are different. The LSTM method is likely to predict a message similar to a normal one will arrive next, and the malicious messages wrongly assigned with low maliciousness scores. Compared to the LSTM method which directly uses n-bits binary values, our proposed method is based on a feature extraction; our well-designed features would contribute to outperform the LSTM method in such a difficult case.

Table 2. AUC values for speedometer attack case

	Small1	Small2	Small3	Small4	Large1	Large2	Large3	Large4	Large5
LSTM-max	0.146	0.105	0.183	0.131	0.194	0.339	0.218	0.071	0.061
LSTM-min	0.571	0.461	0.543	0.561	0.224	0.286	0.399	0.582	0.587
LSTM-mean	0.058	0.004	0.004	0.007	0.004	0.000	0.000	0.000	0.004
All-fields	0.535	0.261	0.541	0.537	0.822	0.622	0.570	0.536	0.538
Checksum-history	0.767	0.783	0.809	0.685	0.566	0.552	0.511	0.798	0.764
All-field-history	0.676	0.690	0.672	0.577	0.481	0.274	0.308	0.742	0.718

Our final experiment targets ID-3, which is for gears. Table 3 shows AUC values and it is seen that All-fields demonstrated the highest detection accuracy among other methods.

Table 3. AUC values for gear attack case

	Large1	Large2
LSTM-max	0.939	0.835
LSTM-min	0.603	0.445
LSTM-mean	0.668	0.358
All-fields	1.000	0.997
Checksum-history	0.922	0.904
All-field-history	0.707	0.542

4   Conclusion

In this paper, we proposed a anomaly detection method for the CAN bus network. Our method focuses on injection attacks, which can control several unintended operations of a vehicle while one can easily implement the attacks. We proposed three types of features and introduced the statistical method for detecting a malicious message. We conducted experiments to investigate the effectiveness of the proposed features for anomaly detection in CAN data and to compare the detection accuracy of the proposed method with that of the LSTM-based method. We used nine types of real datasets including attack messages collected from a moving car; some of them include a large number of injected malicious messages and the others include a small number of malicious messages. The results showed that the proposed method successfully detected the injected malicious messages in most of the datasets and were competitive with or sometimes outperformed the LSTM-based method.

References

1. Cho, K.T., Shin, K.G.: Fingerprinting electronic control units for vehicle intrusion detection. In: Proceedings of the 25th USENIX Security Symposium (USENIX), pp. 911–927 (2016)
2. Hamada, Y., Inoue, M., Horihata, S., Kamemura, A.: Intrusion detection by density estimation of reception cycle periods for in-vehicle networks: a proposal. In: Proceedings of the Embedded Security in Cars Conference (ESCAR) (2016)
3. Koscher, K., et al.: Experimental security analysis of a modern automobile. In: Proceedings of the IEEE Symposium on Security and Privacy, pp. 447–462 (2010)
4. Marchetti, M., Stabili, D., Guido, A., Colajanni, M.: Evaluation of anomaly detection for in-vehicle networks through information-theoretic algorithms. In: Proceedings of the 2016 IEEE 2nd International Forum on Research and Technologies for Society and Industry Leveraging a Better Tomorrow (RTSI), pp. 1–6 (2016)
5. Markovitz, M., Wool, A.: Field classification, modeling and anomaly detection in unknown CAN bus networks. In: Proceedings of the Embedded Security in Cars Conference (ESCAR) (2015)
6. Miller, C., Valasek, C.: Adventures in automotive networks and control units (2013). http://www.ioactive.com/pdfs/IOActive_Adventures_in_Automotive_Networks_and_Control_Units.pdf

7. Müter, M., Asaj, N.: Entropy-based anomaly detection for in-vehicle networks. In: Proceedings of the IEEE Intelligent Vehicles Symposium, pp. 1110–1115 (2011)
8. Taylor, A., Japkowicz, N., Leblanc, S.: Frequency-based anomaly detection for the automotive CAN bus. In: 2015 World Congress on Industrial Control Systems Security (WCICSS) (2015)
9. Taylor, A., Leblanc, S., Japkowicz, N.: Anomaly detection in automobile control network data with long short-term memory networks. In: 2016 IEEE International Conference on Data Science and Advanced Analytics (DSAA), pp. 130–139 (2016)

Biologically-Inspired Techniques for Knowledge Discovery and Data Mining (BDM)

Discovering Strong Meta Association Rules Using Bees Swarm Optimization

Youcef Djenouri[1][(✉)], Asma Belhadi[2], Philippe Fournier-Viger[3],
and Jerry Chun-Wei Lin[4]

[1] IMADA, SDU, Odense, Denmark
`djenouri@imada.sdu.dk`
[2] RIMA, USTHB, Algiers, Algeria
`abelhadi@usthb.dz`
[3] School of Social Sciences and Humanities, Harbin Institute of Technology
(Shenzhen), Shenzhen, China
`philfv8@yahoo.com`
[4] Department of Computing, Mathematics, and Physics, Western Norway University
of Applied Sciences (HVL), Bergen, Norway
`jerrylin@ieee.org`

Abstract. For several applications, association rule mining produces an extremely large number of rules. Analyzing a large number of rules can be very time-consuming for users. Therefore, eliminating irrelevant association rules is necessary. This paper addresses this problem by proposing an efficient approach based on the concept of meta association rules. The algorithm first discovers dependencies between association rules called meta association rules. Then, these dependencies are used to eliminate association rules that can be replaced by a more general rule. Because the set of meta-rules can be very large, a bee swarm optimization approach is applied to quickly extract the strongest meta-rules. The approach has been applied on a synthetic dataset and compared with a state-of-the-art algorithm. Results are promising in terms of number of rules found and their quality.

Keywords: Meta rule extraction · Association rule
Bio-inspired methods

1 Introduction

Association Rule Mining ($\mathcal{ARM}$) is the process of extracting useful rules from a transactional database. ARM is widely used in many application such as Business Intelligence [6,9], Constraint Programming [13,15], and Information Retrieval [5,7]. Many $\mathcal{ARM}$ mining algorithms have been designed (e.g. Apriori [1], FPgrowth [20], and SSFIM [8]). Nevertheless, applying these algorithms on large datasets obtained from the Web, sensor networks and other applications may generate a huge amount of association rules. Thus, interpreting these

© Springer Nature Switzerland AG 2018
M. Ganji et al. (Eds.): PAKDD 2018, LNAI 11154, pp. 195–206, 2018.
https://doi.org/10.1007/978-3-030-04503-6_21

rules can be difficult and very time-consuming for users. To address this problem, a promising solution is to prune irrelevant association rules. Three main approaches have been studied. First, end-users may select rules by hand or provide constraints to filter rules based on their expertise or background knowledge [21]. However, applying this approach to large databases is difficult and time consuming. The second approach relies on the calculation of statistical properties of rules [27]. Algorithms that belong to this category performs additional statistical calculations during the mining process, which may compromise the quality of the final solution. The third approach is to extract meta association rules [2] to identify dependencies between association rules. However, the set of meta-rules may still be very large. Thus, it is desirable to eliminate some rules. This paper addresses this issue by proposing a pruning approach. It first prunes an initial set of association rules to keep only the most representative rules. Then, meta-rules are pruned. This is done using a bio-inspired algorithm, named BSO-MR, which uses a simple neighborhood search and an efficient strategy to select the search area.

The rest of the paper is organized as follows. Section 2 reviews studies on rule pruning. Section 3 presents a meta rule extraction algorithm, which influenced the design of the proposed algorithm for pruning rules, called Pruning Rules. Section 4 presents an experimental evaluation of the two algorithms (Meta-rules and Pruning Rules) on a large dataset, and a comparison with previous work. Section 5 concludes this paper by some remarks and suggestions.

2   Related Work

Studies on association rule pruning can be classified into three categories: end-user based, statistics-based, and meta rule based.

End User Based Approaches. In these approaches, the end user eliminates rules as a postprocessing step, by relying on his background knowledge. This step was introduced by Klemettinen et al. [21] in the *Visualizer* algorithm. The user must specify a set constraints to select relevant rules. Then, an hyper-graph is built to show the selected rules to the user. Although, the information provided by the user is used to reduce the number of association rules, it remains an inefficient process because $\mathcal{ARM}$ is first applied and then rules are selected as a post-processing step. But the user has no control over the $\mathcal{ARM}$ process.

An interactive approach was proposed [23] where the user must specify three sets of knowledge using different levels. Then, association rules are discovered and matched to these sets, and then partitioned into several classes. Then, the rules in each class are stored using semantic measure values in order to find interesting rules. The CAP algorithm [26] was proposed, based on the Apriori principle to select relevant rules and speed up association rule discovery. To apply this approach, the user must provide a set of constraints in first order logic. Then, anti-monotonicity and succinctness properties of the constraints are used to prune rules during the $\mathcal{ARM}$ process. The particularity of this work is that it is interactive. During each iteration of the pattern mining process, the user can delete, remove or modify constraints. A few rule pruning approaches rely on knowledge represented as ontologies to prune rules. An ontology is a formal, explicit specification of a shared conceptualization. An ontology is typically built by domain experts or end users. A post-processing approach [24] uses an ontology to prune rules. First, an ontology must be provided by a domain expert. Then, the ontology is transformed to expression logic, and for each rule, its reliability is calculated and matched to the set of expression logic. Low-reliability rules that do not belong to this set are pruned. An efficient postprocessing and interactive method [25] was also proposed to reduce the number of association rules. First, an ontology is built by the user. Then, some filters are applied to prune uninteresting rules such as the Minimum improvement constraint filter (MICF) and the Item-Relatedness Filter (IRF). The MICF selects rules having a confidence that is greater than a threshold, and the confidence of any of its simplifications. IRF measures the *relatedness* between items, that is their semantic distance in item taxonomies. The main limitation of ontology-based approaches is the generally high cost of constructing a domain ontology. Moreover, it is often difficult to make several users agree on the design of an ontology.

Statistics-Based Approaches. Berrado and Runger [2] proposed an approach, which consists of applying the Apriori algorithm to extract association rules and meta-rules. The information provided by these meta-rules is then used to prune association rules by relying on the concept of rule-transaction (a set of rules having the same consequent and supported by the same transactions in the original database). Dimitrijević and Bošnjak [3] proposed an algorithm for pruning rules found in web log data. It first discards trivial rules (rules having a confidence of 100%). Then, the rules sharing a same subset of items and having approximately the same confidence are replaced by this subset of items. Fernandes and Garcia [16] proposed to reduce the number of rules using a dual scale strategy. It clusters items using the AKMS algorithm. Then, pruning is done by considering only the clusters having a minimum number of items. Liu et al. [22] developed a rule pruning algorithm, which first generates meta-rules from the set of association rules. Then, the algorithm uses these meta-rules to select interesting rules. Watanabe [28] proposed to prune redundant rules during the extraction process using the confidence measure. A rule is considered as redundant if its confidence is less than a rule having the same consequent. In another study, Stumme et al. [27] designed two subsets of association rules called the Guigues-Duquenne sand Luxenburger basis. These sets can be much smaller than the set of all association rules, and can be used to derive information about all association rules (they are a lossless representation of all rules). Another concise representation of association rules based on the Guigues-Duquenne sand Luxenburger basis Zaki [29] proposed a new set of association rules called closer rules. The closer rules reduce largely the rules space. In another study, the StatApriori algorithm was proposed [19]. It is a modified version of the Apriori algorithm, which calculates statistical information to prune redundant and spurious rules. Another algorithm called PDS [22] first prunes insignificant rules. Then, it calculates the correlation between the antecedent and consequent of each significant rule and applies the Chi-square test. Then, the Guigues-Duquenne basis is used to summarize rules.

Meta-rules Based Approaches. The idea of these approaches is to find hidden patterns and relationships between rules. Berrado *et al.* [2] proposed to represent the set of association rules as transactions and to extract meta-rules from this data, where a meta rule is a relationship between two association rules. The main drawback of this approach is that discovering meta-rules can be very time consuming since the Apriori algorithm is applied, which is an exact algorithm. It is well-known that the number of generated rules can be very large. To find rules more efficiently, a promising solution is to apply approximate algorithms for association rule mining. Besides, another limitation is that meta-rules are restricted to having a single rule in their antecedent and consequent. Thus, some important information may be lost about dependencies between association rules. This paper proposes a new approach to discover a set of important meta-rules, which is presented in the next section.

3   The Proposed Two-Step Pruning Approach

This section presents an overview of the proposed approach. A set of rules is first obtained by applying $\mathcal{ARM}$ on a transactional database. Then, a set of meta-rules is generated by applying $\mathcal{ARM}$ and a first pruning step is done. Afterwards, these meta-rules are analyzed and evaluated to find stronger meta-rules (rules that perfectly describe the rule space). Using these stronger meta-rules, a second pruning step is performed to select the most representative rules.

3.1   The First Pruning Step

The first pruning step is done as follows. To discover meta-rules, the association rules found by $\mathcal{ARM}$ are first transformed into a transactional database. Unlike association rule mining which aims at extracting relationships between items, meta-rules are extracted to find relationships between association rules. Therefore, meta-rules can be viewed as more abstract than association rules. The transformation of association rules to transactions is done as follows. Each rule is considered as an item and the rules that have the same consequent are considered as belonging to a same transaction. Consequently, the number of transactions is equal to the number of rule consequents.

Formally, let $AR = \{r_1, r_2, ..., r_n\}$ be a set of association rules. A transactional database is a set of items $I = \{I_1, I_2, ..., I_n\}$ and a set of transactions $T = \{T_1, T_2, ..., T_m\}$. The transformation from the set AR to a transactional database is done as follows:

$$\begin{cases} I = AR \\ x \Rightarrow y \in AR, \ \Rightarrow y \in T. \end{cases} \tag{1}$$

Algorithm 1 explains how the transformation process is performed.

Algorithm 1. Transformation of association rules to a transactional database

Initialize T to $\emptyset$.
    for each rule r: $X \Rightarrow Y$ in AR **do**
3:    exist=false.
        for each t in T and exist=false **do**
          if $Y = t$ **then**
6:            exist=true.
            insert r in t.
          end if
9:      **end for**
    end for
    if exist=false **then**
12:   create a new row in T labeled Y.
    end if

Example 1. Let I be the set of items $\{i_1, i_2, i_3, i_4\}$ and AR be a set of six associ-
ation rules: r_1: $i_1, i_2 \Rightarrow i_3$; r_2: $i_4 \Rightarrow i_3$; r_3: $i_1 \Rightarrow i_2$; r_4: $i_4 \Rightarrow i_1$; r_5: $i_4 \Rightarrow i_1, i_2, i_3$;
r_6: $i_4 \Rightarrow i_2, i_3$.

From these rules, seven consequents (i_1), (i_2), (i_3), (i_1, i_2), (i_1, i_3), (i_2, i_3),
(i_1, i_2, i_3) are identified. Thus, these rules are transformed into seven transactions: $T = \{t_1, t_2, t_3, t_4, t_5, t_6, t_7\}$. The transaction t_1: r_4, r_5 is created because
r_4 and r_5 have the same consequent (i_1). The transaction t_2: r_3, r_5, r_6 is created
because r_3, r_5 and r_6 have the same consequent (i_2). The other transactions t_3
to t_7 are obtained in a similar way.

Meta-rules Discovery. After the transformation of association rules to transactions, an association rule mining algorithm is applied on the resulting transactional base T. Due to the potentially large size of T, applying an exact association rule mining algorithm can be time-consuming. To address this issue, an
approximate bio-inspired algorithm, named $(BSO\text{-}MR)$, is developed for discovering meta-rules using a swarm intelligence approaches [11,17,18]. It extends
the BSO-ARM algorithm [10,14]. Initially, a reference solution is generated randomly or using a heuristic. To avoid considering a same solution multiple times,
in each iteration the current Sref is inserted in a Tabou list. Then, from the current Sref, a search area is assigned to each bee. Each bee explores its assigned
area using a neighborhood search. The neighborhood search consists of modifying the given solution S by flipping a random bit. By this simple operation, n
neighborhoods are created where n is the solution size. After that, bees communicate to select the best solution. This latter becomes the next Sref if it does not
exist in the Tabou list. Otherwise, a solution is randomly chosen in the search
space to continue searching. The best solution of each iteration is kept in memory and the overall best solution is finally selected. This process is repeated until
a maximum number of iterations is reached.

3.2   The Second Pruning Step

A too large number of association rules is difficult to analyze for users. Reducing the number of rules is thus an interesting challenge. Based on the meta rule extraction algorithm proposed in Sect. 3.1, a new algorithm for pruning association rules.

Formulation of Pruning Rules Problem. Let RS be the set of all rules. We define a measure called *Relevance* to evaluate how interesting a rule is. It is a function from RS to R, where R is the set of real values. This function measures the quality of a given rule r and is defined as follows:

Relevance$(r)=\alpha$ Lift$(r)+\beta$ Leverage$(r)+ \gamma$ Conviction(r). The reason for defining this measure is that the traditional support and confidence measures used in association rule mining are often viewed as inappropriate. For this reason, other measures have been proposed such as the lift, leverage, and conviction. The *Relevance* measure combines these measures using weights α, β, and γ which are set to zero or one.

We also use a *Distinct* measure, defined as a function from RS^n to R as: Distinct$(r_1, r_2, ..., r_n) = \sum_{i=1}^{n} \sum_{j=1}^{n} D(r_i, r_j)$, where $D(r_i, r_j)$ measures the distance between the rules r_i and r_j and is calculated by adapting the interestingness measure developed in [4,12].

To find interesting rules that well represent the rule space, the problem of pruning rules can be defined as follows. Find a subset of l rules $\{r_1, r_2, r_3, ..., r_l\}$ denoted as $R_{sub} \subseteq RS$ that maximizes the Distinct function and the Relevance function of each rule in R_{sub}. In other word, maximizes the function f such that:

$$f_{max}(R_{sub}) = \frac{Distinct(R_{sub}) + \sum_{i=1}^{l} Relevant(r_i)}{l}. \tag{2}$$

Rule Pruning Algorithm. The proposed approach to prune rules is composed of two steps. First, the algorithm extracts meta-rules to find relationships and dependencies between extracted rules. Second, according to the relationships found (*meta-rules*), the sets of rules are divided into two categories: representative rules and deductive rules. Representative rules are rules that well describe the data. Deductive rules are rules that can be deduced from representative rules. Finally, deductive rules are deleted and only the representative rules are kept.

Proposition 1. *Consider a meta-rule mr: $X \Rightarrow Y$ where X and Y are rules. If the confidence of mr is 100% then we can say that Y is a set of representative rules and X is a set of deductive rules.*

Proof. Let us consider, $mr: X \Rightarrow Y$. If the confidence of mr is 100% then it means that when the set of rules in the antecedent appears then the set of rules in the consequent also appears. Moreover, rules consequents in X include on the rule consequents of Y. Therefore, we can keep the rules composed of Y and prune the rules of X. In other words, Y can be called representative rules and X can be called deductive rules.

Algorithm 2 gives the pseudo-code of the Rule Pruning Algorithm.

Algorithm 2. Pruning Algorithm

The set of Rules RS.
Output: Set of Representative rules R_{sub}.
Meta-Rules().
4: **for** each meta-rules mr found **do**
    if confidence(mr)=100% **then**
        Add the consequent of mr to R_{sub}.
    end if
8: **end for**

This algorithm finds meta-rules from the set of initial rules RS found by $\mathcal{ARM}$. Then, the confidence of each meta-rule is calculated. If the confidence of such meta-rule is equal to 100% then we consider that the consequent of this meta-rule is a representative rules and insert them in R_{sub}. This process is repeated until all meta-rules are processed.

Example 2. To illustrate how the Rule Pruning Algorithm is applied, consider the following five rules. $r_1 : A \Rightarrow B, C, D$, $r_2 : A \Rightarrow D, E$, $r_3 : A \Rightarrow D$, $r_4 : E \Rightarrow B, C$, and $r_5 : E \Rightarrow C$. First, rules are transformed into a transactional database, shown in Table 1. The first transaction represents the rules that share the item B in their consequent. The same process is repeated for the three remaining transactions. After this step, the Meta-rules algorithm is applied to extract meta-rules. The found meta-rules are shown in Table 2 with their confidence. From this table, the meta-rules that have a confidence of 100% are extracted. Let mr_2 and mr_6 be these meta-rules. For mr_2, we keep r_1 and prune r_6 and for mr_2, we keep r_1 and prune r_3. Thus, the rules r_3 and r_4 are pruned.

Table 1. Transformed transactional database

Transaction ID of t_i	Rules in t_i
B	r_1, r_4
C	r_1, r_4
D	r_1, r_2, r_3
E	r_2

4 Experimental Study

The proposed approach has implemented using Java environment on $I7$ processor and 4GB memory. To validate the pruning approach, several experiments have been performed on the set of association rules described in [2].

Table 2. Meta-rules found

Meta-rule ID	The antecedent part	The consequent part	The confidence
mr_1	r_1	r_4	66%
mr_2	r_4	r_1	100%
mr_3	r_1	r_2	33%
mr_4	r_2	r_1	50%
mr_5	r_1	r_3	33%
mr_6	r_3	r_1	100%

Table 3. Number of meta-rules that have 100% of the different approaches

Input rules	BSO-MR algorithm	Berrado's algorithm
371	**702**	517
797	**408**	147
156	**592**	314
1215	**1770**	1421
1442	**3204**	2698
1374	**1214**	953
1707	**1044**	945

Table 4. Number of remaining rules of the different approaches

Input rules	BSO-MR algorithm	Berrado's algorithm
371	**216**	246
797	**300**	316
156	**198**	211
1215	**452**	502
1442	**502**	567
1374	**138**	120
1707	**202**	300

We first present the results of our pruning approach and compare it to the previous work of Berrado *et al.* [2] which, also employs meta-rules principle. The used parameters are fixed as follows:

- The number of iterations IMAX is set to 100.
- Flip parameter is set to 99.
- The number of bees is set to 80.

Tables 3 and 4 show respectively the number of meta-rules and the number of remaining rules obtained by our algorithm (BSO-MR) and by Berrado's algorithm with different number of input rules.

Table 5. The quality of the rules obtained by the proposed algorithm and Berrado's Algorithm

Input rules	BSO-MR algorithm	Berrado's algorithm
1000	**422**	400
2000	**1425**	1365
3000	**2352**	2105
4000	**2880**	2409
5000	**3115**	3110
6000	**3752**	3325
7000	**4403**	4091
8000	**5196**	4895
9000	**5965**	5263
10000	**6208**	6110

From Table 3, note that the number of meta-rules obtained by our algorithm is greater than those obtained by Berrado's algorithm whatever the number of input rules. Indeed, the proposed approach generates all meta-rules having 100% of confidence. That is not the case of Berrado's algorithm in which only one way meta-rules could be generated. These results validate our studies in Sect. 3.1. Moreover, BSO-MR gives the best results compared to Berrado's algorithm for all the input rules. In addition, from Table 4, the number of output rules obtained by Berrado's algorithm is high compared to the number of rules obtained by our algorithm. This can be explained by missing information in Berrado's algorithm which uses only a few number of meta-rules in the pruning step. Also, in this scenario, BSO-MR algorithm outperforms Berrado's algorithm in terms of number of remaining rules. However, no information can be extracted about the quality of the remaining rules obtained by the proposed algorithm. Therefore, we can not conclude if BSO-MR outperforms Berrado's algorithm too in terms of rules quality.

To evaluate the quality of the remaining rules, another experiment has been performed. Table 5 shows the rules quality returned by the proposed algorithm with different number of input rules. This table depicts that BSO-MR algorithm outperforms Berrado's algorithm in terms of the rules quality whatever the number of used input rules. Indeed, in BSO-MR all kind of rules are found, for consequent the search process of interesting rules can include all the rules space. However, on Berrado's work, just a one way association rules are employed which just similar rules are found. Consequently, the search process does not cover all the rules space.

5   Conclusion

An new approach for pruning association rules has been proposed in this paper. The proposed approach augments the usefulness of the resulted rules for the data analyst and the user. To design such an approach, Meta-Rules algorithm is first employed to find some dependencies between the input rules. To efficiently perform meta-rule extraction, bio-inspired approach is proposed called BSO-MR which uses the entire BSO bio-inspired method in the search process. The mining method described above (Meta-Rules algorithm) is then used to prune the irrelevant rules by keeping the representative rules and eliminating the deductive rules, which can be directly deduced by the representative ones. The proposed approach has then been implemented and tested on large data set. To demonstrate the performance of our approach, different evaluation measures have been considered. The distinct measure has also been proposed, which represents the quality of output rules resulting by pruning rules algorithm. The experimental results are very promising. BSO-MR outperforms Berrado's algorithm in terms of the number of output rules. In addition, BSO-MR also outperforms Berrado's algorithm in terms of the quality of output rules. As a perspective to this work, we plan to integrate other data mining techniques such as clustering methods on pruning rules. This can efficiently increase the performance of the present work.

References

1. Agrawal, R., Imieliński, T., Swami, A: Mining association rules between sets of items in large databases. In: ACM SIGMOD Record, vol. 22, pp. 207–216. ACM (1993)
2. Berrado, A., Runger, G.C.: Using metarules to organize and group discovered association rules. Data Min. Knowl. Discov. **14**(3), 409–431 (2007)
3. Dimitrijević, M., Bošnjak, Z.: Discovering interesting association rules in the web log usage data. Interdiscip. J. Inf. Knowl. Manag. **5**, 191–207 (2010)
4. Djenouri, Y., Drias, H., Habbas, Z., Chemchem, A.: Organizing association rules with meta-rules using knowledge clustering. In: 2013 11th International Symposium on Programming and Systems (ISPS), pp. 109–115. IEEE (2013)
5. Djenouri, Y., Belhadi, A., Belkebir, R.: Bees swarm optimization guided by data mining techniques for document information retrieval. Expert. Syst. Appl. **94**, 126–136 (2018)
6. Djenouri, Y., Belhadi, A., Fournier-Viger, P.: Extracting useful knowledge from event logs: a frequent itemset mining approach. Knowl.-Based Syst. **139**, 132–148 (2018)
7. Djenouri, Y., Belhadi, A., Fournier-Viger, P., Lin, J.C.W.: Fast and effective cluster-based information retrieval using frequent closed itemsets. Inf. Sci. **453**, 154–167 (2018)
8. Djenouri, Y., Comuzzi, M., Djenouri, D.: SS-FIM: single scan for frequent itemsets mining in transactional databases. In: Kim, J., Shim, K., Cao, L., Lee, J.-G., Lin, X., Moon, Y.-S. (eds.) PAKDD 2017. LNCS (LNAI), vol. 10235, pp. 644–654. Springer, Cham (2017). https://doi.org/10.1007/978-3-319-57529-2_50
9. Djenouri, Y., Drias, H., Bendjoudi, A.: Pruning irrelevant association rules using knowledge mining. Int. J. Bus. Intell. Data Min. **9**(2), 112–144 (2014)

10. Djenouri, Y., Drias, H., Habbas, Z.: Bees swarm optimisation using multiple strategies for association rule mining. Int. J. Bio-Inspired Comput. **6**(4), 239–249 (2014)
11. Djenouri, Y., Drias, H., Habbas, Z.: Hybrid intelligent method for association rules mining using multiple strategies. Int. J. Appl. Metaheuristic Comput. (IJAMC) **5**(1), 46–64 (2014)
12. Djenouri, Y., Gheraibia, Y., Mehdi, M., Bendjoudi, A., Nouali-Taboudjemat, N.: An efficient measure for evaluating association rules. In: 2014 6th International Conference of Soft Computing and Pattern Recognition (SoCPaR), pp. 406–410. IEEE (2014)
13. Djenouri, Y., Habbas, Z., Djenouri, D.: Data mining-based decomposition for solving the MAXSAT problem: toward a new approach. IEEE Intell. Syst. **32**(4), 48–58 (2017)
14. Djenouri, Y., Habbas, Z., Djenouri, D., Comuzzi, M.: Diversification heuristics in bees swarm optimization for association rules mining. In: Kang, U., Lim, E.-P., Yu, J.X., Moon, Y.-S. (eds.) PAKDD 2017. LNCS (LNAI), vol. 10526, pp. 68–78. Springer, Cham (2017). https://doi.org/10.1007/978-3-319-67274-8_7
15. Djenouri, Y., Habbas,, Z., Djenouri, D., Fournier-Viger, P.: Bee swarm optimization for solving the MAXSAT problem using prior knowledge. Soft Comput. pp. 1–18 (2017)
16. Fernandes, L.A.F., García, A.C.B.: Association rule visualization and pruning through response-style data organization and clustering. In: Pavón, J., Duque-Méndez, N.D., Fuentes-Fernández, R. (eds.) IBERAMIA 2012. LNCS (LNAI), vol. 7637, pp. 71–80. Springer, Heidelberg (2012). https://doi.org/10.1007/978-3-642-34654-5_8
17. Gheraibia, Y., Moussaoui, A., Djenouri, Y., Kabir, S., Yin, P.Y.: Penguins search optimisation algorithm for association rules mining. J. Comput. Inf. Technol. **24**(2), 165–179 (2016)
18. Gheraibia, Y., Moussaoui, A., Djenouri, Y., Kabir, S., Yin, P.-Y., Mazouzi, S.: Penguin search optimisation algorithm for finding optimal spaced seeds. Int. J. Softw. Sci. Comput. Intell. (IJSSCI) **7**(2), 85–99 (2015)
19. Hämäläinen, W.: StatAapriori: an efficient algorithm for searching statistically significant association rules. Knowl. Inf. Syst. **23**(3), 373–399 (2010)
20. Han, J., Pei, J., Yin, Y.: Mining frequent patterns without candidate generation. In: ACM SIGMOD Record, vol. 29, pp. 1–12. ACM (2000)
21. Klemettinen, M., Mannila, H., Ronkainen, P., Toivonen, H., Verkamo, A.I.: Finding interesting rules from large sets of discovered association rules. In: Proceedings of the Third International Conference on Information and Knowledge Management, pp. 401–407. ACM (1994)
22. Liu, B., Hsu, W., Ma, Y.: Pruning and summarizing the discovered associations. In: Proceedings of the fifth ACM SIGKDD International Conference on Knowledge Discovery and Data Mining, pp. 125–134. ACM (1999)
23. Liu, B., Hsu, W., Wang, K., Chen, S.: Visually aided exploration of interesting association rules. In: Zhong, N., Zhou, L. (eds.) PAKDD 1999. LNCS (LNAI), vol. 1574, pp. 380–389. Springer, Heidelberg (1999). https://doi.org/10.1007/3-540-48912-6_52
24. Mansingh, G., Osei-Bryson, K.-M., Reichgelt, H.: Using ontologies to facilitate post-processing of association rules by domain experts. Inf. Sci. **181**(3), 419–434 (2011)
25. Marinica, C., Guillet, F.: Knowledge-based interactive postmining of association rules using ontologies. IEEE Trans. Knowl. Data Eng. **22**(6), 784–797 (2010)

26. Ng, R.T., Lakshmanan, V.S., Han, J., Pang, A.: Exploratory mining and pruning optimizations of constrained associations rules. In: ACM SIGMOD Record, vol. 27, pp. 13–24. ACM (1998)
27. Stumme, G., Taouil, R., Bastide, Y., Pasquier, N., Lakhal, L.: Intelligent structuring and reducing of association rules with formal concept analysis. In: Baader, F., Brewka, G., Eiter, T. (eds.) KI 2001. LNCS (LNAI), vol. 2174, pp. 335–350. Springer, Heidelberg (2001). https://doi.org/10.1007/3-540-45422-5_24
28. Watanabe, T.: An improvement of fuzzy association rules mining algorithm based on redundacy of rules. In: 2010 2nd International Symposium on Aware Computing (ISAC), pp. 68–73. IEEE (2010)
29. Zaki, M.J.: Generating non-redundant association rules. In: Proceedings of the sixth ACM SIGKDD International Conference on Knowledge Discovery and Data Mining, pp. 34–43. ACM (2000)

ParallelPC: An R Package for Efficient Causal Exploration in Genomic Data

Thuc Duy Le[1]([⊠]), Taosheng Xu[2], Lin Liu[1], Hu Shu[1], Tao Hoang[1],
and Jiuyong Li[1]

[1] University of South Australia, Mawson Lakes, SA 5095, Australia
{Thuc.Le,Lin.Liu,Jiuyong.Li}@unisa.edu.au, hushu@iie.ac.cn,
hoatn002@mymail.unisa.edu.au
[2] Institute of Intelligent Machines, Hefei Institutes of Physical Science,
Chinese Academy of Sciences, Hefei 230031, China
xutaosheng@aliyun.com

Abstract. Discovering causal relationships from genomic data is the ultimate goal in gene regulation research. Constraint based causal exploration algorithms, such as PC, FCI, RFCI, PC-simple, IDA and Joint-IDA have achieved significant progress and have many applications. However, their applications in bioinformatics are still limited due to their high computational complexity. In this paper, we present an R package, *ParallelPC*, that includes the parallelised versions of these causal exploration algorithms and 12 different conditional independence tests for each. The parallelised algorithms help speed up the procedure of experimenting large biological datasets and reduce the memory used when running the algorithms. Our experiment results on a real gene expression dataset show that using the parallelised algorithms it is now practical to explore causal relationships in high dimensional datasets with thousands of variables in a personal multicore computer. We present some typical applications in bioinformatics using different algorithms in *ParallelPC*. *ParallelPC* is available in CRAN repository at https://cran.r-project.org/web/packages/ParallelPC/index.html.

Keywords: Causality discovery · Bayesian networks
Parallel computing · Constraint-based methods

1 Introduction

Inferring causal relationships between variables is an ultimate goal of many research areas, e.g. investigating the causes of cancer, finding the factors affecting life expectancy. Therefore, it is important to develop tools for causal exploration from real world datasets.

One of the most advanced theories with widespread recognition in discovering causality is the Causal Bayesian Network (CBN), [1]. In this framework, causal relationships are represented with a Directed Acyclic Graph (DAG). There are

© Springer Nature Switzerland AG 2018
M. Ganji et al. (Eds.): PAKDD 2018, LNAI 11154, pp. 207–218, 2018.
https://doi.org/10.1007/978-3-030-04503-6_22

two main approaches for learning the DAG from data: the search and score approach, and the constraint based approach. While the search and score approach raises an NP-hard problem, the complexity of the constraint based approach is exponential to the number of variables. Constraint based approach for causality discovery has been advanced in the last decade and has been shown to be useful in some real world applications. The approach includes causal structure learning methods, e.g. PC [2], FCI and RFCI [3], causal inference methods, e.g. IDA [4] and Joint-IDA [5], and local causal structure learning such as PC-Simple [6,7]. However, the high computational complexity has hindered the applications of causal discovery approaches to high dimensional datasets, e.g. gene expression datasets where the number of genes (variables) is large and the number of samples is normally small.

In [8], we presented a method that is based on parallel computing technique to speed up the PC algorithm. Here in the *ParallelPC* package, we parallelise a family of causal structure learning and causal inference methods, including PC, FCI, RFCI, PC-simple, IDA, and Joint-IDA. We also collate 12 different conditional independence (CI) tests that can be used in these algorithms. The algorithms in this package return the same results as those in the *pcalg* package [9], but the runtime is much lower depending on the number of cores CPU specified by users. Our experiment results show that with the *ParallelPC* package it is now practical to apply those methods to genomic datasets in a modern personal computer.

2   Contraint Based Algorithms and Their Parallelised Versions

We paralellised the following causal discovery and inference algorithms.

– PC, [2]. The PC algorithm is the state of the art method in constraint based approach for learning causal structures from data. It has two main steps. In the first step, it learns from data a skeleton graph, which contains only undirected edges. In the second step, it orients the undirected edges to form an equivalence class of DAGs. In the skeleton learning step, the PC algorithm starts with the fully connected network and uses the CI tests to decide if an edge is removed or retained. The stable version of the PC algorithm (the Stable-PC algorithm, [10]) updates the graph at the end of each level (the size of the conditioning set) of the algorithm rather than after each CI test. Stable-PC limits the problem of the PC algorithm, which is dependent on the order of the CI tests. It is not possible to parallelise the Stable-PC algorithm globally, as the CI tests across different levels in the Stable-PC algorithm are dependent to one another. In [8], we proposed the Parallel-PC algorithm which parallelised the CI tests inside each level of the Stable-PC algorithm. The Parallel-PC algorithm is more efficient and returns the same results as that of the Stable-PC algorithm (see Fig. 1).

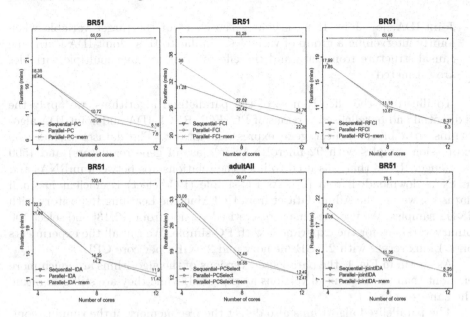

Fig. 1. Runtime of the sequential and parallelised versions (with and without the memory efficient option) of PC, FCI, RFCI, IDA, PC-simple, and Joint-IDA

- FCI, [3]. FCI is designed for learning the causal structure that takes latent variables into consideration. In real world datasets, there are often unmeasured variables and they will affect the learnt causal structure. FCI was implemented in the *pcalg* package and it uses PC algorithm as the first step. The skeleton of the causal structure learnt by PC algorithm will be refined by performing more CI test. Therefore, FCI is not efficient for large datasets.
- RFCI, [3]. RFCI is an improvement of the FCI algorithm to speed up the running time when the underlying graph is sparse. However, our experiment results show that it is still impractical for high dimensional datasets.
- PC-simple, [6]. PC-simple is a local causal discovery algorithm to search for parents and children of the target variable. In dense datasets where the target variable has large number of causes and effects, the algorithm is not efficient. We utilise the idea of taking order-independent approach [8] on the local structure learning problem to parallelise the PC-simple algorithm.
- IDA, [4]. IDA is a causal inference method which infers the causal effect that a variable has on another variable. It firstly learns the causal structure from data, and then based on the learnt causal structure, it estimates the causal effect between a cause node and an effect node by adjusting the effects of the parents of the cause. Learning the causal structure is time consuming. Therefore, IDA will be efficient when the causal structure learning step is improved. Figure 1 shows that our parallelised version of the IDA improves the efficiency of the IDA algorithm significantly.

– Joint-IDA, [5]. Joint-IDA estimates the effect of the target variable when jointly intervening a group of variables. Similar to IDA, Joint-IDA learns the causal structure from data and the effects of intervening multiple variables are estimated.

To illustrate the effectiveness of the parallelised algorithms, we apply the sequential and parallelised versions of PC, FCI, RFCI, IDA and Joint-IDA algorithms to a breast cancer gene expression dataset. The dataset includes 50 expression samples with 92 microRNAs (a class of gene regulators) and 1500 messenger RNAs that was used to infer the relationships between miRNAs and mRNAs downloaded from [11]. As PC-simple (PC-Select) is efficient in small datasets, we use the Adult dataset from UCI Machine Learning Repository with 48842 samples. We use the binary discretised version from [12,13] and select 100 binary variables for the experiment with PC-simple. We run all the experiments on a Linux server with 2.7 GB memory and 2.6 GHz per core CPU.

As shown in Fig. 1, the parallelised versions of the algorithms are much more efficient than the sequential versions as expected, while they are still generating the same results.

The parallelised algorithms also detect the free memory of the running computer to estimate the number of CI tests that will be distributed evenly to the cores. This step is to ensure that each core of the computer will not hold off a big amount of memory while waiting for the synchronisation step. The memory-efficient procedure may consume a little bit more time compared to the original parallel version. However, this option is recommended for computers with limited memory resources or for big datasets.

3   Conditional Independence Tests for Constraint Based Methods

It is shown that different CI tests may lead to different results for a particular constraint based algorithm, and a CI test may be suitable for a certain type of datasets. In this package, we collate 12 CI tests in the *pcalg* [9] and *bnlearn* [14] packages to provide options for function calls of the constraint-based methods. These CI tests can be used separately for the purpose of testing (conditional) dependency between variables. They can also be used within the constraint based algorithms (both sequential and parallelised algorithms) in the *ParallelPC* package. The following codes show an example of running the FCI algorithm using the sequential version in the *pcalg* package, the parallelised versions with or without the memory efficient option, and using a different CI test (mutual information) rather than the Gaussian CI test.

```
## Using the FCI-stable algorithm in the pcalg package
library(pcalg)
data("gmG")
p<-ncol(gmG$x)
```

```
suffStat<-list(C=cor(gmG$x),n=nrow(gmG$x))
fci_stable(suffStat, indepTest=gaussCItest, p=p,
skel.method="stable", alpha=0.01)
## Using fci_parallel without the memory efficient option
fci_parallel(suffStat, indepTest=gaussCItest, p=p,
skel.method="parallel",
alpha=0.01, num.cores=2)
## Using fci_parallel with the memory efficient option
fci_parallel(suffStat, indepTest=gaussCItest, p=p,
skel.method="parallel",
alpha=0.01, num.cores=2, mem.efficient=TRUE)
## Using fci_parallel with mutual information test
fci_parallel(gmG$x, indepTest=mig, p=p,
skel.method="parallel",
alpha=0.01, num.cores=2, mem.efficient=TRUE)
```

4   Finding a Group of Genes Directly Connected to a Gene of Interest

Given a gene of interest, we may want to see all genes that have direct causal relationship with the interest gene. The following workflow shows the convenient approach to achieve the analysis target by using the *pcSelect_parallel()* function in the *ParallelPC* package. We take the TP53 gene as an example gene of interest. We aim to infer the parent and children genes of TP53 within the top 40 differentially expression genes in the TCGA BRCA dataset.

```
## Select a gene of interest
index3=which(rownames(Tumor_Exp)=="BRCA1")
BRCA1_exp=Tumor_Exp[index3 ,]

## Use pcSelect() function to find the set of parents and
## children of the gene of interest.
pcSelect_Result<- pcSelect(BRCA1_exp,BRCA_mRNA,alpha=0.01)

##Select the causal effect features
index4=which(pcSelect_Result$G==TRUE)
pcSelect_Result_1=data.frame(
"mRNA"=names(pcSelect_Result$G)[index4]
,"zMin"=pcSelect_Result$zMin[index4])

##        mRNA       zMin
##1  PAFAH1B3  3.735917
##2     FXYD1  2.788022
##3      NEK2  5.868730
##4   ADAMTS5  3.216070
```

5   Predicting miRNA Targets Using a Causal Inference Method

The regulatory relationship between an miRNA and an mRNA means that a change in the expression level of the miRNA results in the change in the expression level of the mRNA. Causal inference methods allow us to estimate the change of a gene when manipulating another and therefore applicable for identifying the causal effects that the miRNAs have on the mRNAs [15–17]. The estimated causal effect is different from the correlation coefficient as in traditional correlation-based methods. In this section, we present the usage of the *ParallelPC* package in predicting the targets of a miRNA with a causal inference method called IDA.

5.1   Obtaining a Matched miRNA, mRNA Gene Expression Dataset

Firstly we retrieve the mRNA and miRNA expression data from TCGA BRCA dataset.

```
rm(list = ls())
####Retrieve TCGA mRNA and miRNA expression dataset
library("RTCGA.mRNA")
library("RTCGA.miRNASeq")
#### Extract the mRNA expression data of breast cancer
data(BRCA.mRNA)
mRNA=t(as.matrix(BRCA.mRNA[,-1]))
colnames(mRNA)=BRCA.mRNA[,1]
#### data imputation for the missing measurements
library("impute")
mRNA=impute.knn(mRNA)$data
###Split the normal and tumor samples
index=which(as.numeric(substr(colnames(mRNA),14,15))>9)
Normal_Exp=mRNA[,index]
Tumor_Exp=mRNA[,-index]
### Remove the duplicated tumor samples
index1=which(as.numeric(substr(colnames(Tumor_Exp),14,15))>1)
Tumor_Exp=Tumor_Exp[,-index1]
colnames(Tumor_Exp)=substr(colnames(Tumor_Exp),1,12)

#### Extract the miRNA expression data of breast cancer
data(BRCA.miRNASeq)
BRCA.miRNASeq=
BRCA.miRNASeq[seq(2, nrow(BRCA.miRNASeq), 3),-c(1,2)]
miRNASeq=apply(as.matrix(BRCA.miRNASeq),1,as.numeric)
rownames(miRNASeq)=colnames(BRCA.miRNASeq)
###Split the normal and tumor samples
index3 =which(as.numeric(substr(colnames(miRNASeq),14,15))>9)
Normal_miRNA=miRNASeq[,index3]
Tumor_miRNA=miRNASeq[,-index3]
### Remove the duplicated samples
```

```
index4=
which(as.numeric(substr(colnames(Tumor_miRNA),14,15))>1)
Tumor_miRNA=Tumor_miRNA[,-index4]
colnames(Tumor_miRNA)=substr(colnames(Tumor_miRNA),1,12)
```

5.2 Finding the Differentially Expressed miRNAs and mRNAs

```
library(limma)
c1=ncol(Normal_Exp)
c2=ncol(Tumor_Exp)
mR=cbind(Normal_Exp, Tumor_Exp)
design1=cbind(Normal=c(rep(1,c1), rep(0,c2)),
Cancer=c(rep(0,c1), rep(1,c2)))
##In order to return the index of features ,
##set the same name for two feature
rownames(mR)[1]="repeat"
rownames(mR)[2]="repeat"
mRfit=lmFit(mR, design1)
contrast.matrix=makeContrasts(NormalvCancer=Normal - Cancer ,
levels=design1)
mRfit1=contrasts.fit(mRfit , contrast.matrix)
mRfit1=eBayes(mRfit1)
##Choose the top 100 differentially expression genes
##for experiment analysis
mRresults=
topTable(mRfit1 , number= 100, sort.by="p", adjust="BH")
index2= as.numeric(row.names(mRresults))
##Extract the experiment dataset for downstream analysis
BRCA_mRNA_100=t(Tumor_Exp[index2 ,])

##Extract the most different expression miRNAs
c1=ncol(Normal_miRNA)
c2=ncol(Tumor_miRNA)
miRNA=cbind(Normal_miRNA, Tumor_miRNA)
design2=cbind(Normal=c(rep(1,c1), rep(0,c2)),
Cancer=c(rep(0,c1), rep(1,c2)))
###In order to return the index of features,
###set the same name for two feature
rownames(miRNA)[1]="repeat"
rownames(miRNA)[2]="repeat"
miRNA <- voom(miRNA, design2, plot=TRUE)
mRfit=lmFit(miRNA, design2)
contrast.matrix=makeContrasts(NormalvCancer=Normal - Cancer,
levels=design2)
mRfit2=contrasts.fit(mRfit, contrast.matrix)
mRfit2=eBayes(mRfit2)
##Choose the top 10 differentially expression miRNAs
##for experiment analysis
```

```
mRresults=
topTable(mRfit2, number= 10, sort.by="p", adjust="BH")
index5 = as.numeric(row.names(mRresults))
##Get the final experiment dataset.
BRCA_miRNA_10=t(Tumor_miRNA[index5,])
```

5.3   Extract the Matched Samples with mRNA and miRNA

```
mRNA_samples=rownames(BRCA_mRNA_100)
miRNA_samples=rownames(BRCA_miRNA_10)
intersect_samples=intersect(mRNA_samples,miRNA_samples)
```

```
index6 =match(intersect_samples,mRNA_samples)
index7=match(intersect_samples,miRNA_samples)
BRCA_mRNA_100_matched=BRCA_mRNA_100[index6,]
BRCA_miRNA_10_matched=BRCA_miRNA_10[index7,]
##Test all samples are matched or not
all(rownames(BRCA_mRNA_100_matched)==
rownames(BRCA_miRNA_10_matched))

BRCA_matched_miRNA10_mRNA100=cbind(BRCA_miRNA_10_matched,
BRCA_mRNA_100_matched)
write.csv(BRCA_matched_miRNA10_mRNA100,
file = "BRCA_matched_miRNA10_mRNA100.csv",
row.names = FALSE)
```

5.4   Applying Parallel IDA for Inferring miRNA-mRNA Causal Effects Without Using Target Binding Information

```
library("ParallelPC")
library("pcalg")
library("parallel")
library("miRLAB")
miRNAFrom=1
miRNATo=10
mRNAFrom=11
mRNATo=110
IDAResult_noTargetBinding =IDA_parallel(
"BRCA_matched_miRNA10_mRNA100.csv",
miRNAFrom:miRNATo,mRNAFrom:mRNATo,"parallel",0.01, 2, TRUE)
##Extract the top 20 targets of hsa-mir-139 as an example
library("miRLAB")
miRTop20_1=bRank(IDAResult_noTargetBinding, 2,20, TRUE)
##          miRNA     mRNA    Causal effects
##51 hsa-mir-139    PDE2A   0.4176818
```

```
##15  hsa-mir-139     BTNL9   0.3658065
##4   hsa-mir-139       CA4   0.3655496
##98  hsa-mir-139    AQP7P2   0.3512459
##79  hsa-mir-139     LYVE1   0.3499394
##7   hsa-mir-139   CD300LG   0.3488489
##31  hsa-mir-139     AVPR2   0.3420044
##68  hsa-mir-139      NPR1   0.3372393
##91  hsa-mir-139   MAP1LC3C  0.3324548
##12  hsa-mir-139     ATOH8   0.3232092
##78  hsa-mir-139    KCNIP2   0.3205980
##87  hsa-mir-139      PCK1   0.3198771
##44  hsa-mir-139    ACVR1C   0.3163645
##21  hsa-mir-139   HSD17B13  0.3156995
##13  hsa-mir-139  LOC387911  0.3138352
##33  hsa-mir-139      FIGF   0.3131615
##20  hsa-mir-139   C1QTNF9   0.3109739
##57  hsa-mir-139      TNMD   0.3081662
##28  hsa-mir-139      SDPR   0.3051225
##37  hsa-mir-139     ITIH5   0.3050095
```

5.5 Applying IDA for Inferring miRNA-mRNA Causal Effects with Target Binding Information

We use the IDA() function from *miRLAB* package [18] to infer the miRNA-mRNA causal effects with the TargetScan 7 as the target binding information. The TargetScan7.csv file can be downloaded from (nugget.unisa.edu.au/Thuc/TargetScan7.csv).

```
library("miRLAB")
IDAResult_TargetBinding =
IDA("BRCA_matched_miRNA10_mRNA100.csv",
miRNAFrom:miRNATo, mRNAFrom:mRNATo,
targetbinding = "TargetScan7.csv")
```

5.6 Creating an Ensemble Method by Combining IDA and Lasso

```
### Ensemble method
library(miRLAB)
IDA_Ensemble =
IDA_parallel("BRCA_matched_miRNA10_mRNA100.csv",
cause = miRNAFrom:miRNATo,
effect = mRNAFrom:mRNATo,
"parallel",0.01, 2, TRUE)

Lasso_Ensemble = Lasso("BRCA_matched_miRNA10_mRNA100.csv",
cause=miRNAFrom:miRNATo,
```

```
effect=mRNAFrom:mRNATo)
Borda_Ensemble = Borda(list(Lasso_Ensemble, IDA_Ensemble))
miRTop20_2= bRank(Borda_Ensemble,2,20,TRUE )

## Extract the top 20 targets of hsa-mir-139 as an example
##          miRNA      mRNA   Ranking Score
##51 hsa-mir-139      PDE2A   100.000000
##98 hsa-mir-139     AQP7P2   28.571429
##4  hsa-mir-139        CA4   28.571429
##7  hsa-mir-139     CD300LG  25.000000
##15 hsa-mir-139      BTNL9   18.181818
##79 hsa-mir-139      LYVE1   15.384615
##68 hsa-mir-139       NPR1   15.384615
##78 hsa-mir-139     KCNIP2   11.111111
##44 hsa-mir-139     ACVR1C   10.526316
##87 hsa-mir-139       PCK1   9.090909
##91 hsa-mir-139    MAP1LC3C  8.333333
##21 hsa-mir-139    HSD17B13  8.000000
##31 hsa-mir-139      AVPR2   7.407407
##57 hsa-mir-139       TNMD   6.250000
##6  hsa-mir-139      GLYAT   6.060606
##33 hsa-mir-139       FIGF   5.714286
##19 hsa-mir-139       AQP7   5.128205
##13 hsa-mir-139    LOC387911  5.128205
##61 hsa-mir-139      GPR146  5.000000
##12 hsa-mir-139      ATOH8   5.000000
```

6   Identifying Double Intervention Effects from Expression Data

While IDA estimates the causal effect of one variable on the other, joint_IDA estimate the joint causal effect of multiple variables on the other. In this scenario, we show the example of calculating the joint causal effect of 2 miRNAs on a target gene. The joint causal effect here is similar to the effect of knocking down the two miRNAs at the same time.

6.1   Identify Differentially Expressed Genes for the TCGA Breast Cancer Dataset

We extract the top 40 differentially expressed mRNAs and 10 differentially expressed miRNAs for the analysis.

```
BRCA_mRNA_40_matched=BRCA_mRNA_100_matched[,1:40]
BRCA_matched_miRNA10_mRNA40=cbind(BRCA_miRNA_10_matched,
BRCA_mRNA_40_matched)
write.csv(BRCA_matched_miRNA10_mRNA40,
```

```
file = "BRCA_matched_miRNA10_mRNA40.csv",
row.names = FALSE)
```

6.2   Identify Join-Effect of 2 miRNAs on a Gene

We find the joint causal effect of *hsa-mir-592* and *hsa-mir-139* on *RDH5*.

```
causal_coefficient=
jointIDA_parallel("BRCA_matched_miRNA10_mRNA40.csv", 1:2,40,
pcmethod="parallel", 0.01, 2, technique="RRC")
##hsa.mir.592  -0.1999714
##hsa.mir.139   0.2391669

#Let X1 be the average expression level of  hsa-miR-592
#and X2 be the average expression level of hsa-miR-139
X1=colMeans(BRCA_miRNA_10_matched)[1]# hsa_mir_592
X2=colMeans(BRCA_miRNA_10_matched)[2]# hsa_mir_139

#The joint causal effect of the two miRNAs on RDH5 is:
Joint_effect = -X1*( -0.1999714)-X2*0.2391669
##The result is -10.17413.
```

7   Conclusion

In this paper, we present a software package, *ParallelPC*, for efficient causal exploration using genomic data. We present some use cases of the package including (i) inferring gene regulatory networks, (ii) finding a set of genes having causal relationship with a gene of interest, (iii) predicting miRNA targets with causal inference methods, and (iv) identifying the joint causal effect of multiple miRNAs on a given gene. The package will offer a set of useful causal exploration tools for novel applications.

Funding
Thuc Duy Le was supported by NHMRC [grant ID 1123042]. This work has been partly supported by ARC Discovery Project DP170101306.

References

1. Pearl, J.: Causality. Cambridge University Press, Cambridge (2009)
2. Spirtes, P., Glymour, C.N., Scheines, R.: Causation, Prediction, and Search, vol. 81. MIT Press, Cambridge (2000)
3. Colombo, D., Maathuis, M.H., Kalisch, M., Richardson, T.S., et al.: Learning high-dimensional directed acyclic graphs with latent and selection variables. Ann. Stat. **40**(1), 294–321 (2012)

4. Maathuis, M.H., Kalisch, M., Bühlmann, P., et al.: Estimating high-dimensional intervention effects from observational data. Ann. Stat. **37**(6A), 3133–3164 (2009)
5. Nandy, P., Maathuis, M.H., Richardson, T.S.: Estimating the effect of joint interventions from observational data in sparse high-dimensional settings. Ann. Stat. **45**(2), 647–674 (2017)
6. Bühlmann, P., Kalisch, M., Maathuis, M.H.: Variable selection in high-dimensional linear models: partially faithful distributions and the PC-simple algorithm. Biometrika **97**(2), 261–278 (2010)
7. Li, J., Liu, L., Le, T.D.: Practical Approaches to Causal Relationship Exploration. SECE. Springer, Cham (2015). https://doi.org/10.1007/978-3-319-14433-7
8. Le, T.D., Hoang, T., Li, J., Liu, L., Liu, H., Hu, S.: A fast PC algorithm for high dimensional causal discovery with multi-core PCs. IEEE/ACM Trans. Comput. Biol. Bioinform. (2016)
9. Kalisch, M., Mächler, M., Colombo, D., Maathuis, M.H., Bühlmann, P.: Causal inference using graphical models with the R package pcalg. J. Stat. Softw. **47**(11), 1–26 (2012)
10. Colombo, D., Maathuis, M.H.: Order-independent constraint-based causal structure learning. J. Mach. Learn. Res. **15**(1), 3741–3782 (2014)
11. Le, T.D., Liu, L., Zhang, J., Liu, B., Li, J.: From miRNA regulation to miRNA-TF co-regulation: computational approaches and challenges. Brief. Bioinform. **16**(3), 475–496 (2015)
12. Li, J., Le, T.D., Liu, L., Liu, J., Jin, Z., Sun, B.: Mining causal association rules. In: 2013 IEEE 13th International Conference on Data Mining Workshops (ICDMW), pp. 114–123 (2013)
13. Li, J., Le, T.D., Liu, L., Liu, J., Jin, Z., Sun, B., Ma, S.: From observational studies to causal rule mining. ACM Trans. Intell. Syst. Technol. **7**(2), 14 (2016)
14. Scutari, M.: Learning Bayesian networks with the bnlearn R package. arXiv preprint arXiv:0908.3817 (2009)
15. Le, T.D., et al.: Inferring microRNA–mRNA causal regulatory relationships from expression data. Bioinformatics **29**(6), 765–771 (2013). https://doi.org/10.1093/bioinformatics/btt048
16. Zhang, J., et al.: Inferring condition-specific miRNA activity from matched miRNA and mRNA expression data. Bioinformatics **30**(21), 3070–3077 (2014)
17. Zhang, J., et al.: Identifying direct miRNA–mRNA causal regulatory relationships in heterogeneous data. J. Biomed. Inform. **52**, 438–447 (2014)
18. Le, T.D., Zhang, J., Liu, L., Liu, H., Li, J.: miRLAB: an R based dry lab for exploring miRNA-mRNA regulatory relationships. PLoS One **10**(12), e0145386 (2015)

Citation Field Learning by RNN with Limited Training Data

Yiqing Zhang[1]([✉]), Yimeng Dai[1], Jianzhong Qi[1], Xinxing Xu[2], and Rui Zhang[1]

[1] School of CIS, The University of Melbourne, Parkville, Australia
{yiqingz2,yimengd}@student.unimelb.edu.au,
{jianzhong.qi,rui.zhang}@unimelb.edu.au
[2] Institute of High Performance Computing, A*STAR, Singapore, Singapore
xxxing1987@gmail.com

Abstract. Citation field learning is to segment a citation string into fields of interest such as *author*, *title*, and *venue* from plain text. We are interested in citation field learning from researchers' homepages. This task is challenging due to the free citation styles used by different creators of the homepages. We aim to address the challenge by neural network based approaches which learn the citation field styles automatically. Neural network based approaches are data-hungry, but manually labeled training data is expensive to obtain. Therefore, we propose a novel framework that utilizes auto-generated training data and domain adaptation to enhance a manually labeled training dataset of limited size. At the same time, we design an adaptive Recurrent Neural Network (RNN) to learn citation styles from the enhanced training data effectively. Extensive experiments show that the proposed methods outperform state-of-the-art methods for citation field learning.

1 Introduction

Webpages about academic researchers such as researchers' academic homepages are growing rapidly. Such webpages are important channels for learning researchers' profiles. In this paper, we focus on researchers' publication records, as these records contain rich information, such as researchers' areas of interest, usual venues of publication, co-author networks, etc. Specifically, we aim to learn citation fields including *author*, *title*, *venue*, and *year* from each citation item contained in a researcher's homepage. For example, given a citation item as shown in Fig. 1, we aim to extract the author ("Robert Ross, Robert Latham, William Gropp, Ewing Lusk and Rajeev Thakur"), title ("Processing MPI Datatypes Outside MPI"), venue ("EuroPVMMPI"), and year ("09").

Citation field learning from researchers' homepages is much more challenging than that from well-formatted bibliographies in research papers. The bibliography in a research paper is often produced by software, such as LATEX with one of the predefined referencing styles (e.g., the Harvard referencing style), whereas publication lists in researchers' homepages have free styles and are more error-prone, making them more difficult to analyze. For example, the owners of homepages may write their names in full, but use initials for the co-author names.

M. Ganji et al. (Eds.): PAKDD 2018, LNAI 11154, pp. 219–232, 2018.
https://doi.org/10.1007/978-3-030-04503-6_23

Also, citations in homepages may contain extra information that does not usually appear in references of research papers, such as acceptance rate or awards. In Fig. 1, the citation item extracted from an academic homepage (http://wgropp. cs.illinois.edu/) contains abbreviated venue and year as well as extra award information, which makes learning the fields from it difficult.

Commercial systems such as Google Scholar collect citation information by purchasing proprietary bibliography databases from publishers. These commercial systems have high-quality citation information but are very expensive to maintain. In the research community, researchers have attempted to build systems, such as CiteSeer [7], ParsCit [4], and ArnetMiner [20], which collect large numbers of academics' research information. Their citation field learning algorithms rely on a large amount of manually labeled data. We aim at an algorithm that requires only a limited amount of manually labeled data for citation field learning without sacrificing the learning accuracy.

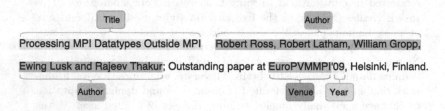

Fig. 1. A citation item example and its labels.

Existing approaches for citation field learning fall into two categories: rule-based approaches and sequence labeling approaches. Rule-based approaches [2, 6] require significant human efforts to create a large number of rules to cover various citation styles. Traditional sequence labeling approaches such as *Hidden Markov Model* (HMM) and *Conditional Random Fields* (CRF) require careful feature engineering, which also means non-trivial human efforts.

We take a sequence labeling approach and make use of the *Recurrent Neural Network* (RNN), which has shown promising results in other sequence labeling tasks. RNN is a deep learning model, which requires a large amount of training data. For our task, large bibliography dataset, such as DBLP and PubMed Central, may be helpful. However, these datasets only cover well-formatted citation styles, and they cannot be used as training data directly, since the metadata in the datasets are in the format of XML or BibTex, which are not presented in a naturally readable citation format (e.g., Fig. 1).

To exploit the large publicly available bibliography datasets, we propose to render the metadata into naturally readable citations with the aid of a large online knowledge base of citation styles, *CSL Style Repository*[1], which contains over 8,000 citation styles. These allow us to generate a much larger training dataset. We refer to this dataset the *generated dataset* or the *generated data*.

---

[1] http://citationstyles.org/styles/.

There are still limitations with the generated dataset. While CSL Style Repository helps cope with the various styles of different fields (e.g., author names in full or in initials), it does not cover free writing styles such as adding extra award information or typos. To address this limitation, we collect 3,000 citation items from academic homepages in a range of disciplines and manually labeled the fields of these items. We call this dataset *Cita*.

We find that training the model on a straightforward combination of the generated dataset and Cita does not offer better learning result due to the unbalanced sizes of the two datasets. To address this limitation, we utilize domain adaptation techniques and propose an adaptive Bi-directional LSTM-CRF model. This model adds constraints at the fine-tuning stage during the training process to prevent over-fitting and preserve information learned from the generated data. As a result, the adaptive model can handle citations items with different styles from various disciplines with high precision and recall, requiring only a limited amount of manually labeled training data.

This paper makes the following contributions:

- We address the problem of citation field learning by an RNN approach. To obtain sufficient data for the RNN approach, we propose a training data enhancement method that makes use of publicly available citation data and citation styles, and use it together with a small manually labeled dataset.
- We propose an adaptive Bi-LSTM-CRF model that can combine a small manually labeled training dataset with a large generated dataset to achieve high model accuracy that using either dataset alone would not achieve.
- We conduct an extensive experimental study. The results validate the effectiveness of the proposed framework, which outperforms the state-of-the-art by up to 5.2% in F1-score.

The rest of the paper is organized as follows. Section 2 reviews related work and Sect. 3 presents preliminaries. Section 4 details our methods. Section 5 reports the results of the experimental study and Sect. 6 concludes the paper.

2   Related Work

To extract citation fields, various rule-based [2,3,9] and sequence labeling based [18,21] methods have been developed. Rule-based methods require extensive engineering of rules while the rules may target at limited styles. For example, Day et al. [6] only consider six citation styles. Rule-based methods do not suit our needs to extract citation fields from homepages, where there are a large variety of citation styles.

From the machine learning perspective, citation field learning can be viewed as a sequence labeling problem. HMM and CRF have shown good results [17,19,22] in sequence labeling tasks such as *Part-of-speech* (POS) tagging and *Named-entity Recognition* (NER). They have also been applied to citation field learning [18,21]. Particularly, CRF is the state-of-the-art method [4,23] for this task. However, it still suffers from relying on hand engineered feature functions which require significant human efforts.

Recently, Huang et al. [10] propose to combine *Long Short-Term Memory* (LSTM), a variant of RNN, with CRF for sequence labeling. LSTM is trained to represent the context of a word in a sequence of tokens (e.g., a sentence), and CRF is stacked on the top of the output of LSTM to represent the structural relationship between tokens and labels. The LSTM-CRF model outperforms the CRF model and has been successfully applied in various tasks, such as NER [12,14,15]. However, the LSTM-CRF model requires a large labeled dataset to train, while existing citation field learning public datasets such as Cora [16] are relatively small and may not suit the need.

To overcome the limitation on training dataset size, we propose to generate more training data from various publicly available sources. This idea is inspired by Jaderberg et al. [11] who propose to generate text in natural scene images to form a training dataset. While this is a known technique in the area of scene text recognition, training data enhancement has not been widely used in natural language processing, probably because generated text can hardly resemble real-world text to a satisfactory level.

To overcome the lack of free styles in the generated data, we borrow an idea of *fine-tuning* from the area of object recognition [1]. The fine-tuning method inspires us to train our Bi-LSTM-CRF model with a large generated dataset first, and then adapt it with a manually labeled dataset. This adaptation allows the model to take advantage of both a large amount of generated data and the resemblance of the manually labeled data to the target data.

3   Preliminaries

We present four basic models upon which the proposed models are built. These basic models will also serve as baseline methods in the experimental study.

3.1   CRF

CRF [13] is often used for sequence labeling. It captures the correlations between an observation (e.g., a token) and its label, and the correlations between labels in a neighborhood. For example, in citation field learning, a token "William" is likely to have an *author* label since the token usually represents a name; an *author* label is likely to be followed by another *author* label, since a citation usually has multiple author tokens. These intuitions are captured by the formula below:

$$p(\boldsymbol{y}|\boldsymbol{x}, \lambda) = \frac{1}{Z(\boldsymbol{x})} exp(\sum \lambda_j F_j(\boldsymbol{y}, \boldsymbol{x})),$$

where $\boldsymbol{x}$ is an observation (token) sequence, $\boldsymbol{y}$ is the predicted label sequence, $Z(\boldsymbol{x})$ is for normalization, λ are parameters to be learned, and $F(\boldsymbol{y}, \boldsymbol{x})$ are *potential functions*.

To predict labels of the tokens, we search for the label sequence y^* with the maximum predicted probability:

$$y^* = \underset{y \in Y(x)}{\arg\max}\, p(y|x, \lambda).$$

Applying CRF for citation field learning requires a training dataset where every token (e.g., a word) in a citation item is labeled with a citation filed label. Then the model can learn and predict the most likely label sequence y^* given a token sequence x (i.e., a citation item).

3.2   Bi-directional LSTM

RNN is powerful in capturing long-term dependency but it suffers from the vanishing gradient problem. *Long Short-Term Memory* (LSTM) is a variant of RNN that addresses this problem. The main difference between LSTM and a basic RNN is that LSTM has a memory cell, which remembers the current network state. Since LSTM only captures information flow from one direction, it is common to use two LSTMs, each trained from one direction, to capture the context of a word from its both sides. This results in a Bi-LSTM model. To get a single output at time step t, we concatenate the forward pass output h_t^f and the backward pass output h_t^b from the two LSTMs as $[h_t^f, h_t^b]$.

3.3   Bi-LSTM-CRF

CRF and Bi-LSTM can be combined together to form a Bi-LSTM-CRF model [10]. Such a model can take advantage of CRF to capture word-label relationship information and Bi-LSTM to capture word context information. Applying Bi-LSTM-CRF for citation field learning requires first getting word embeddings (e.g., using GloVe) for the tokens in a citation item. These word embeddings are then fed into Bi-LSTM as input. The output of Bi-LSTM is further fed into a linear chain CRF layer, which predicts the field labels of the tokens. For regularization, dropout is applied on all the LSTM layers.

3.4   Input Augmentation for Bi-LSTM-CRF

As discussed earlier, we may obtain training data from publicly available bibliography datasets. However, these datasets are imbalanced. Training a Bi-LSTM-CRF model on a straightforward combination of the data from multiple datasets may not yield the best citation field learning result. Daumé [5] proposes an input augmentation method for domain adaptation to handle training data from different datasets. This method augments a training data point to differentiate it from data points of other datasets. To apply this idea to train the Bi-LSTM-CRF model described above, we augment the input via extending the d-dimensional word embedding vector x of each token t to an $(m + 1) \cdot d$ dimensional vector x_e, where m is the total number of datasets from which the training data is obtained. The first d dimensions and the $k \cdot d$ to $(k+1) \cdot d$ dimensions of x_e are two copies of x, while the rest of the dimensions are all 0s, assuming that t is from the kth dataset.

4 Proposed Methods

We propose a citation field learning framework that utilizes training data from various sources to learn to label the different fields (i.e., *author, title, venue,* and *year*) of citation items from researchers' homepages. Figure 2 illustrates the framework. We first collect citation metadata from various public bibliography datasets and use the styles in the CSL Style Repository to generate a large number of citation items with a large variety of styles. Meanwhile, we manually label a small set of citation items from different disciplines. The generated data and the manually labeled data are then fed into an adaptive Bi-LSTM-CRF model for citation field learning. The aim of using both generated data and manually labeled data is to take advantage of both the large size of the generated data and the large variety of styles of the manually labeled data. To make the most of the two types of data, we propose multiple models to learn from the two types of data. In what follows, we first present our data enhancement method in Sect. 4.1, and then describe our models in Sect. 4.2.

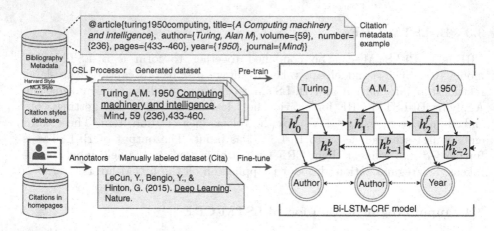

Fig. 2. Citation field learning using enhanced training data via adaptive Bi-LSTM-CRF

4.1 Training Data Enhancement

The main challenge of applying the Bi-LSTM-CRF model to the citation field learning task lies in the need for a large amount of training data. To address this challenge, we design a training data generator. This data generator takes a seed citation metadata and a set of κ citation styles as the input, where κ is a system parameter. It then generates κ citation items with the same content but in κ different styles. The seed citation metadata are obtained from publicly available bibliography datasets (which are in XML or BibTex format). In the experiments, we use three different bibliography datasets with different discipline distributions, one in computer science, one in medical, and one in 22 different disciplines. We will detail them in Sect. 5.1. The citation styles are obtained from the CSL Style Repository, which contains over 8,000 citation styles. Each citation style is described with the *Citation Style Language* (CSL), which is an XML based language. We use a CSL Processor called *citeproc-py*[2] to generate the citations.

To simulate the progress of writing citation items in researchers' homepages, we add random noise and make adjustments to the generated citation items as follows: (1) we randomly remove and alter some letters from the generated citation items; (2) we alter the content of citations following patterns which are observed in citation data from researchers' homepages, e.g., adding award or conference ranking information; (3) we adjust the discipline distribution of the generated citation items via sampling from the three datasets based on a discipline distribution obtained from a manually labeled dataset named Cita. The Cita dataset is created by ourselves and it will be detailed in Sect. 5.1. It covers more than 20 disciplines. We simply use the "Fields of Research" in the Australian and New Zealand Standard Research Classification[3] as the disciplines.

4.2   Adaptive Bi-LSTM-CRF Models

Using the data enhancement method described above, we can easily generate a large number (e.g., 100,000) of labeled citation items. Meanwhile, we can also manually label a small number of (e.g., 3,000) citation items extracted from researchers' homepages. Next, we discuss how to train a Bi-LSTM-CRF model that can learn the most from the generated data and manually labeled data together.

Bi-LSTM-CRF with Enhanced Data. Let the set of generated data be $\mathcal{S}_a$ and the set of manually labeled data be $\mathcal{S}_h$. A straightforward approach is to train a Bi-LSTM-CRF on the union of the two datasets, i.e., $\mathcal{S}_a \cup \mathcal{S}_h$. Let function $f(\boldsymbol{W})$ be the optimization objective function of the Bi-LSTM-CRF model, where $\boldsymbol{W}$ is the trainable parameter set of the model. Then training the model is to obtain the parameter values $\boldsymbol{W}^*$ that minimizes $f(\boldsymbol{W})$, i.e.,

$$\boldsymbol{W}^* = min_{\boldsymbol{W}} f(\boldsymbol{W})$$

When training the model, we use *Stochastic Gradient Descent* (SGD) to update model parameters iteratively.

---

[2] https://github.com/brechtm/citeproc-py.
[3] http://www.arc.gov.au/rfcd-seo-and-anzsic-codes.

Bi-LSTM-CRF with Fine-Tuning. Simply merging the two datasets does not take the unbalanced data sizes into account. Inspired by the fine-tuning methods used in the object recognition community, we first train the Bi-LSTM-CRF model on the large generated dataset $\mathcal{S}_a$ and save the trained model parameters as $\mathcal{M}_a$. We then initialize another model $\mathcal{M}_h$ with $\mathcal{M}_a$. Model $\mathcal{M}_h$ is later trained on $\mathcal{S}_h$ with a lower learning rate.

The intuition is that we can first train the model to learn general structures of citations and words used in citations from generated data. When the model is fine-tuned on the manually labeled data, the model can then learn the specific knowledge about citation items in homepages. If trained only on generated data, the model may fail to capture features of citation items in researchers' homepages, while trained only on manually labeled data, it may suffer from overfitting due to the limited dataset size.

Adaptive Bi-LSTM-CRF. To avoid overfitting, we further propose a method called *adaptive fine-tuning*, which adds a regularization term to the objective function:

$$W^* = min_W f(W) + \frac{\gamma}{2} \|W - W_0\|^2$$

Here, W is the parameter set of $\mathcal{M}_h$; W_0 is the parameter set of the pre-trained model $\mathcal{M}_a$; γ is a hyper-parameter to control how strong the regularization term will influence the adaptive fine-tuning results.

MA-Bi-LSTM-CRF. Since the generated dataset contain citation items from different sources, we propose another adaptive fine-tuning method that first train multiple Bi-LSTM-CRF models, one on a dataset generated from a different source. Then we fine-tune the Bi-LSTM-CRF model by training it on the manually labeled dataset $\mathcal{S}_h$. The training goal now becomes as follows.

$$W^* = min_W f(W) + \sum_{k=1}^{K} \frac{\gamma}{2K} \|W - W_0^k\|^2$$

Here, K denotes the total number of data sources, e.g., $K = 3$ is used in our experiments, and W_0^k is the parameter set of model pre-trained on the dataset generated from the kth source. We call this model *Multi-Adaptive Bi-LSTM-CRF* (MA-Bi-LSTM-CRF). During adaptive fine-tuning of MA-Bi-LSTM-CRF, the primary model parameters W are initialized with the average of all W_0^k.

5   Experiments

In this section, we report experimental results of our models compared with five baseline methods in citation field learning.

5.1   Settings

Datasets. We crawled 16,400 researchers' homepages from university websites worldwide. A total of 67,243 citation items are extracted from these pages. This dataset is denoted as χ, but there are no field labels for these citation items. Thus, we randomly sample 3,000 citation items from χ for manual labeling. We ensure that at most one item is extracted from each homepage, and there are no duplicates in the sample data. We call this sampled dataset Cita, and more than 20 disciplines are covered in Cita.

Six annotators label ground truth labels to all citation items in Cita. Labels used include *author, title, year, venue*, and *other*. The labeling is time-consuming. On average, it takes about an hour to label 100 items. The annotators are encouraged to consult Google Scholar and online bibliography databases to ensure the label quality. Among the 3,000 labeled citation items, 60% are used for training and validation, and 40% are used for testing. We denote the subsets of Cita used for training and testing as S_{gt} and S_{test}, respectively.

An additional public dataset named Cora [16] is used in the experiments. This dataset contains 500 citation items in computer science. A 70-30 split is used for training and testing on this dataset, following previous studies [2,18]. We denote the subsets used for training and testing as S_{cgt} and S_{ctest}, respectively. The two datasets S_{gt} and S_{cgt} will serve as the manually labeled dataset S_h used in the discussions in Sect. 4.2.

For training data enhancement, we use three bibliography data sources: (1) $\mathcal{D}_1$: DBLP hosts more than 3.9 million computer science citation items from 1.9 million authors. (2) $\mathcal{D}_2$: PubMed Central is an online archive of research in life science and biomedical that contains more than 4.5 million papers, among which a subset of over one million papers are publicly available. (3) $\mathcal{D}_3$: Queensland University of Technology ePrints is an online bibliography library that hosts around 80 thousand citation items. Although relatively small, this dataset covers 22 different disciplines.

Table 1. Summary of datasets: $\natural S_i$ means sampled data from dataset S_i according to the discipline distribution in Cita dataset.

Generated datasets					Manually labeled datasets				
Dataset	Source	Size	Discipline	Usage	Dataset	Source	Size	Discipline	Usage
S_1	$\mathcal{D}_1$	100,000	Computer Science	Pre-train	S_{cgt}	Cora	350	Computer Science	Training
S_2	$\mathcal{D}_2$	100,000	Medical	Pre-train	S_{ctest}	Cora	150	Computer Science	Testing
S_3	$\mathcal{D}_3$	100,000	Cross-discipline	Pre-train	S_{gt}	Cita	1,800	Cross-discipline	Training
S_a	$\cup_{i=1}^{3}\natural S_i$	100,000	Cross-discipline	Pre-train	S_{test}	Cita	1,200	Cross-discipline	Testing

Following the method described in Sect. 4.1, we generate three training datasets[4] S_1, S_2, and S_3 based on metadata obtained from the three sources $\mathcal{D}_1$, $\mathcal{D}_2$, and $\mathcal{D}_3$, respectively. We further sample citation items from S_1, S_2, and S_3 according to the discipline distribution of Cita to form an additional dataset S_a. We limit the size of all these generated datasets to 100,000 to constrain the training time. A summary of all the datasets used in experiments is given in Table 1.

---

[4] Datasets available at: https://github.com/yiqingzhang/citationlearn.

Parameters. In data generation process, we use $\kappa = 4$ to generate $\mathcal{S}_1$, $\mathcal{S}_2$, and $\mathcal{S}_3$. Before training, each input citation item is tokenized into words and punctuations. We then use GloVe, a pre-trained 100-dimensional word embedding, to map the tokens into vectors before feeding them into the Bi-LSTM-CRF model. During the training and fine-tuning process, Adagrad with momentum is used. We use 0.01 as the initial learning rate, and 32 as the batch size. We use 20% of the training set as the validation set for early stopping and hyperparameter tuning. The optimal hyperparameters are obtained with a standard grid search procedure on the validation set.

Measures. To make a fair comparison, models trained with Cita data are tested with the Cita testing dataset $\mathcal{S}_{test}$; models trained with Cora data are tested with the Cora testing dataset $\mathcal{S}_{ctest}$. We report both field-level matching and token-level matching precision, recall, and F1-score, following previous studies [8,12].

5.2 Methods Evaluated

We evaluate five baseline methods and four variants of the proposed framework. **The baseline methods used are**:

- **BibPro**: A rule-based citation parser algorithm proposed by Chen et al. [2].
- **ParsCit**: A citation string parsing software developed by Councill et al. [4]. We use the above tools as off-the-shelf tools without changing any parameters.
- **CRF**: We implement a CRF model as described in Sect. 3.1 and train it with the manually labeled dataset ($\mathcal{S}_{gt}$ and $\mathcal{S}_{cgt}$ when evaluated on Cita and Cora, respectively. We denote these datasets as $\mathcal{S}_{(c)gt}$).
- **Bi-LSTM-CRF**: We also implement a straightforward Bi-LSTM-CRF model as described in Sect. 3.3 and train it with the manually labeled data ($\mathcal{S}_{(c)gt}$).
- **Bi-LSTM-CRF with Input Augmentation**: This method also uses Bi-LSTM-CRF as its main model, but adds an input augmentation layer, as described in Sect. 3.4. This model is trained on four datasets, i.e., $\mathcal{S}_1$, $\mathcal{S}_2$, $\mathcal{S}_3$, and $\mathcal{S}_{gt}$, when evaluated on Cita. The model is trained on two datasets $\mathcal{S}_1$ and $\mathcal{S}_{cgt}$ when evaluated on Cora.

The four variants of the proposed framework tested are:

- **Bi-LSTM-CRF with Enhanced Data**: This model is trained on a straightforward combination of generated data and manually labeled data, i.e. $\mathcal{S}_a \cup \mathcal{S}_{gt}$ when evaluated on Cita and $\mathcal{S}_1 \cup \mathcal{S}_{gt}$ when evaluated on Cora.
- **Bi-LSTM-CRF with Fine-tuning**: This model is trained on generated data ($\mathcal{S}_a$ when evaluated on Cita and $\mathcal{S}_1$ when evaluated on Cora) first, and then fine-tuned on manually labeled data ($\mathcal{S}_{(c)gt}$).
- **Adaptive Bi-LSTM-CRF**: This model is similar to Bi-LSTM-CRF with Fine-tuning, but an adaptive constraint term γ is used to prevent the model from overfitting the manually labeled data. We use $\gamma = 0.001$ in the experiments.

– **MA-Bi-LSTM-CRF**: Cita contains citation items from multiple disciplines. We can train a multi-adaptive Bi-LSTM-CRF model based on multiple generated datasets covering different disciplines for testing on the Cita dataset. In particular, we use S_1, S_2, and S_3 independently to train three Bi-LSTM-CRF models first, and fine-tune a Bi-LSTM-CRF model using S_{gt} following the method described in Sect. 4.2. For the Cora dataset, we do not apply this model, since Cora only contains citation items from a single discipline.

Table 2. Performance on the Cita dataset.

Methods	Token-level			Field-level		
	Precision	Recall	F1	Precision	Recall	F1
BibPro [2]	93.60	69.87	80.01	60.76	50.92	55.41
ParsCit [4]	88.23	82.04	85.02	76.84	72.84	74.78
CRF [13]	96.05	96.57	96.31	85.29	80.06	82.59
Bi-LSTM-CRF [10]	97.51	96.42	96.96	80.30	79.34	79.82
Bi-LSTM-CRF with Input Augmentation [5]	98.09	95.71	96.88	83.89	83.51	83.70
Bi-LSTM-CRF with Enhanced Data (Proposed)	96.60	95.35	95.97	81.80	81.43	81.61
Bi-LSTM-CRF with Fine-tuning (Proposed)	98.21	97.38†	97.79†	84.17	83.88	84.03
Adaptive Bi-LSTM-CRF (Proposed)	**98.36**	**97.58**†	**97.97**†	86.09†	85.83†	85.96†
MA-Bi-LSTM-CRF (Proposed)	98.27	97.44†	97.85†	**86.27**†	**86.16**†	**86.21**†

Table 3. Performance on the Cora dataset.

Methods	Token-level			Field-level		
	Precision	Recall	F1	Precision	Recall	F1
BibPro [2]	97.38	87.96	92.43	74.34	73.98	74.16
ParsCit [4]	90.85	97.25	93.94	75.93	76.47	76.20
CRF [13]	93.81	93.08	93.44	88.40	82.17	85.18
Bi-LSTM-CRF [10]	95.84	94.16	94.99	81.21	81.64	81.42
Bi-LSTM-CRF with Input Augmentation [5]	96.43	96.86	96.64	87.83	88.77	88.30
Bi-LSTM-CRF with Enhanced Data (Proposed)	93.81	94.77	94.29	80.24	82.53	81.37
Bi-LSTM-CRF with fine-tuning (Proposed)	97.61	96.36	96.98	89.45†	89.13†	89.29†
Adaptive Bi-LSTM-CRF (Proposed)	**98.02**†	**97.58**	**97.80**†	**93.42**†	**93.58**†	**93.50**†

5.3   Results

Tables 2 and 3 present experimental results of the models. They show that the proposed framework (with different model variants) consistently outperforms all the baselines, with a statistical significant margin[5]. The improvements achieved by the proposed framework over the baseline models is up to 5.2% in field-level

---

[5] The $\dagger$ symbol in the tables denotes that the result is statistically significantly better than all the baselines, with $p < .01$ based on t-test.

F1-score (i.e., 93.50% for our Adaptive Bi-LSTM-CRF model compared with 88.3% for the best baseline model on the Cora dataset). The experiments are run on NVIDIA Tesla K40 GPUs.

Effect of Generated Data. An important observation from the two tables is that a straightforward combination of the generated data and manually labeled data does not lead to optimal results (the Bi-LSTM-CRF with Enhanced Data model), even if the generated dataset contains rich information. This justifies the proposal of adaptive Bi-LSTM-CRF models, which involve a fine-tuning process.

Effect of Fine-Tuning. The results also show that fine-tuning helps increase the overall model performance (Bi-LSTM-CRF with Fine-tuning). Adding an adaptive term (adaptive Bi-LSTM-CRF and MA-Bi-LSTM-CRF) further improves the performance, which justifies the benefits of forcing the model not to deviate too far from the model trained from the generated training data, since this generalizes the model to avoid overfitting.

5.4   Discussion

Traditional citation field learning methods require feature engineering, which is a manual process to incorporate human knowledge into machine learning models. The proposed framework uses many different citation styles to generate training data for Bi-LSTM-CRF models. The CSL Style Repository that we use covers over 8,000 styles, which are created and maintained by over 500 contributors. This is a large collection of knowledge of citation styles, which requires many man hour to generate. The proposed framework can learn from such knowledge to solve the citation field learning problem with lightweight manual labeling efforts. However, the proposed framework has limitations of its reliance on generated data. Designing a data generation method without the need for the CSL Style Repository would be an interesting problem for future work.

6   Conclusion

We studied citation field learning and proposed a training data enhancement method that takes advantage of a large citation style knowledge base. We trained a Bi-LSTM-CRF model for the learning task over a combination of generated and manually labeled training data. We proposed an adaptive Bi-LSTM-CRF model to overcome the problem of imbalance size of generated and manually labeled data. Using the proposed model to extract citations fields requires little human-engineering effort and limited manually labeled data. Experimental results show that the proposed framework outperforms the state-of-the-art methods by 5.2% in field-level F1-score.

Acknowledgments. We gratefully acknowledge the support of NVIDIA Corporation with the donation of the Tesla K40 GPU used for this research.

References

1. Agrawal, P., Girshick, R., Malik, J.: Analyzing the performance of multilayer neural networks for object recognition. In: Fleet, D., Pajdla, T., Schiele, B., Tuytelaars, T. (eds.) ECCV 2014. LNCS, vol. 8695, pp. 329–344. Springer, Cham (2014). https://doi.org/10.1007/978-3-319-10584-0_22

2. Chen, C.C., Yang, K.H., Chen, C.L., Ho, J.M.: BibPro: a citation parser based on sequence alignment. TKDE **24**(2), 236–250 (2012)

3. Cortez, E., da Silva, A.S., Gonçalves, M.A., Mesquita, F., de Moura, E.S.: FLUX-CIM: flexible unsupervised extraction of citation metadata. In: JCDL, pp. 215–224 (2007)

4. Councill, I.G., Giles, C.L., Kan, M.: ParsCit: an open-source CRF reference string parsing package. In: LREC, pp. 661–667 (2008)

5. Daumé III, H.: Frustratingly easy domain adaptation. In: ACL, pp. 256–263 (2007)

6. Day, M.Y., et al.: Reference metadata extraction using a hierarchical knowledge representation framework. Decis. Support. Syst. **43**(1), 152–167 (2007)

7. Giles, C.L., Bollacker, K.D., Lawrence, S.: CiteSeer: an automatic citation indexing system. In: ACM DL, pp. 89–98 (1998)

8. Hetzner, E.: A simple method for citation metadata extraction using hidden Markov models. In: JCDL, pp. 280–284 (2008)

9. Huang, I.-A., Ho, J.-M., Kao, H.-Y., Lin, W.-C.: Extracting citation metadata from online publication lists using BLAST. In: Dai, H., Srikant, R., Zhang, C. (eds.) PAKDD 2004. LNCS (LNAI), vol. 3056, pp. 539–548. Springer, Heidelberg (2004). https://doi.org/10.1007/978-3-540-24775-3_64

10. Huang, Z., Xu, W., Yu, K.: Bidirectional LSTM-CRF models for sequence tagging. arXiv preprint arXiv:1508.01991 (2015)

11. Jaderberg, M., Simonyan, K., Vedaldi, A., Zisserman, A.: Synthetic data and artificial neural networks for natural scene text recognition. arXiv preprint arXiv:1406.2227 (2014)

12. Jagannatha, A., Yu, H.: Structured prediction models for RNN based sequence labeling in clinical text. In: EMNLP, pp. 856–865 (2016)

13. Lafferty, J.D., McCallum, A., Pereira, F.C.N.: Conditional random fields: probabilistic models for segmenting and labeling sequence data. In: ICML, pp. 282–289 (2001)

14. Lample, G., Ballesteros, M., Subramanian, S., Kawakami, K., Dyer, C.: Neural architectures for named entity recognition. In: NAACL HLT, pp. 260–270 (2016)

15. Ma, X., Hovy, E.H.: End-to-end sequence labeling via bi-directional LSTM-CNNs-CRF. In: ACL, pp. 1064–1074 (2016)

16. McCallum, A.: Andrew McCallum data (2005). https://people.cs.umass.edu/~mccallum/data.html

17. McCallum, A., Li, W.: Early results for named entity recognition with conditional random fields, feature induction and web-enhanced lexicons. In: NAACL HLT, pp. 188–191 (2003)

18. Peng, F., McCallum, A.: Information extraction from research papers using conditional random fields. IJIPM **42**(4), 963–979 (2006)

19. Sha, F., Pereira, F.C.N.: Shallow parsing with conditional random fields. In: NAACL HLT, pp. 134–141 (2003)

20. Tang, J., Yao, L., Zhang, D., Zhang, J.: A combination approach to web user profiling. TKDD **5**(1), 1–38 (2010)

21. Yin, P., Zhang, M., Deng, Z., Yang, D.: Metadata extraction from bibliographies using bigram HMM. In: ICADL, pp. 310–319 (2004)
22. Zhou, G., Su, J.: Named entity recognition using an HMM-based chunk tagger. In: ACL, pp. 473–480 (2002)
23. Zhu, J., Zhang, B., Nie, Z., Wen, J.R., Hon, H.W.: Webpage understanding: an integrated approach. In: SIGKDD, pp. 903–912 (2007)

Affine Transformation Capsule Net

Runkun Lu, Jianwei Liu[⊠], Siming Lian, and Xin Zuo

Department of Automation, China University of Petroleum,
Beijing Campus (CUP), Beijing 102249, China
zsylrk@gmail.com, liujw@cup.edu.cn

Abstract. CapsNet is a great attempt to relieve the drawback of CNN, where the routing by agreement method is tolerant to small changes in the viewpoint on one entity inside an image. This ground breaking has attracted the attention of many researchers, however, original CapsNet only utilizes the length of digit capsules in the classification task, which ignores the information of orientation. Based on this, we propose an Affine Transformation Capsule Net (AT-CapsNet) which we leverage both of the length and orientation information of digit capsules by adding a single-layer perceptron substitutes for the operation of computing length of vectors. In addition, we explain AT-CapsNet model's architecture from five perspectives and further analyse model complexity and the difference between dynamic routing and attention mechanism. The experimental results outperform the efficiency of our proposed algorithm in real world data sets.

Keywords: CapsNet · AT-CapsNet · Perceptron · Affine transformation

1 Introduction

Convolutional neural network (CNN) has strong, powerful capability for handling the image information with the multi-layer hierarchical network, the vision system in CNN utilizes the shared knowledge at all locations in the image. In this way, the weights of features filters are shared so that features learned at one location can be applied in other locations. CNN also reduces the number of training parameters by mining the spatial correlation in date set, so as to improve the efficiency of algorithm. However, with the properties of limiting the recognition on the spatial relationship between objects and ignoring the viewpoint variation while processing the images, CNN only causes complex impacts on pixel intensities but simple effects on the pose matrix which represents the relationship between target object and the viewer [1]. Motivated by this, Capsules Network (CapsNet) is proposed by the Hinton et al. [2], which more focus on a single fixation rather than a single identified object and its properties, and meanwhile these single fixations are coordinated over multiple fixations with the multi-layer visual system based on the CapsNet.

The capsules are generally constructed with the group of neurons, and these activities of the neurons within an active capsule represent the diverse properties of the specific entity that is present in the image [3]. Furthermore, the properties of the images include various types of the instantiation parameters as the inputs of the capsules

M. Ganji et al. (Eds.): PAKDD 2018, LNAI 11154, pp. 233–242, 2018.
https://doi.org/10.1007/978-3-030-04503-6_24

network. The specific component of the CapsNet can be divided into several parts: stride convolutional net [4], PrimaryCaps, DigitCaps and the routing-by-agreement. And so far, there exist some problems of CapsNet as follows:

(1) The choice of different routing mechanisms: the dynamic routing adopts the simple iteration rule to update the routing parameters $(\widehat{u}_{j|i}, r, l)$, However, the EM routing also proposed to replace the dynamic routing in the routing-by-agreement [1], but there is no qualitative or quantitative analysis of routing mechanisms from primary capsule to digit capsule;

(2) The length and orientation of the outputs of DigitCaps should be exploited to represent the properties of the entity, and the length of vectors indicate the probability that entity exists but not make use of the orientation of the outputs vectors. In this way, some of the relevant information may be discarded;

(3) Compare with the CNN approach, the performance of the CapsNet has been improved only in MNIST, but there are too many weight parameters need to be updated. Thus, the algorithm is computational expensive, which is not more readily satisfied for larger scale data sets;

(4) On some real data sets, such as CIFAR-10, the CapsNet achieved 10.6% error with the ensemble of 7 models. The result is easily achieved by standard convolutional nets [5]. However, if we adopt a single model to conduct the experiments, the classification performance is poor; but only if to reduce the error rate, the ensemble model will lead to the high computational complexity.

The specific analysis of the above viewpoints is discussed in the following section. In particular, we mainly aim at addressing the second problem to give the corresponding solution, named as Affine Transformation based Capsule Networks (AT-CapsNet). Instead of abandoning the orientation information of vectors, we utilize the perceptron to learn the mapping relationship from digit capsule to target tasks. The AT-CapsNet has better results than the original CapsNet in MNIST. We summarize the contributions of the AT-CapsNet as follows:

(1) We elaborated the detailed architecture of the CapsNet model, and the distinctions between the AT-CapsNet and the CapsNet has been pointed out;

(2) We analyse the problems existing in the CapsNet model with a practical example, such as the computational complexity and the difference dynamic routing and attention mechanism;

(3) An improved strategy has been proposed for the decision process, which utilize the length and orientation of digit capsules outputs;

(4) The experimental results demonstrate that our ATCapsNet is feasible on MNIST and Fashion MNIST, and we also make an analysis of the reason that CapsNet is difficult to achieve good results on CIFAR-10.

2 Related Work

CNN has brought about widespread attention because its powerful ability to process the large scales images [6]. Many researchers are not satisfied with the fact that only use it to process the ordinary images, hence, they devise the capsules and routing-by-agreement built in to CNN to process the shift and rotational images and capture the spatial relationship between objects, called CapsNet. The CapsNet is first introduced into the machine learning to address the above limitations of CNN [2], allowing networks to automatically learn the spatial relationships with the transformation matrix in a novel architecture when face up to the viewpoint variation problem.

The transformation matrix is also considered to be added to the spatial transformer networks [7], which also adopt the idea that the viewpoint invariant by changing the sampling progress of CNNs. And then De Brabandere et al. extend spatial transformer networks where filters are adapted during inference depending on the input [8]. Besides, [9] also uses the transformation matrix to deal with the images under different viewpoint variations. [10–12] also motivated by the CNNs and extend a novel architecture to process the images, in particular, [13] highlights the results of the single model on the CIFAR-10 data set that is hardly to reach the average level, and they conjecture the gap between MNIST and CIFAR-10 may possibly be attributed to the reconstruction method. The recent approaches focus on the capsules network to process the rotational images and capture the spatial relationships by using the capsules and returning the length of the vectors which indicate the probability of the entity exists, which share the similar idea of capsules but with an incremental advance in capsules structure [1, 13].

Our proposed method makes use of both the length and the orientation of digit capsule outputs to process the images from the view of the perceptron, and the details of ATCapsNet are described in Sect. 3, which also involves stride convolutional net, PrimaryCaps, and digit capsule. In addition, model complexity is a key problem of CapsNet, we also give a detailed analysis of it. The routing algorithm could not simply be seen as an attention mechanism, and we also have an analysis of the discrepancy between the routing-by-agreement and attention mechanism. Finally, the experimental results show that compared with CapsNet our proposed method makes an improvement on MNIST.

3 Framework

In this section, we will give an explicit description of ATCapsNet with an example of training on MNIST data set [14], and the detailed training processes are shown in Fig. 1 which can be divided into five stages: A, B, C, D, E.

Stage A: In Fig. 1, the image is the input of the system, and its shape is $[28, 28, 1]$. Further consideration, Conv1 is a shallow feature of raw image with a convolutional operation which has 256, 9×9 convolution kernels with a stride of 1 and ReLu activation.

Stage B: We view Conv1 as the input of another convolution layer which has 32, 32, $9 \times 9 \times 256$ convolution kernels with a stride of 2 and ReLu activation (note that with the ReLu activation the classification performance gets better). And we do this operation for 8 times which are shown in Fig. 2 in details. The second column of Fig. 2 comprises 8 blocks and each of them is composed of 32, 6×6 matrixes. We combine the same layers in each block to form the Primary Capsule, which is shown in the third column of Fig. 2. A key concept of Primary Capsule is that of each capsule is an 8-dimensional vector and totally there are $6 \times 6 \times 32$ Primary Capsules. As a consequence, it can be reshaped into a matrix with its size of $[6 \times 6 \times 32, 8]$, which means there are $6 \times 6 \times 32$ inputs of the stage C as shown in the 4-th column in Fig. 1.

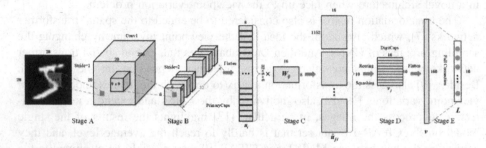

Fig. 1. Frame work of AT-CapsNet

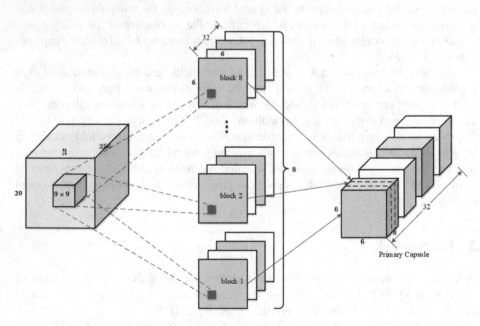

Fig. 2. Explanation of primary capsule

Stage C: W_{ij} is a transform matrix which project each 8-dimensional primary capsule u_i into a 16-dimensional vector which is defined as prediction vectors $\hat{u}_{j|i}$, which can be formulated as follows:

$$\hat{u}_{j|i} = W_{ij}u_i \tag{1}$$

It is a very significant concept that each primary capsule need to be transformed 10 times because $j \in \{1, 2, \cdots, 10\}$ represents the j-th category of the data set. Note that the model complexity of CapsNet mainly comes from this stage, and the details will be discussed in Sect. 4.

Stage D: This stage describes the routing mechanism, which is the core idea of CapsNet. Note that v_j is the digit capsule and it comes from s_j by using a non-linear squashing function, and s_j is defined as follows:

$$s_j = \sum_i c_{ij}\hat{u}_{j|i} \tag{2}$$

where c_{ij} is the coupling coefficients that are determined by dynamic routing process, and it is determined by the log prior probabilities b_{ij} that capsule i should be coupled to capsule j. The functional relationship between c_{ij} and b_{ij} and the updating rule of b_{ij} are shown as follows:

$$c_{ij} = \frac{\exp(b_{ij})}{\sum_k \exp(b_{ik})}, b_{ij} \leftarrow b_{ij} + \hat{u}_{j|i} \cdot v_j \tag{3}$$

Equation (3) only shows the relationship between c_{ij} and b_{ij}, however, this is an iterative routing process and its details are shown in [1]. Interestingly, we may find the dynamic routing process is similar to attention mechanism, where they all consider the inner product between two vectors that come from two layers. However, they are essentially different, and we will discuss it in details in Sect. 5.

Stage E: Obviously, this stage corresponds to a classification task. And in the stage D, we have obtained 10 digit capsules v_j, $j \in \{1, 2, \cdots, 10\}$ as the input of stage E, and v_j is advanced feature of each category. Unlike the traditional classification methods, the output is not a set of conditional probability but a set of vectors. As we known, a vector involves richer information than a scalar, which is the great advantage of capsule that it may have better prediction performance as well as the ability of generalization. In [3], they predict class labels only using the length of each digit capsule but ignore the orientation of the vector. However, the orientation of the vector represents the properties of the entity, which is meaningful and useful. The stage E in this paper as shown in Fig. 1 is pretty different from the way proposed in [3]. We flatten the digit capsule matrix from a 10 by 16 matrix into a 160-dimensional vector, and we use a single layer perceptron with sigmoid activation function to map it into a 10-dimensional vector. The perceptron corresponds to a non-linear affine transformation with the form as follows:

$$L = sigmoid(f), \ f = wv + b \tag{4}$$

where $v \in \Re^{160 \times 1}$ is defined as the concatenation of v_j, and w is regression coefficient as well as b is the bias. Unlike the way to only compute the length of each vector v_j, which is also an affine transformation but its form is certain and unconstrained, we get the affine transformation through the constraint of loss function and make use of the relationships between each digit capsules which is discussed in details in Subsect. 6.1.

4 Model Complexity Analysis

The model complexity of CapsNet mainly comes from $W_{ij} \in \Re^{8 \times 16}$, which transform the low-level capsules into high level ones. In the training process for only a picture coming from MNIST, we find that the number of the elements in W_{ij} is more than a million $(32 \times 6 \times 6 \times 8 \times 16 \times 10 = 1474560 = O(10^6))$, because $i \in \{1, 2, \cdots, 32 \times 6 \times 6\}$, and $j \in \{1, 2, \cdots, 10\}$). Note that the training set comprises 55000 pictures (the remaining 5000 pictures are validation set), which means in the training stage nearly one hundred billion parameters will be updated only about $W_{ij}(55000 \times 1474560 \approx 8 \times 10^{10} \approx O(10^{11}))$.

In practical operations, an acceptable CNN with the accuracy 99.1% only needs to be running for 2 min on a 1080Ti GPU, and a CapsNet with the accuracy about 9.5% to spend at least 4 h. The discussion above is based on MNIST, and its picture size is $28 \times 28 \times 1$. However, almost all RGB image data sets have a bigger volume than MNIST, which requires highly computational costs. Simultaneously, with the increase of channel numbers of image, CapsNet suffers from poor performance problem which is discussed in details in Sect. 6.2. As a consequence, it is very necessary and impendency to improve the problem of model and computational complexity. In spite of the presence of complexity makes it difficult to apply capsule net in practical project, CapsNet is still a great attempt to improve CNN and has overcame some defects of CNN.

One way to reduce the complexity of the model is to share the weights which exist in Stage C of Fig. 2, and we call it the transform matrix W_{ij}, where $i \in \{1, 2, \cdots, 32 \times 6 \times 6\}$, and $j \in \{1, 2, \cdots, 10\}$. By drawing on the idea of CNN, we let all the primary capsules share the same transform matrix W_j, and $j \in \{1, 2, \cdots, 10\}$, and the structure is shown in Fig. 3. In this way, the numbers of transform matrix

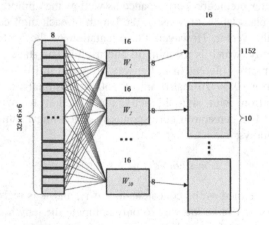

Fig. 3. The structure of sharing the weights

reduce from 11520 to 10, and the complexity of the model decrease three percentage points. However, with the decrease of transform matrix, the classification performance must be influenced. Fortunately, the impact is small and acceptable, and the comparison of classification result is shown in Table 1.

Table 1. Table captions should be placed above the tables.

Method	MNIST	Fashion-MNIST
CapsNet	99.4391%	90.8454%
CapsNet-shared	99.3928%	90.0234%

5 The Difference Between Dynamic Routing and Attention Mechanism

We may find that the updating rule of b_{ij} in dynamic routing is similar to soft attention [15], but they are actually different. In soft attention mechanism, assumption that h is the state of encoder and s is state of decoder, the computation rule of context vector c_i is defined as:

$$c_i = \sum_j \alpha_{ij} h_j \tag{5}$$

where the weight α is defined as follows:

$$\alpha_{ij} = \frac{\exp(e_{ij})}{\sum_k \exp(e_{ik})}, \quad e_{ij} = s_{i-1} \cdot h_j \tag{6}$$

Compared with the soft attention mechanism discussed above, one of the differences is that the routing mechanism in CapsNet is an iterative algorithm [3]. On the other hand, according to the routing mechanism which is formulated in (1), (2), and (3), it's not hard to see that the dynamic routing actually is the way to determine the low-level capsules are delivered to which high level capsules. On the contrary, attention mechanism is the way that the output in decoder select the fraction of information of input latent vectors in encoder as shown in (5), and (6). Hence the discussion above verify that the dynamic routing is different from soft attention mechanism, but it may draw lessons from the thought of attention mechanism and it has been proven that the dynamic routing is an efficient method.

6 Experiment

6.1 Classification Performance

In this subsection, we mainly compare the classification results between AT-CapsNet and CapsNet on MNIST and Fashion-MNIST data sets [16]. We just use the trick proposed in [3] both on CapsNet and AT-CapsNet, such as reconstruction and mask out. Meanwhile, the numbers and the shapes of each kind of layers in CapsNet and AT-CapsNet are the same and the specific experiment configuration is similar to the method in [3]. The results are shown in Table 2.

Table 2. Table captions should be placed above the tables.

Method	MNIST	Fashion-MNIST
CapsNet	99.4391%	90.8454%
AT-CapsNet	99.5192%	90.6542%
AT-CapsNet-1H	99.3589%	90.3221%
AT-CapsNet-2H	99.2388%	90.3314%
AT-CapsNet-10P	99.2789%	89.6935%

We also consider some additional case of AT-CapsNet. We add additional hidden layers in the stage E in Fig. 1 to improve the classification performance. However, when adding 1 hidden layer with 64 nodes (AT-CapsNet-1H) and adding 2 hidden layers with 64 and 32 nodes (AT-CapsNet-2H), the classification results deteriorate. The reason is that the digit capsules have been pretty good features so that we do not need to add additional hidden layers to further refine features.

Furthermore, we find that a 16-dimensional digit capsule v_j is the feature of category j, however, we flatten the 10 vectors together and make the 160-dimensional vector as the input of perceptron, which means that for category j the other 9 digit capsules also contribute to its classification task. Is that right? To validate our guess, we construct 10 perceptrons respectively (AT-CapsNet-10P), the j-th perceptron's input is v_j and its output is the probability of the image belongs to category j. Interestingly, the poor classification performance occurred, which means the for category j the other 9 digit capsules have positive contribution on it.

The results of the above discussion are listed in Table 1, and we find that the AT-CapsNet is both efficient and effective. On MNIST, we get a better result than the baseline method of CapsNet, and on Fashion-MNIST the performance is acceptable. Note that the source code of CapsNet is come from https://github.com/naturomics/ CapsNet-Tensorflow, and we modify it to form the version which is suitable for our purpose.

6.2   Classification Performance

In [3], they achieve 10.6% error with an ensemble of 7 models on CIFAR-10 data set, however, without ensemble, the single CapsNet model can only achieve about 40% error according to our experimental results. In [13], they attempt to find the best configuration set that yield the optimal test error on CIFAR-10 data set. In their experiment, by using a single baseline CapsNet model, they achieve 31.07% error. This is not an acceptable result because the start-of-art result on CIFAR-10 is 4.50%. As a result, the failure on CIFAR-10 means that we face great challenge on complex data sets.

On the other hand, the reconstructed image of CIFAR-10 is bad by using the CapsNet as shown in Fig. 4. Unlike digits on MNIST, objects on CIFAR10 have more than one viewpoint for each class due to changes in viewpoint. Applying a 2-dimensional reconstruction regularization method on 3-dimensional data may cause inaccurate reconstruction regularization values, which may be a contributing factor to the subpar performance of capsule net on complex data [13].

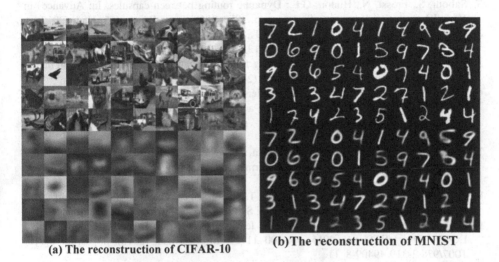

(a) The reconstruction of CIFAR-10   (b) The reconstruction of MNIST

Fig. 4. The reconstruction of CIFAR-10 and MNIST

7 Conclusion and Future Work

It is undeniable that CapsNet is a great attempt to improve CNN, however, we have a long way to go if we really want to put it into practical use. Our proposed AT-CapsNet model considers both of the length and orientation of digit capsules, which makes full use of the advantage of vector neurons. However, we have not solved some problems existing in CapsNet, such as poor performance on complex data set, and the existing capsule net cannot solve it too. The future works may focus on it, but it is only one of the research directions of CapsNet, there are still plenty of fields can be attempted. Considering the structure of model, the routing mechanism, squashing function, and the selection of the dimension of capsules may also the key direction to improve the performance of CapsNet. And in application, how to transfer CapsNet into other fields is necessary and interesting, such as GANs [17], what will happen if we use CapsNet to replace the discriminator (CNN) in DCGAN [18]. On the other hand, to apply CapsNet into industrial scene, we also need to find a better way to improve model complexity, our proposed method in Sect. 5 is a way, but there must be other more effective methods.

References

1. Frosst, N., Hinton, G.E., Sabour, S.: Matrix capsules with em routing. In: International Conference on Learning Representations, accepted as poster, Vancouver, BC, Canada (2018)
2. Hinton, G.E., Krizhevsky, A., Wang, S.D.: Transforming auto-encoders. In: Honkela, Timo, Duch, W., Girolami, M., Kaski, S. (eds.) ICANN 2011. LNCS, vol. 6791, pp. 44–51. Springer, Heidelberg (2011). https://doi.org/10.1007/978-3-642-21735-7_6
3. Sabour, S., Frosst, N., Hinton, G.E.: Dynamic routing between capsules. In: Advances in Neural Information Processing Systems, pp. 3859–3869, Long Beach, CA, USA (2017)
4. Springenberg, J.T., Dosovitskiy, A., Brox, T., Riedmiller, M.: Striving for simplicity: The all convolutional net. arXiv preprint arXiv:1412.6806 (2014)
5. Zeiler, M.D., Fergus, R.: Stochastic pooling for regularization of deep convolutional neural networks. arXiv preprint arXiv:1301.3557 (2013)
6. LeCun, Y., Boser, B.E., Denker, J.S., Henderson, D., Howard, R.E., Hubbard, W.E., Jackel, L.D.: Handwritten digit recognition with a back-propagation network. In: Advances in Neural Information Processing Systems, pp. 396–404, Morgan Kaufmann, Denver, Colorado, USA (1990)
7. Jaderberg, M., Simonyan, K., Zisserman, A., Kavukcuoglu, K.: Spatial transformer networks. In: Advances in Neural Information Processing Systems, pp. 2017–2025, Montreal, Quebec, Canada (2015)
8. Jia, X., De Brabandere, B., Tuytelaars, T., Gool, L.V.: Dynamic filter networks. In: Advances in Neural Information Processing Systems, pp. 667–675, Barcelona, Spain (2016)
9. Lenc, K., Vedaldi, A.: Learning covariant feature detectors. In: Hua, G., Jégou, H. (eds.) ECCV 2016. LNCS, vol. 9915, pp. 100–117. Springer, Cham (2016). https://doi.org/10.1007/978-3-319-49409-8_11
10. Cohen, T., Welling, M.: Group equivariant convolutional networks. In: International Conference on Machine Learning, pp. 2990–2999, New York City, NY, USA (2016)
11. Dieleman, S., De Fauw, J., Kavukcuoglu, K.: Exploiting cyclic symmetry in convolutional neural networks. arXiv preprint arXiv:1602.02660 (2016)
12. Oyallon, E., Mallat, S.: Deep roto-translation scattering for object classification in CVPR. In: IEEE Conference on Computer Vision and Pattern Recognition, pp. 2865–2873, Boston, MA, USA (2015)
13. Xi, E., Bing, S., Jin, Y.: Capsule network performance on complex data. arXiv preprint arXiv:1712.03480 (2017)
14. LeCun, Y.: The mnist database of handwritten digits. http://yann.lecun.com/exdb/mnist/ (2008)
15. Yao, L., et al.: Video description generation incorporating spatio-temporal features and a soft-attention mechanism. arXiv preprint arXiv:1502.08029 (2015)
16. Xiao, H., Rasul, K., Vollgraf, R.: Fashionmnist: a novel image dataset for benchmarking machine learning algorithms. arXiv preprint arXiv:1708.07747 (2017)
17. Goodfellow, I.J., et al.: Generative adversarial nets. In: International Conference on Neural Information Processing Systems MIT Press, pp. 2672–2680, Montreal, Quebec, Canada (2014)
18. Radford, A., Metz, L., Chintala, S.: Unsupervised representation learning with deep convolutional generative adversarial networks. arXiv preprint arXiv:1511.06434 (2015)

A Novel Algorithm for Frequent Itemsets Mining in Transactional Databases

Huan Phan[1,2(✉)] and Bac Le[3]

[1] Division of IT, University of Social Sciences and Humanities, VNU-HCM,
Ho Chi Minh City, Vietnam
huanphan@hcmussh.edu.vn
[2] Faculty of Mathematics and Computer Science, University of Science,
VNU-HCM, Ho Chi Minh City, Vietnam
[3] Faculty of IT, University of Science, VNU-HCM, Ho Chi Minh City, Vietnam
lhbac@fithcmus.edu.vn

Abstract. Since the era of data explosion, data mining in transactional databases has become more and more important. There are many data mining techniques like association rule mining, the most important and well-researched one. Furthermore, frequent itemset mining is one of the fundamental but time-consuming steps in association rule mining. Most of the algorithms used in literature find frequent itemsets on search space items having at least a *minsup* and are not reused for mining next time. To deal with this problem, NOV-FI algorithms are proposed as a new approach in order to quickly detect frequent itemsets from transactional databases using an array of co-occurrences and occurrences of kernel item in at least one transaction. NOV-FI algorithms are easily expanded in distributed systems. Finally, the experimental results show that the proposed algorithms perform better than other existing algorithms.

Keywords: Association rules · Co-occurrence items · Frequent itemsets

1 Introduction

Mining frequent itemsets is a fundamental and essential problem in many data mining applications such as the discovery of association rules, strong rules, correlations, multi-dimensional patterns, and many other important discovery tasks. The problem is formulated as follows: Given a large database of set of items transactions, find all frequent itemsets, where a frequent itemset is one that occurs in *at least a user-specified percentage* of transaction database [7].

Most of the mining algorithms for frequent itemsets, proposed by various authors around the world, are based on Apriori [1, 2] and FP-Tree [3, 6]. Apriori generated candidate using a breadth-first search. Besides, the algorithms improve upon the Apriori using a depth-first search. The main limitation of algorithms based on Apriori, a major advance developed using pattern-growth based on FP-Tree. In recent years, Deng et al., proposed the PrePost [6] algorithm for constructing a *FP-Tree-like* to mining frequent itemsets from a database and shows the better performance result. In

© Springer Nature Switzerland AG 2018
M. Ganji et al. (Eds.): PAKDD 2018, LNAI 11154, pp. 243–255, 2018.
https://doi.org/10.1007/978-3-030-04503-6_25

this paper, we propose a NOV-FI algorithm that mines frequent itemsets, and easily expanded in distributed systems.

- **Algorithm 1**: Computing Kernel_COOC array of co-occurrences and occurrences of kernel item in at least one transaction;
- **Algorithm 2**: Generating all frequent itemsets based on Kernel_COOC array;

This paper is organized as follows: in Sect. 2, we describe the basic concepts for mining frequent itemsets and data structure for transaction databases. Some theoretical aspects of our approach relies, are given in Sect. 3. Besides, we describe our NOV-FI algorithm to compute frequent itemsets based on Algorithm 1 and Algorithm 2. Details on implementation and experimental tests are discussed in Sect. 4. Finally, we conclude with a summary of our approach, perspectives and extensions of this future work.

2   Background

2.1   Frequent Itemset Mining

Let $I = \{i_1, i_2, \ldots i_m\}$ be a set of m distinct items. A set of items $X = \{i_1, i_2, \ldots i_k\}, \forall i_j \in I (1 \leq j \leq k)$ is called an itemset, an itemset with k items is called a $k - itemset$. **D** be a dataset containing n transaction, a set of transaction $T = \{t_1, t_2, \ldots t_n\}$, and each transaction $t_j = \{i_{k_1}, i_{k_2}, \ldots i_{k_l}\}, \forall i_{k_l} \in I (1 \leq k_l \leq m)$.

Definition 1. The support of an itemset X is the number of transaction in which occurs as a subset, denoted as $sup(X)$.

Definition 2. Let $minsup$ be the threshold minimum support value specified by user. If $sup(X) \geq minsup$, itemset X is called a frequent itemset, denoted FI is the set of all the frequent itemset.

Property 1. $\forall X \subseteq Y$, If $sup(Y) \geq minsup$ then $sup(X) \geq minsup$;

Property 2. $\forall X \subset Y$, If $sup(X) < minsup$ then $sup(Y) < minsup$;
See an example transaction database $\mathcal{D}$ in Table 1.

Table 1. The transaction database $\mathcal{D}$ used as our running example.

TID	Items						
t1	A		C		E	F	
t2	A		C				G
t3					E			H
t4	A		C	D		F	G
t5	A		C		E		G
t6					E		
t7	A	B	C		E		
t8	A		C	D			
t9	A	B	C		E		G
t10	A		C		E	F	G

Table 2. Mining frequent itemsets.

k-itemset	FI (*minsup = 2*)	FI (*minsup = 3*)	FI (*minsup = 5*)
1	D, B, F, G, E, A, C	F, G, E, A, C	G, E, A, C
2	BE, BA, BC, DA, DC, FE, FG, FA, FC, GE, GA, GC, EA, EC, AC	FA, FC, GE, GA, GC, EA, EC, AC	GA, GC, EA, EC, AC
3	BAC, BEA, DAC, FEA, BEC, FEC, FGA, CFG, FAC, GEA, GEC, EAC, GAC	FAC, GEA, GEC, GAC, EAC	GAC, EAC
4	BEAC, FGAC, FEAC, GEAC	GEAC	

Example 1. See Table 1. There are eight different items $I = \{A, B, C, D, E, F, G, H\}$ and ten transactions $T = \{t1, t2, t3, t4, t5, t6, t7, t8, t9, t10\}$. Table 2 shows the frequent itemsets at three different minsup values – 2 (20%), 3 (30%) and 5 (50%) respectively.

2.2 Data Structure for Transaction Database

The binary matrix is an efficient data structure for mining frequent itemsets [4, 5]. The process begins with the transaction database transformed into a binary matrix BiM, in which each row corresponds to a transaction and each column corresponds to an item. Each element in the binary matrix BiM contains 1 if the item is presented in the current transaction; otherwise it contains 0 (Fig. 1).

TID	A	B	C	D	E	F	G	H
t1	1	0	1	0	1	1	0	0
t2	1	0	1	0	0	0	1	0
t3	0	0	0	0	1	0	0	1
t4	1	0	1	1	0	1	1	0
t5	1	0	1	0	1	0	1	0
t6	0	0	0	0	1	0	0	0
t7	1	1	1	0	1	0	0	0
t8	1	0	1	1	0	0	0	0
t9	1	1	1	0	1	0	1	0
t10	1	0	1	0	1	1	1	0

Fig. 1. A binary matrix BiM representation of example transaction database.

3 The Proposed Algorithms

3.1 Generating Array of Co-occurrence Items of Kernel Item

In this part, we illustrate the framework of the algorithm generating co-occurrence items of items in transaction database.

Definition 3. Project set of item i_k on database $\mathcal{D}$: $\pi(i_k) = \{t_j \in \mathcal{D}|i_k \subseteq t_j\}$ is set of transaction contain item i_k (π–*decreasing monotonic*). According to Definition 1:

$$\sup(i_k) = |\pi(i_k)| \tag{1}$$

Example 2. See Table 1. Consider item B, we detect project set of item B on database $\mathcal{D}$: $\pi(B) = \{t_7, t_9\}$ then $sup(B) = |\pi(B)| = 2$.

Definition 4. Project set of itemset $X = \{i_1, i_2, \ldots i_k\}, \forall i_{j=\overline{1,k}} \in I$: $\pi(X) = \pi(i_1) \cap \pi(i_2) \ldots \pi(i_k)$.

$$\sup(X) = |\pi(X)| \tag{2}$$

Example 3. See Table 1. Consider item E, we detect project set of item E on database $\mathcal{D}$: $\pi(E) = \{t_1, t_3, t_5, t_6, t_7, t_9, t_{10}\}$ then $sup(BE) = |\pi(BE)| = |\pi(B) \cap \pi(E)| = |\{t_7, t_9\} \cap \{t_1, t_3, t_5, t_6, t_7, t_9, t_{10}\}| = |\{t_7, t_9\}| = 2$.

Definition 5. Let $i_k \in I$ is called a kernel item. Itemset $X_{cooc} \subseteq I$ is called co-occurrence items with kernel item i_k, so that satisfy $\pi(i_k) \equiv \pi(i_k \cup X_{cooc})$. Denoted as $cooc(i_k) = X_{cooc}$.

Example 4. See Table 1. Consider item B as kernel item, we detect co-occurrence items with item B as $cooc(B) = \{A, C, E\}$ and $sup(B) = sup(BACE) = 2$.

Definition 6. Let $i_k \in I$ is called a kernel item. Itemset $Y_{looc} \subseteq I$ is called occurrence items with kernel item i_k in as least one transaction, but not co-occurrence items, so that satisfy $1 \leq |\pi(i_k \cup i_{looc})| < |\pi(i_k)|, \forall i_{looc} \in Y_{looc}$. Denoted as $looc(i_k) = Y_{looc}$.

Example 5. See Table 1. Consider item B as kernel item, we detect occurrence items with item B in as least one transaction $looc(B) = \{G\}$ and $\pi(BG) = \{t_9\} \subset \pi(B) = \{t_7, t_9\}$.

Algorithm Generating Array of Co-occurrence Items

This algorithm is generating co-occurrence items of items in transaction database and archive into the *Kernel_COOC* array. Each element within the *Kernel_COOC*, 4 fields (Fig. 2):

- Kernel_COOC[k].*item* : kernel item k;
- Kernel_COOC[k].*sup* : support of kernel item k;
- Kernel_COOC[k].*cooc* : co-occurrence items with kernel item k;
- Kernel_COOC[k].*looc* : occurrence items kernel item k in least one transaction.

The framework of Algorithm 1 is as follows:

Algorithm 1. Generating Array of Co-occurrence Items

    Input : Dataset $\mathcal{D}$

    Output: *Kernel_COOC array*

1:    **foreach** Kernel_COOC[k] **do**

2:        Kernel_COOC[k].item = i_k

3:        Kernel_COOC[k].sup = **0**

4:        Kernel_COOC[k].cooc = $2^m - 1$

5:        Kernel_COOC[k].looc = 0

6:    **foreach** $t_j \in T$ **do**

7:        **foreach** $i_k \in t_j$ **do**

8:            Kernel_COOC[k].sup ++

9:            Kernel_COOC[k].cooc = Kernel_COOC[k].cooc **AND** vectorbit(t_j)

10:       Kernel_COOC[k].looc = Kernel_COOC[k].looc **OR** vectorbit(t_j)

11:   sort Kernel_COOC array in ascending by support

We illustrate Algorithm 1 on example database in Table 1.

Initialization of the Kernel_COOC array, number items in database m = 8;

Item	A	B	C	D	E	F	G	H
sup	0	0	0	0	0	0	0	0
cooc	11111111	11111111	11111111	11111111	11111111	11111111	11111111	11111111
looc	00000000	00000000	00000000	00000000	00000000	00000000	00000000	00000000

Read once of each transaction from *t1* to *t10*

Transaction $t1 = \{A, C, E, F\}$ has vector bit representation **10101100**;

Item	A	B	C	D	E	F	G	H
sup	1	0	1	0	1	1	0	0
cooc	10101100	11111111	10101100	11111111	10101100	10101100	11111111	11111111
looc	10101100	00000000	10101100	00000000	10101100	10101100	00000000	00000000

Transaction $t2 = \{A, C, G\}$ has vector bit representation **10100010**;

Item	A	B	C	D	E	F	G	H
sup	2	0	2	0	1	1	1	0
cooc	**10100000**	11111111	**10100000**	11111111	10101100	10101100	**10100010**	11111111
looc	**10101110**	00000000	**10101110**	00000000	10101100	10101100	**10100010**	00000000

Transaction $t3 = \{E, H\}$ has vector bit representation **00001001**;

Item	A	B	C	D	E	F	G	H
sup	2	0	2	0	2	1	1	1
cooc	10100000	11111111	10100000	11111111	**00001000**	10101100	10100010	**00001001**
looc	10101110	00000000	10101110	00000000	**10101101**	10101100	10100010	**00001001**

Transaction $t4 = \{A, C, D, F, G\}$ has vector bit representation **10110110**;

Item	A	B	C	D	E	F	G	H
sup	3	0	3	1	2	2	2	1
cooc	**10100000**	11111111	**10100000**	**10110110**	00001000	**10100100**	**10100010**	00001001
looc	**10111110**	00000000	**10111110**	**10110110**	10101101	**10111110**	**10110110**	00001001

Transaction $t5 = \{A, C, E, G\}$ has vector bit representation **10101010**;

Item	A	B	C	D	E	F	G	H
sup	4	0	4	1	3	2	3	1
cooc	**10100000**	11111111	**10100000**	10110110	**00001000**	10100100	**10100010**	00001001
looc	**10111110**	00000000	**10111110**	10110110	**10101111**	10111110	**10111110**	00001001

Transaction $t6 = \{E\}$ has vector bit representation **00001000**;

Item	A	B	C	D	E	F	G	H
sup	4	0	4	1	4	2	3	1
cooc	10100000	11111111	10100000	10110110	**00001000**	10100100	10100010	00001001
looc	10111110	00000000	10111110	10110110	**10101111**	10111110	10111110	00001001

Transaction $t7 = \{A, B, C, E\}$ has vector bit representation **11101000**;

Item	A	B	C	D	E	F	G	H
sup	5	1	5	1	5	2	3	1
cooc	**10100000**	**11101000**	**10100000**	10110110	**00001000**	10100100	10100010	00001001
looc	**11111110**	**11101000**	**11111110**	10110110	**11101111**	10111110	10111110	00001001

Transaction $t8 = \{A, C, D\}$ has vector bit representation **10110000**;

Item	A	B	C	D	E	F	G	H
sup	6	1	6	2	5	2	3	1
cooc	**10100000**	11101000	**10100000**	**10110000**	00001000	10100100	10100010	00001001
looc	**11111110**	11101000	**11111110**	**10110110**	11101111	10111110	10111110	00001001

Transaction $t9 = \{A, B, C, E, G\}$ has vector bit representation **11101010**;

Item	A	B	C	D	E	F	G	H
sup	7	2	7	2	6	2	4	1
cooc	**10100000**	**11101000**	**10100000**	10110000	**00001000**	10100100	**10100010**	00001001
looc	**11111110**	**11101010**	**11111110**	10110110	**11101111**	10111110	**11111110**	00001001

The last, transaction $t10 = \{A, C, E, F, G\}$ has vector bit representation **10101110**;

Item	A	B	C	D	E	F	G	H
sup	8	2	8	2	7	3	5	1
cooc	10100000	11101000	10100000	10110000	00001000	10100100	10100010	00001001
looc	11111110	11101010	11111110	10110110	11101111	10111110	11111110	00001001

After the processing of Algorithm 1, the Kernel_COOC array as follows:

Table 3. Kernel_COOC array are ordered in support ascending order.

Item	H	B	D	F	G	E	A	C
sup	1	2	2	3	5	7	8	8
cooc	E	A, C, E	A, C	A, C	A, C	Ø	C	A
looc	Ø	G	F, G	D, E, G	B, D, E, F	A, B, C, F, G, H	B, D, E, F, G	B, D, E, F, G

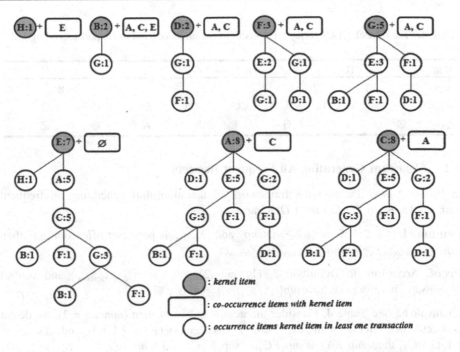

Fig. 2. The pattern-tree of occurrence items with kernel item in as least one transaction.

See Table 3, we have $cooc(A) = \{C\}$ and $cooc(C) = \{A\}$. In this case, the frequent itemset generated from A and C items will be duplicated. We provide a Definitions 7, 8 to eliminate the duplication when generating frequent itemsets.

Definition 7. Let $i_k \in I(i_1 \prec i_2 \prec \ldots \prec i_m)$ items are ordered in support ascending order, i_k is called a kernel item. Itemset $X_{lexcooc} \subseteq I$ is called co-occurrence items with the kernel item i_k, so that satisfy $\pi(i_k) \equiv \pi(i_k \cup i_j), i_k \prec i_j, \forall i_j \in X_{lexcooc}$. Denoted as $lexcooc(i_k) = X_{lexcooc}$.

Definition 8. Let $i_k \in I(i_1 \prec i_2 \prec \ldots \prec i_m)$ items are ordered in support ascending order, i_k is called a kernel item. Itemset $Y_{lexlooc} \subseteq I$ is called occurrence items with kernel item i_k in as least one transaction, but not co-occurrence items, so that satisfy $1 \leq |\pi(i_k \cup i_{lexlooc})| < |\pi(i_k)|, \forall i_{lexlooc} \in Y_{lexlooc}$. Denoted as $lexlooc(i_k) = Y_{lexlooc}$.

Additional command line 12, 13 and 14 into Algorithm 1 (Fig. 3):

12:	**foreach** $i_k \in t_j$ **do**
13:	Kernel_COOC[k].cooc = lexcooc(i_k)
14:	Kernel_COOC[k].looc = lexlooc(i_k)

According to Definition 7, we have $cooc(C) = \{A\}$, where $A \prec C$ so $lexcooc(C) = \{\varnothing\}$. Similary, according to Definition 8, we have $looc(G) = \{B, D, E, F\}$, where $B, D \prec F \prec G \prec E$, so $lexlooc(G) = \{E\}$. Execute command line 12, 13 and 14 has result on Table 4.

Table 4. The Kernel_COOC array are co-occurrence items ordered in support ascending order.

Item	H	B	D	F	G	E	A	C
sup		1	2	2	3	5	7	8	8
cooc	E	A, C, E	A, C	A,C	A, C	$\varnothing$	C	$\varnothing$
looc	$\varnothing$	G	F, G	G, E	E	A, C	$\varnothing$	$\varnothing$

3.2   Algorithm Generating All Frequent Itemsets

In this part, we illustrate the framework of the algorithm generating all frequent itemsets bases on the *Kernel_COOC* array.

Lemma 1. $\forall i_k \in I$, if $sup(i_k) \geq minsup$ and $X_{lexcooc}$ is powerset of $lexcooc(i_k)$ then $sup(i_k \cup x_{lexcooc}) \geq minsup, \forall x_{lexcooc} \in X_{lexcooc}$.

Proof. According to Definition 8, (1) and (2): $\pi(i_k) \equiv \pi(i_k \cup x_{lexcooc})$ and $sup(i_k) \geq minsup$. Therefore, we have $sup(i_k \cup x_{lexcooc}) \geq minsup$ ∎.

Example 6. See Table 4. Consider the item F as kernel item ($minsup = 2$), we detect co-occurrence items with the item F as $lexcooc(F) = \{A, C\}$ and $X_{lexcooc} = \{A, C, AC\}$, then $sup(FA) = sup(FC) = sup(FAC) = 3 \geq minsup$.

Lemma   2. $\forall i_k \in I, Y_{lexlooc}$ is powerset of $lexlooc(i_k), \forall y_{lexlooc} \in Y_{lexlooc}$, if $sup(i_k \cup y_{lexlooc}) \geq minsup$ and $X_{lexcooc}$ is powerset of $lexcooc(i_k)$ then $sup(i_k \cup y_{lexlooc} \cup x_{lexcooc}) \geq minsup, \forall x_{lexcooc} \in X_{lexcooc}$.

Proof. According to Definitions 7, 8: $|\pi(i_k \cup y_{lexlooc})| < |\pi(i_k)| = |\pi(i_k \cup x_{lexcooc})|$ and $sup(i_k \cup y_{lexlooc}) \geq minsup$. Therefore, we have $sup(i_k \cup y_{lexcooc} \cup x_{lexcooc}) \geq minsup$, $\forall x_{lexcooc} \in X_{lexcooc}, \forall y_{lexlooc} \in Y_{lexlooc}$■.

Example 7. See Table 4. Consider the item G as kernel item ($minsup = 2$), we detect co-occurrence items with item G as $lexcooc(G) = \{A, C\}$, $X_{lexcooc} = \{A, C, AC\}$; $lexlooc(G) = \{E\}$ and $sup(GE) = 3 \geq minsup$ then $sup(GEA) = sup(GEC) = sup(GEAC) = 3 \geq minsup$.

The framework of Algorithm 2 is presented as follows:

Algorithm 2. Generating all frequent itemsets satisfy *minsup*

    Input   : *minsup, Kernel_COOC array,* Dataset $\mathcal{D}$

    Output: FI consists all frequent itemsets

1:    **foreach** Kernel_COOC[k].sup $\geq$ *minsup* **do** //*property 2*

2:        FI[k] = i_k

3:        **if** (Kernel_COOC[k].sup = *minsup*) **then**

4:            Co $\leftarrow$ GenSub(Kernel_COOC[k].cooc)//*generating noempty subsets of cooc*

5:            **foreach** $is_j \in Co$ **do**

6:                FI[k] = FI[k] $\cup \{i_k \cup is_j\}$ //*lemma 1*

7:        **else**

8:            **if** (Kernel_COOC[k].cooc = $\varnothing$) **then**

9:                Lo $\leftarrow$ GenSub(Kernel_COOC[k].looc)//*generating noempty subsets of looc*

10:                **foreach** $is_j \in Lo$ **do**

11:                    FI[k] = FI[k] $\cup \{i_k \cup is_j\}$

12:            **else**

13:                Co $\leftarrow$ GenSub(Kernel_COOC[k].cooc)

14:                **foreach** $is_j \in Co$ **do**

15:                    $F_t = F_t \cup \{i_k \cup is_j\}$

16:                Lo $\leftarrow$ GenSub(Kernel_COOC[k].looc)

17:                **foreach** $is_j \in Lo$ **do**

18:                    $F_k = F_k \cup \{i_k \cup is_j\}$

19:                **foreach** $f_i \in F_t$ **do**

20:                    **foreach** $is_j \in Lo$ **do**

21:                        FI[k] = FI[k] $\cup \{f_i \cup is_j\}$ //*lemma 2*

22:                FI[k] = FI[k] $\cup F_t$

23:    sort **FI** in descending by support

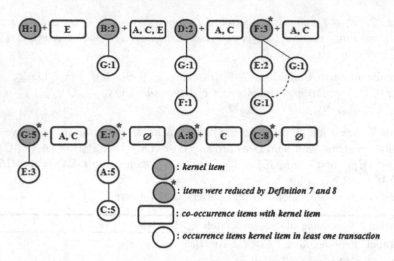

Fig. 3. The pattern-tree was reduced by Definitions 7 and 8.

We illustrate Algorithm 2 on example database in Table 1, and *minsup* = 3. After the processing Algorithm 1, the Kernel_COOC array in Table 4 is showed.

Line 3, consider items satisfying *minsup* as kernel items {F, G, E, A, C};

Consider *kernel item* F, $sup(F) = 3 = minsup$ (Lemma 1- *form line 5 to 6*) generating all frequent with *kernel item F* as $FI_{[F]} = \{(\underline{F}, 3), (\underline{F}A, 3), (\underline{F}C, 3), (\underline{F}AC, 3)\}$ (Tables 5 and 6).

Consider the *kernel item* G *(from line 12 to 21)*: the powerset of co-occurrence items of *kernel item* G as set Co = {A, C, AC}, generating frequent itemsets $F_t = \{(\underline{G}A, 5), (\underline{G}A, 5), (\underline{G}AC, 5)\}$; line 16 – generating noempty subsets of *looc* field Lo = {E}, $F_k = \{\underline{G}E\}$ – generating frequent itemsets $FI_{[G]} = \{(\underline{G}, 5), (\underline{G}A, 5), (\underline{G}C, 5), (\underline{G}AC, 5), (\underline{G}E, 3), (\underline{G}EA, 3), (\underline{G}EC, 3), (\underline{G}EAC, 3)\}$.

Consider the *kernel item* E *(from line 8 to 11)*: generating noempty subsets of *looc* field Lo = {A, C, AC}, line 10 and 11 – generating frequent itemsets $FI_{[E]} = \{(\underline{E}, 7), (\underline{E}A, 5), (\underline{E}C, 5), (\underline{E}AC, 5)\}$.

Consider the *kernel item* A (similary *kernel item* G): Co = {C}, $F_t = \{(AC, 8)\}$, Lo = {∅}, $F_k = \{∅\}$ – generating frequent itemsets $FI_{[A]} = \{(\underline{A}, 8), (\underline{A}C, 8)\}$.

Consider the *kernel item* C (similary *kernel item* E): Lo = {∅} – generating frequent itemsets $FI_{[C]} = \{(\underline{C}, 8)\}$ (Figs. 4 and 5).

Table 5. All frequent itemsets satisfy *minsup* = 3 (example database in Table 1).

Kernel item	Frequent itemsets - FI							
F	(F,3)	(FA,3)	(FC,3)	(FAC,3)				
G	(GE,3)	(GEA,3)	(GEC,3)	(GEAC,3)	(GA,5)	(GC,5)	(GAC,5)	(G,5)
E	(EC,5)	(EA,5)	(EAC,5)	(E,7)				
A	(A,8)	(AC,8)						
C	(C,8)							

Table 6. Datasets description in experiments.

Name	#Trans	#Items	#Avg. length	Type	Density (%)
Chess	3,196	75	37	Dense	49.3
Mushroom	8,142	119	23	Dense	19.3
T10I4D100K	100,000	870	10	Sparse	1.1
T40I10D100K	100,000	942	40	Sparse	4.2

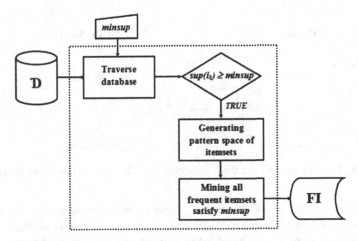

Fig. 4. The diagram general algorithm for frequent itemsets mining.

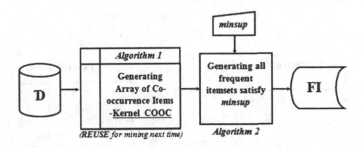

Fig. 5. The diagram NOV-FI algorithm for frequent itemsets mining.

4 Experiments

All experiments were conducted on a PC with a CPU 2.0 GHz, 4 Gb main memory, running Microsoft Windows 7 Ultimate. All codes were compiled using C#, Microsoft Visual Studio 2010, .Net Framework 4.

We experimented on two instance types of datasets:

- Two real datasets are both dense form of UCI Machine Learning Repository [http://archive.ics.uci.edu/ml] as **Chess** and **Mushroom** datasets.

– Two synthetic sparse datasets are generated by software of IBM Almaden Research Center [http://www.almaden.ibm.com] as **T10I4D100K** and **T40I10D100K** datasets.

We have compared the NOV-FI algorithm with two algorithms: the first algorithm is Eclat [2] improved upon the Apriori; the second algorithm is PrePost [6] constructive a *FP-tree-like*. In recent years, PrePost algorithm shows the better performance result on dense datasets.

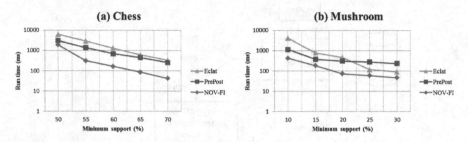

Fig. 6. Running time of the three algorithms on **Chess** and **Mushroom** datasets.

Figure 6(a) and (b) show the running time of the compared algorithms on real datasets Chess and Mushroom. NOV-FI runs faster among two algorithms Eclat and PrePost under all minimum supports; Fig. 6(a) shows PrePost runs faster and more scalable than Eclat. However, Fig. 6(b) shows PrePost runs slower than Eclat with minimum support threshold as 25% and 30% (*Mushroom dataset has low density as 19.3%*).

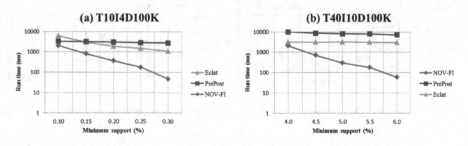

Fig. 7. Running time of the three algorithms on **T10I4D100K** and **T40I10D100K** datasets.

Figure 7(a) and (b) show the running time of the compared algorithms on synthetic sparse datasets T10I4D100K and T40I10D100K. NOV-FI runs faster among two algorithms Eclat and PrePost under all minimum supports. However, Fig. 7 shows PrePost inefficient with sparse datasets (*T10I4D100K, T40I10D100K dataset have low density as 1.1% and 4.2%*).

In summary, experimental results suggest the following ordering of these algorithms as running time is concerned: NOV-FI runs faster among two algorithms Eclat and PrePost under all minimum supports on sparse and dense datasets.

5 Conclusion

In this paper, we have proposed a sequential architecture mining frequent itemsets on transaction databases, consisting of two phases: *the first phase*, quickly detect a the Kernel_COOC array of co-occurrences and occurrences of kernel item in at least one transaction; *the second phase*, the algorithm is proposed for fast mining all frequent itemset based on Kernel_COOC array. Besides, when using mining frequent itemsets with *other minsup value* then the proposed algorithm only performs mining frequent itemsets based on the Kernel_COOC array that is calculated previously, reducing the significant processing time. The experimental results show that the proposed algorithms perform better than other existing algorithms.

The results from the algorithm proposed: In the future, we will expand the algorithm to be able to mining frequent itemsets on weighted transaction databases, as well as to expand the NOV-FI algorithm on distributed computing systems.

Acknowledgements. This work was supported by University of Social Sciences and Humanities; University of Science, VNU-HCM, Vietnam.

References

1. Agrawal, R., Imilienski, T., Swami, A.: Mining association rules between sets of large databases. In: Proceedings of the ACM SIGMOD International Conference on Management of Data, Washington, DC, pp. 207–216 (1993)
2. Zaki, M.J.: Scalable algorithms for association mining. IEEE Trans. Knowl. Data Eng. **12**(3), 372–390 (2000)
3. Han, J., Pei, J., Yin, Y.: Mining frequent patterns without candidate generation. In: Proceedings of 2000 ACM-SIGMOD International Conference on Management of Data (SIGMOD 2000), Dallas, TX, pp. 1–12 (2000)
4. Dong, J., Han, M.: BitTableFI: an efficient mining frequent itemsets algorithm. Knowl.-Based Syst. **20**(4), 329–335 (2007)
5. Song, W., Yang, B.: Index-BitTableFI: an improved algorithm for mining frequent itemsets. Knowl.-Based Syst. **21**, 507–513 (2008)
6. Deng, Z.H., Wang, Z.H., Jiang, J.J.: A new algorithm for fast mining frequent itemsets using N-lists. Sci. China Inf. Sci. **55**(9), 2008–2030 (2012)
7. Philippe, F.V., Jerry, C.W.L., Bay, V., Tin, C.T., Ji, Z., Bac, L.: A survey of itemset mining. Wiley Interdiscip. Rev.: Data Min. Knowl. Discov. **7**(4) (2017)

Rare-Events Classification: An Approach Based on Genetic Algorithm and Voronoi Tessellation

Abdul Rauf Khan[1]([⊠]), Henrik Schiøler[1], Mohamed Zaki[2],
and Murat Kulahci[3,4]

[1] Department of Electronic Systems, Aalborg University, Aalborg, Denmark
arkh@dtu.dk, henrik@es.aau.dk
[2] Department of Engineering, Institute for Manufacturing,
University of Cambridge, Cambridge, UK
mehyz2@cam.ac.uk
[3] Department of Applied Mathematics and Computer Science,
Technical University of Denmark, Kongens Lyngby, Denmark
muku@dtu.dk
[4] Luleå University of Technology Sweden, Luleå, Sweden

Abstract. Classification is a major constituent of the data mining tool kit. Well-known methods for classification are either built on the principle of logic or on statistical reasoning. For imbalanced and noisy cases, classification may however fail to deliver on basic data mining goals, i.e., identifying statistical dependencies in data. In this article, we propose a novel strategy for data mining based on partitioning of the feature space through Voronoi tessellation and Genetic Algorithm, where the latter is applied to solve a combinatorial optimization problem. We apply the suggested methodology to a range of classification problems of varying imbalance and noise and compare the performance of the suggested method with well-known classification methods such as (SVM, KNN, and ANN). The results obtained indicate the proposed methodology to be well suited for data mining tasks in case of highly imbalanced classes and significant noise.

Keywords: Data mining · Classification · Imbalance · Noisy data
Voronoi tessellation · Genetic algorithm

1 Introduction

Classification methods can be grouped into two broad categories; Logic based algorithms and Statistical learning algorithms [10]. Classification trees and Decision trees are prominent members of the logic based algorithms. Classification and regression trees (CART) [2], ID3 [16], C4.5 [15], SLIQ [12] and PUBLIC Algorithm [17] are some of the popular tree based algorithms. Most of these algorithms are designed to solve specialized problems, e.g., ID3 is limited to discrete attributes whereas C4.5 does not have this restriction and SLIQ especially

© Springer Nature Switzerland AG 2018
M. Ganji et al. (Eds.): PAKDD 2018, LNAI 11154, pp. 256–266, 2018.
https://doi.org/10.1007/978-3-030-04503-6_26

handles scalability and flexibility issues [13]. Another renowned family of methods is perceptron [19] based algorithms. The well-known artificial neural network (ANN) can be seen as an extension of single layer perception into multilayer perceptron [23]. Artificial neural network and deep learning are some of the rapidly growing research areas in the field of machine learning. In contrast to the logical and the perceptron based methods, statistical learning approaches are based on the underlying probability model [10]. Discriminant analysis and Bayes rules can be considered as distinguished members of this class. In case of linear classification and continuous data, linear discriminant analysis (LDA) or fisher linear discriminant analysis (an extension of LDA for more than two classes) [7] are widely used in literature, whereas in case of categorical data discriminant correspondence analysis [20] is more relevant. Secondly, for nonlinear classification problems, methods such as Naive Bayes classifier, Bayesian Networks [22] and instance based learning principles (e.g. k-nearest neighbour) [5] belongs to the statistical learning class along with the relatively newer addition in supervised learning approaches; Support vector machines (SVMs) [21]. The motivational problem for this work is a classification of *pass* and *fail* categories on the basis of quality control data in highly optimized industrial processes. The case is highly *imbalanced*, i.e. the pass class contains 99% of the data whereas the fail class only the remaining 1%, which may lead to a degenerate *all pass* classification. Classification methods may therefore fail to reveal existing statistical dependencies in data, in case of imbalance. As demonstrated below another cause of this phenomenon is *overlap*, where there are no clear classification boundaries. We say that data are *noisy*. Since the identification of statistical dependencies is the main outcome of data mining, we may conclude, that imbalance and noise may render classification methods unsuitable for such data mining tasks. In this article we propose a methodology for data mining in case of imbalanced and noisy data. The structure of this article is as follows; we will first give mathematical preliminaries. Next we present the methodology for data mining in imbalanced and noisy data followed by a section presenting results from applying the proposed methodology to a range of different problems. In the same section we will present the results of comparison with well-known classification methods (SVM, KNN and ANN). We will conclude this article with a discussion on the data mining capabilities of the proposed methodology and possible extensions of this work.

2   Mathematical Preliminaries

Generally we may define a classifier by a discriminating function F mapping a feature space X (most often a real space) onto a set of classes $\mathcal{C} = \{1, .., N\}$ such that inverse images $\{F^{-1}(i)\}$ constitute a partition $\{A_1, .., A_N\}$ on X. Data $\{(x_i, c_i)\}$ are considered drawn independently from a (typically) unknown distribution P on $X \times \mathcal{C}$. Successful classification is when class can be predicted from feature, i.e.

$$F(x_i) = c_i \tag{1}$$

A basic classification objective is therefore to minimize the probability of classification failure, i.e.

$$min_F P(F(x_i) \neq c_i) \tag{2}$$

More generally a positive N by N cost matrix λ may be defined leading to the optimization problem

$$min_F E(\lambda(F(x_i), c_i)) \tag{3}$$

where (2) is the special case of (3) for $\lambda(k, j) = 1 - \delta(k, j)$ (δ is the Kronecker delta).

The solution to (3) is given by

$$F(x) = argmin_k \sum_{j=1}^{N} \lambda(k, j) f(x|j) P(j) \tag{4}$$

where $f(x|j)$ denotes the class conditional feature density and $P(j) = P(c_i = j)$ the *a priori* class probability.

For the basic classification problem given in (2) with uniform prior probabilities, i.e. $P(j) = 1/N$ (the exact balanced situation), the solution to (2) is given by

$$F(x) = argmax_j f(x|j) \tag{5}$$

Now consider the conditionals

$$P(c_i = k | x \in A_j) = \frac{1}{N} \frac{\int_{A_j} f(x|k) dx}{\int_{A_j} f(x) dx} \tag{6}$$

such that $P(c_i = j | x \in A_j) > P(c_i = j | x \in A_k)$ for all $k \neq j$. Thus, for the balanced case, optimal classification yields a partition on X revealing probabilistic *dependency rules*, i.e.

$$x_i \in A_j \text{ is a stronger cause for } c_i = j \text{ than } x_i \in A_k$$

For some cases (as exemplified below) the underlying distribution is highly imbalanced, i.e.

$$P(p) >> P(k) \text{ for } k \neq p \tag{7}$$

which for moderate cost λ is likely to render the solution of (3) constant, i.e.

$$F(x) = p \; \forall x \tag{8}$$

such that $A_p = X$ and $A_j = \emptyset$ for all $j \neq p$.

Even for such cases, another partition of X $B_1, .., B_N$ may exist, such that conditionals

$$P(c_i = k | x \in B_j) = \frac{\int_{B_j} f(x|k) P(k) dx}{\int_{B_j} f(x) dx} \tag{9}$$

exhibit significant differences which may underpin dependency rules, i.e. we may find for some j that

$$P(c_i = j | x \in B_j) >> P(c_i = j | x \in B_k) \; \forall k \neq j \tag{10}$$

and state that

$x_i \, in \, B_j \, is \, a \, stronger \, cause \, for \, c_i = j \, than \, x_i \, in \, B_k$

Even if such a partition $B_1, .., B_N$ exists, it may not be revealed through classification due to imbalance.

As a practical example x_i may comprise test measurements from earlier stages in a multi-stage manufacturing process and c_i the binary outcome (pass,fail) of the final quality test. For well managed processes we should have $P(pass) >> P(fail)$, hence the case is high imbalance. However identifying a partition B_{pass}, B_{fail} such that

$$P(c_i = fail | x_i \in B_{fail}) >> P(c_i = fail | x_i \in B_{pass}) \tag{11}$$

may bring valuable insights into the manufacturing process and finally lead to optimizations and competitive margin.

For the *pass/fail* case we have (for uniform cost) $F(x) = pass$ whenever

$$f(x|pass)P(pass) > f(x|fail)P(fail) \tag{12}$$

With high imbalance and *overlap*, where

$$f(x|pass)f(x|fail) > 0 \; \forall x \tag{13}$$

even substantial variations in $f(x|fail)$ may lead to the degenerate classifier $F(x) = pass \; \forall x$. Substantial variations in $f(x|fail)$ may through (9) result in (11). Altogether, imbalance and overlap may lead to a situation, where valuable statistical structure is missed by classification.

One direct way of dealing with imbalance is to introduce *balancing* costs, i.e.

$$\lambda(k, j) = \frac{1}{P(j)} \tag{14}$$

This through (4), instead of (12), gives

$$f(x|pass) > f(x|fail) \tag{15}$$

which seemingly solves the problem. However a statistical problem remains, namely when

$$P(pass) >> P(fail)$$

estimating $f(x|fail)$ and $P(fail)$ for solving (14) and (15) will be based on scarce data, leading to significant relative uncertainty. The methodology presented in the sequel is specifically designed to deal with high imbalance and overlap and to produce statistically robust results accompanied with comparable quality labels. It relies on a *non-parametric* discretization of the feature space and subsequent combinatorial optimization of a specially designed separation statistic. The separation statistic is designed to be sensitive to estimation uncertainty and serves finally as a suitable quality measure.

3 Methods

3.1 Proposed Methodology

Fundamentally, our data mining strategy is built on the following three pillars; selection of seed (for tessellation) by Learning vector quantization (LVQ) [9,18], Voronoi tessellation [11] based on selected seed and optimization through genetic algorithm. We first divide (or tessellate) the feature space X into a pre-selected number M of tiles $\{T_1, .., T_M\}$, where $T_j \subseteq X$ and then group these tiles into two groups defining a binary partition $\mathcal{B} = \{B_{pass}, B_{fail}\}$ of the feature space. $\mathcal{B}$ is obtained through maximization of an appropriately designed objective function. As presented below the defined maximization problem is of combinatorial nature deeming stochastic optimization methods such as Genetic Algorithms (GA) [6] suitable for the task.

A similar approach was pursued in [8], where GA was used to group categorical feature values into a binary partition $\mathcal{B} = \{B_{pass}, B_{fail}\}$. In this work the feature space is discretized through Voronoi tessellation, transforming continuous feature values into their categorical counterparts and as such feasible for the grouping approach presented in [8]. One important difference is however that, while in [8] the number of categorical values was fixed, the number of tiles M is selectable and as such a target for estimation.

Suppose we have a multi-stage (N stages) process yielding, for each product item i number of quality control measurements $x_i \in X = R^N$ as well as the binary outcome of a final acceptance test $c_i \in \{pass, fail\}$. After having collected an entire sample $\{(x_i, c_i)\}$ we wish to reveal statistical dependencies between measurements x_i and test outcome c_i. More formally we may express the data mining task as identifying a partition $\mathcal{B} = \{B_{pass}, B_{fail}\}$ such that

$$P(fail|B_{fail}) >> P(fail|B_{pass}) \tag{16}$$

where $P(fail|B_{fail})$ is shorthand for $P(c_i = fail|x \in B_{fail})$.

Identification of $\mathcal{B}$ could be arranged as an optimization problem, i.e.

$$argmax_\mathcal{B} P(fail|B_{fail}) - P(fail|B_{pass}) \tag{17}$$

which for all partitions require the, rather unlikely, knowledge of the involved conditionals. In a practical situation, recorded data could for every partition be used to generate the contingency table (Table 1) and in turn maximum likelihood estimates of the class conditional probabilities of the following form

$$\hat{P}_{fail|B_{fail}} = \frac{X_{fail, B_{fail}}}{X_{.B_{fail}}} \quad and \quad \hat{P}_{fail|B_{pass}} = \frac{X_{fail, B_{pass}}}{X_{.B_{pass}}} \tag{18}$$

applicable for the optimization defined in (17). Such a strategy would however be rather sensitive to estimation uncertainty, which is emphasized by imbalance. Therefore we seek a more robust strategy.

Table 1. Contingency table for partition $\{B_{pass}, B_{fail}\}$

Status	Class		
	B_{pass}	B_{fail}	
Pass	$X_{pass,B_{pass}}$	$X_{pass,B_{fail}}$	$X_{pass.}$
Fail	$X_{fail,B_{pass}}$	$X_{fail,B_{fail}}$	$X_{fail.}$
Total	$X_{.,B_{pass}}$	$X_{.,B_{fail}}$	$X_{..}$

3.2   Objective/Fitness Function

The objective function presented in [8] is based on the maximization of the difference between the two confidence bounds. Let U_i and L_i, for $i \in \{fail, pass\}$ be the upper and lower Clopper-Pearson confidence bounds [1,4] of $\hat{P}_{fail|B_{fail}}$ and $\hat{P}_{fail|B_{pass}}$, and the objective function presented in [8] is

$$L_{fail} - U_{pass} \qquad (19)$$

For imbalanced situations, we are likely to have $X_{fail,B_{fail}} \ll X_{B_{fail}}$ and $X_{fail,B_{pass}} \ll X_{B_{pass}}$ such that uncertainties of estimates $\hat{P}_{fail|B_{fail}}$ and $\hat{P}_{fail|B_{pass}}$ are dominated by the uncertainties of $X_{fail,B_{fail}}$ and $X_{fail,B_{pass}}$. Therefore approximate confidence intervals can be computed on the assumption that table counts $X_{fail,B_{fail}}$ and $X_{fail,B_{pass}}$ are conditionally binomial given $X_{B_{pass}.}$ and $X_{B_{fail}.}$ respectively. Lower and upper for the confidence intervals are calculated through the Beta transformation of the Binomial distribution [1].

The main motivation for the objective function presented in (4) is to counteract the variability in the estimates. Although it served such a purpose in [8] it was accompanied by auxiliary constraints impairing its autonomy, i.e. some specialist intervention was needed during the inference process. To remove the need for auxiliary constraints, in this study we suggest a slightly modified objective function I defined by

$$I = \frac{L_{fail} - U_{pass}}{(U_{pass} - L_{pass}) + (U_{fail} - L_{fail})} \qquad (20)$$

where the numerator is maximized for maximum separation of confidence intervals and the denominator yields preference for narrow confidence intervals when maximizing I. As a result I is designed to render statistically significant results by preventing B_{fail} or B_{pass} being *too small*.

3.3   Genetic Algorithm

Summarizing the arguments of the previous subsection we examine the following optimization problem

$$argmax_{\{B_{fail}, B_{pass}\}} I \qquad (21)$$

Since $B_{fail} = \cup_{j \in J_f} T_j$ and $B_{pass} = B_{fail}^C = \cup_{j \in J_f^c} T_j$, where $T_j \subseteq X$ are Voronoi tiles, the optimization problem expressed in (21) is conducted in the power set $2^{\{1,...,M\}}$, where "M" refers to the number of seed points. This yields $\#(2^{\{1,...,M\}}) = 2^M$ distinct possibilities. Even for moderate cases, e.g., $M = 30$, this gives approximately $1e9$ possibilities. Thus, without an efficient solution this calls for non-exhaustive heuristics for even moderate values of M.

We may encode the problem into binary strings (with a proper identification)

$$B_{fail} = b_0, b_1, .., b_{M-1}$$

such that $b_i = 1$ *iff* $T_i \subseteq B_{fail}$ and

$$B_{pass} = (1 - b_0), (1 - b_1), .., (1 - b_{M-1})$$

which deems the problem suitable for stochastic optimization via *evolutionary methods* such as *Genetic Algorithms* (GA) as presented in [8].

3.4   Cross Validation

The number of Voronoi tiles (or Tessellations) M is an important parameter in our learning strategy. If M chosen too small, descriptiveness is impaired and a satisfactory level for the I-statistic in (20) can not be achieved. If on the other hand M is chosen too large, results may not generalize from the data used for estimation (training set) to an independent data set (validation set), i.e., *problem of over fitting*. In the extreme case every feature sample x_i in the training set may define a Voronoi tile. Choosing $B_{fail} = \cup_{\{i|c_i=fail\}} T_i$ and $B_{pass} = B_{pass}^c$ could produce very large values for the I-statistic. Thus, even with the I-statistic designed with robustness to statistical variation in mind, it is by itself, is no guarantee against over fitting.

In order to select a suitable number of tiles we apply a Monte Carlo cross validation strategy, where the available data set for estimation is divided randomly a 100 times (trials) into a training- and test-set of equal sizes. For each trial a variety of values for M are used for estimating $\mathcal{B}$ based on the training set. For each value of M the I-statistic is measured for the test set. I-statistic values are finally averaged over all trials and maximum value for M is chosen. Ultimately I-statistic is measured along with a number of other performance statistics for an independent validation set. In addition to I-statistic we measure *Markedness* [14], *F-inverse* and position in *ROC-space* for the validation set for the proposed method as well as a selection of classification methods: *K Nearest Neighbour* (KNN), *Artificial Neural Network* (ANN) and *Support Vector Machine* (SVM) for comparison.

4   Results

In order to test the classification/data mining capabilities of the proposed methodology, it is applied to four different classification problems as presented

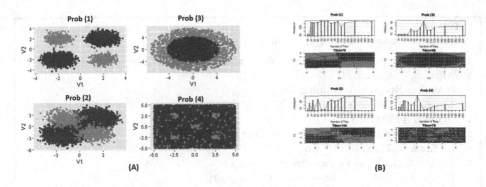

Fig. 1. Classification results (Color figure online)

in Fig. 1. These four test cases include different degree of imbalance as well as noise (overlap).

Problem-1 No Overlap, Moderate Imbalance: Problem 1 shown in the top left representation of Fig. 1(A) is well known XOR problem with nice separation between two classes. But classes here are imbalanced in such a way that the counts for *pass* and *fail* are 9000 and 1000 respectively.

Problem-2 Significant Overlap, Moderate Imbalance: In bottom left representation of Fig. 1(A) an XOR problem with significant overlap is presented, where ratio of imbalance is as it was in problem 1.

Problem-3 No Overlap, Moderate Imbalance: The top right plot in Fig. 1(A) is a classification problem where one class is surrounded by another class. The imbalance between classes is same as in previous two problems.

Problem-4 Significant Overlap, Highly Imbalanced: Problem 4 in Fig. 1(A) represents the highly imbalanced (10000 counts for *pass* and 200 counts for *fail*). As observed from the figure, there are areas with a relatively high concentration of *fail* (red dots). This problem is key to the comparison as the classification algorithms are prone to be degenerate although an observable statistical dependence may exist between feature and class.

Figure 1(B) represents the classification boundaries computed through the proposed methodology. The matrix of plots in Fig. 1(B) obey the order of Fig. 1(A) top left is the XOR problem with well separated classes and on the right bottom corner we have significantly overlapped classes. We observe that the proposed methodology is working fairly well for both rather easy (the left hand column of Fig. 1(A) and (B) and reasonably complex (right hand column of Fig. 1(A) and (B) classification/data mining problems.

We observe that for first three cases I-statistic values are rather high, $I > 10$ compared to the highly imbalanced case with overlap where $I \in [0.2, 0.8]$. We note that positive values for I corresponds to non overlapping confidence intervals $[L_{fail}, U_{fail}]$ and $[L_{pass}, U_{pass}]$, whereas $I = -0.5$ is equivalent to identical confidence intervals.

```
Input   : D_{m,n} = Data {x_i, c_i}
Output: Class Labels for B_{pass} and B_{fail}
1   Divide D_{m,n} into three disjoint sets: training, validation and testing ;
2   Initialize T[2, 3, ......M] ;
    Training Step: On Training Data
3   Adjust selected seeds through LVQ;
4   foreach #T[M] do
5   |   Tessellate information space w.r.t #T[M];
6   |   for k ← 1 to #T[M] do
7   |   |   X_{pass}[M, k] ← PassCount(Tile_{M,k});
8   |   |   X_{fail}[M, k] ← RepairCount(Tile_{M,k});
9   |   |   GA(X_{pass}, X_{fail});
10  |   |   return(B_{pass}, B_{fail})  eq(23) is fulfilled;
11  |   |   Remember(ClassificationRule);
12  |   end
    |   Cross validation Step: On validation data
13  |   for j ← 1 to 100 do
14  |   |   Calculate I for each j by previously learned classification rule;
15  |   |   Calculate M_I = avg(I_j);
16  |   |   Select I #T[s] with max(M_I);
17  |   end
18  end
```

Algorithm 1: Complete Learning Strategy

Table 2. Comparison with existing methods

	F-Inverse								Markedness							
	Miss-classification cost = $\frac{1}{ClassSize}$				SMOTE Treatment for Imbalance				Miss-classification cost = $\frac{1}{ClassSize}$				SMOTE Treatment for Imbalance			
	SVM	NNET	KNN	Vor_{ga}	SVM	NNET	KNN	Vor_{ga}	SVM	NNET	KNN	Vor_{ga}	SVM	NNET	KNN	Vor_{ga}
Prob1	0.999	1	1	0.999	0.999	1	1	0.999	0.999	1	1	1	0.999	1	1	1
Prob2	0.812	0.842	0.865	0.703	0.830	0.817	0.798	0.703	0.909	0.906	0.836	0.547	0.889	0.859	0.860	0.547
Prob3	0.952	0.998	0.983	0.765	0.959	0.983	0.969	0.765	0.988	0.999	0.976	0.621	0.990	0.988	0.988	0.621
Prob4	0.044	0.038	0.052	0.091	0.054	0.075	0.078	0.091	0.073	0.015	0.028	0.355	0.256	0.328	0.344	0.355

We compare the performance of the proposed algorithm with R-implementations of existing methodologies SVM, ANN and KNN. Comparison is made on the basis of 3 well known classification performance measures Markedness and F-inverse (Table 2 and Fig. 2(A)), ROC-space (Fig. 2(B)) as well as the I statistics defined in (20) (Fig. 2(A)). All results represent averages over 100 independent test sets.

4.1   Comparison with Existing Classification Methods

Table 2 presents a detailed comparison of the proposed algorithm (Vor_{ga}) with some of the other classification methods. Two different techniques (Inverse class proportion and SMOTE [3]) are used to deal with the issue of class imbalance. Rows in Table 2 represent four cases (problems), whereas columns compare the results of comparison measures (F-inverse and Markedness) for all four methods

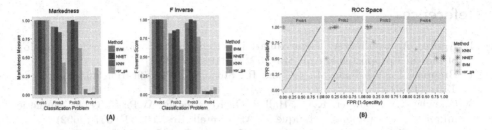

Fig. 2. Comparison with existing classification methods

(SVM, NNET, KNN and Vor_{ga}). The results of Table 2 suggests a superiority of the proposed method on other three methods for both comparison measures.

Figure 2(A) and (B) visualize the same results (only for the case where imbalanced is controlled through Inverse class proportion). The left hand plot in Fig. 2(A) represents the value of the markedness measure for all four problems (on-x-axis) and for all four methods (differentiated by colors). We have observed that vor_{ga} algorithm (proposed method) prevails for Problem-4 but other methods are performing better for Problem-1, Problem-2 and Problem-3. The right hand plot in Fig. 2(A) (F-Inverse score) supports the trend observed for markedness. Figure 2(B) represents the position of proposed method along with SVM, ANN and KNN classifier for all four problems (see Fig. 1(A)). The black line in the plots represents the random guess and colors are used to distinguish methodologies. ROC-space results also support the trend that the proposed method (purple colour) prevails in the case of high imbalance and significant overlap (Problem-4), whereas other methods perform better for Problem-1, Problem-2 and Problem-3.

5   Conclusion

Comparison results suggest that the proposed methodology performs reasonably well on moderately balanced classification problems, however not as well as existing classification methods. For highly imbalanced and overlapping settings however, the proposed method outperforms existing classification techniques. In most of the real life problems imbalance is a consequence of the nature of the data generating process (e.g. optimized manufacturing) along with a high degree of overlap, which justifies the relevance of the proposed method. One application of this work can be in quality and reliability engineering as it is demonstrated in this article. In this work our comparison strategy is mainly based on the cost sensitive strategies for the existing methodologies (SVM, NNET and KNN) to deal with imbalance. This comparison can be extended to other strategies such as ensemble methods. Another possible extension of this work is to implement this methodology to solve the multi-class classification problem.

References

1. Agresti, A., Coull, B.A.: Approximate is better than exact for interval estimation of binomial proportions. Am. Stat. **52**(2), 119–126 (1998)
2. Breiman, L., Friedman, J., Olshen, R., Stone, C.: Classification and Regression Trees (1984)
3. Chawla, N.V., Bowyer, K.W., Hall, L.O., Kegelmeyer, W.P.: SMOTE: synthetic minority over-sampling technique. J. Artif. Intell. Res. **16**, 321–357 (2002)
4. Clopper, C.J., Pearson, E.S.: The use of confidence or fiducial limits illustrated in the case of the binomial. Biometrika **26**(4), 404–413 (1934)
5. Cover, T.M., Hart, P.E.: Nearest neighbor pattern classification. IEEE Trans. Inf. Theory **13**(1), 21–27 (1967)
6. Fidelis, M.V., Lopes, H.S., Freitas, A.A.: Discovering comprehensible classification rules with a genetic algorithm. In: Proceedings of the 2000 Congress on Evolutionary Computation, vol. 1, pp. 805–810. IEEE (2000)
7. Friedman, J.H.: Regularized discriminant analysis. J. Am. Stat. Assoc. **84**(405), 165–175 (1989)
8. Khan, A.R., Schioler, H., Knudsen, T., Kulahci, M.: Statistical data mining for efficient quality control in manufacturing. In: 2015 IEEE 20th Conference on Emerging Technologies & Factory Automation (ETFA), pp. 1–4. IEEE (2015)
9. Kohonen, T.: The self-organizing map. Proc. IEEE **78**(9), 1464–1480 (1990)
10. Kotsiantis, S.B., Zaharakis, I., Pintelas, P.: Supervised machine learning: a review of classification techniques. Emerg. Artif. Intell. Appl. Comput. Eng. **160**, 3–24 (2007)
11. Lee, D.-T., Schachter, B.J.: Two algorithms for constructing a delaunay triangulation. Int. J. Comput. Inf. Sci. **9**(3), 219–242 (1980)
12. Mehta, M., Agrawal, R., Rissanen, J.: SLIQ: a fast scalable classifier for data mining. In: Apers, P., Bouzeghoub, M., Gardarin, G. (eds.) EDBT 1996. LNCS, vol. 1057, pp. 18–32. Springer, Heidelberg (1996). https://doi.org/10.1007/BFb0014141
13. Niuniu, X., Yuxun, L.: Review of decision trees. In: 2010 3rd IEEE International Conference on Computer Science and Information Technology (ICCSIT), pp. 105–109 (2010)
14. Powers, D.M.: Evaluation: from precision, recall and f-measure to roc, informedness, markedness and correlation. J. Mach. Learn. Technol. (2011)
15. Quinlan, J.: Programs for Machine Learning (1993)
16. Quinlan, J.R.: Induction of decision trees. Mach. Learn. **1**(1), 81–106 (1986)
17. Rastogi, R., Shim, K.: PUBLIC: a decision tree classifier that integrates building and pruning. VLDB **98**, 24–27 (1998)
18. Ripley, B., Venables, W.: Package class. CRAN R Project (2015)
19. Rosenblatt, F.: Principles of Neurodynamics (1962)
20. Scholkopft, B., Mullert, K.-R.: Fisher discriminant analysis with kernels. Neural Netw. Signal Process. IX **1**(1), 1 (1999)
21. Vladimir, V.N., Vapnik, V.: The Nature of Statistical Learning Theory (1995)
22. Wan, E.A.: Neural network classification: a bayesian interpretation. IEEE Trans. Neural Netw./A Publ. IEEE Neural Netw. Counc. **1**(4), 303–305 (1989)
23. Williams, D.R.G.H.R., Hinton, G.: Learning representations by back-propagating errors. Nature **323**, 533–536 (1986)

Particle Swarm Optimization-Based Weighted-Nadaraya-Watson Estimator

Jie Jiang[1,2], Yulin He[1,2(✉)], and Joshua Zhexue Huang[1,2]

[1] Big Data Institute, College of Computer Science and Software Engineering, Shenzhen University, Shenzhen 518060, Guangdong, China
[2] National Engineering Laboratory for Big Data System Computing Technology, Shenzhen University, Shenzhen 518060, Guangdong, China
2172272027@email.szu.edu.cn, {yulinhe,zx.huang}@szu.edu.cn

Abstract. This paper proposes a **P**article **S**warm **O**ptimization-based **W**eighted-**N**adaraya-**W**atson (PSO-WNW) estimator which uses the standard PSO algorithm to choose the optimal weights for WNW estimator. PSO-WNW estimator gives up the weight constraints of the classical WNW estimator, which makes PSO algorithm to find the more appropriate weights. The estimation performances of PSO-WNW estimator are tested based on 6 well-known testing functions. The experimental results show that PSO-WNW estimator can further reduce the training and testing **R**oot **M**ean **S**quare **E**rrors (RMSEs) of WNW estimator for any given bandwidth parameter and demonstrate the feasibility and effectiveness of PSO-WNW estimator.

Keywords: Weighted-Nadaraya-Watson estimator
Particle swarm optimization · Kernel regression · Kernel function
Bandwidth parameter

1 Introduction

In statistics, the kernel regression [7,16] is a non-parametric technique which uses the kernel density estimation [11,20] to determine the non-linear relation between a pair of input $\mathcal{X}$ and output $\mathcal{Y}$, where $\mathcal{X}$ is either a scalar [1] or a vector. This kind of non-linear relation can be expressed as the conditional expectation $E(\mathcal{Y}|\mathcal{X})$ of random variable $\mathcal{Y}$ relative to random variable $\mathcal{X}$. The kernel regression technique has been widely applied to the fields of pattern recognition, such as image processing [19,23], signal processing [12,24] and time series prediction [5,6].

The kernel regression estimates the output of unknown input with the locally weight average of outputs corresponding to the training inputs near the unknown input. Nadaraya-Watson estimator [2,10,22] is a typical kernel regression method, which uses the kernel as the weighting function. There is a

---

[1] This paper focuses on the case of scalar-input.

© Springer Nature Switzerland AG 2018
M. Ganji et al. (Eds.): PAKDD 2018, LNAI 11154, pp. 267–279, 2018.
https://doi.org/10.1007/978-3-030-04503-6_27

very important bandwidth parameter which has a great impact on the performance of Nadaraya-Watson estimator. The larger bandwidth usually leads to the smoother regression estimation, while the smaller bandwidth causes the rougher regression estimation. How to select a proper bandwidth parameter is the key point of Nadaraya-Watson estimator construction. There are many works which have studied the problem of bandwidth selection. The representative studies include the consistent bandwidth selection [1], local bandwidth choice [8], adaptive bandwidth choice [15], locally adaptive bandwidth choice [3], plug-in bandwidth selector [17], etc. Up to now, there is still no such a bandwidth selection method which has been widely recognized by academia and industry.

The Weighted-Nadaraya-Watson (WNW) estimator is a variant of classical Nadaraya-Watson estimator, which was firstly proposed by Cai [4] and then extended in [14,18,21]. WNW estimator tries to improve the prediction capability of Nadaraya-Watson estimator by taking both advantages of Nadaraya-Watson estimator and local polynomial fitting [4] rather than tuning the complex bandwidth parameter. The Newton-Raphson iteration method was recommended in [4] to determine the optimal weights for WNW estimator. The further analysis shows that the weight constraints of WNW estimator may limit the selection of optimal weights. In addition, the optimization performance of Newton-Raphson iteration method heavily depends on the selection of initial value. The improper initial value will lead to the slow convergence or non-convergence of Newton-Raphson iteration method. Hence, this paper proposes a Particle Swarm Optimization-based WNW (PSO-WNW) estimator which gives up the weight constraints of WNW estimator and uses the standard PSO algorithm [9,13] to choose the optimal weights for WNW estimator. We test the estimation performances of PSO-WNW estimator based on 6 well-known testing functions. The experimental results show that PSO-WNW estimator can further reduce the training and testing Root Mean Square Errors (RMSEs) of WNW estimator for any given bandwidth parameter and demonstrate the feasibility and effectiveness of PSO-WNW estimator.

The remainder of this paper is organized as follows. In Sect. 2, we provide a brief introduction to WNW estimator. In Sect. 3, we analyze the shortcomings of WNW estimator and present the proposed PSO-WNW estimator. In Sect. 4, we report experimental comparisons that demonstrate the feasibility and effectiveness of PSO-WNW estimator. Finally, we give our conclusions in Sect. 5.

2   Weighted-Nadaraya-Watson Estimator

For the given data set $D = \{(x_i, y_i) | x_i, y_i \in R, i = 1, 2, \cdots, \mathcal{N}\}$ including $\mathcal{N}$ training data, where x_i and y_i, $i = 1, 2, \cdots, \mathcal{N}$ can be treated as $\mathcal{N}$ observations of random variables $\mathcal{X}$ and $\mathcal{Y}$, respectively. Nadaraya-Watson estimator uses the conditional expectation $E(\mathcal{Y}|\mathcal{X})$ of $\mathcal{Y}$ relative to $\mathcal{X}$ to express the non-linear relation $y(x)$ between $\mathcal{X}$ and $\mathcal{Y}$, i.e.,

$$y(x) = \mathrm{E}(\mathcal{Y}|\mathcal{X} = x)$$
$$= \int \mathcal{P}(y|x)\,y\,dy,$$
$$= \int \frac{\mathcal{P}(x,y)}{\mathcal{P}(x)}\,y\,dy \tag{1}$$

where $\mathcal{P}(\mathcal{X})$ is the probability density function ($p.d.f.$) of $\mathcal{X}$ and $\mathcal{P}(\mathcal{X},\mathcal{Y})$ is the joint $p.d.f.$ of $\mathcal{X}$ and $\mathcal{Y}$. The kernel density estimation [11,20] is used to approximate $\mathcal{P}(x)$ and $\mathcal{P}(x,y)$ as

$$\hat{\mathcal{P}}(x) = \frac{1}{\mathcal{N}}\sum_{i=1}^{\mathcal{N}} K_h(x - x_i) \tag{2}$$

and

$$\hat{\mathcal{P}}(x,y) = \frac{1}{\mathcal{N}}\sum_{i=1}^{\mathcal{N}} K_h(x - x_i) K_h(y - y_i), \tag{3}$$

respectively, where $h > 0$ is the bandwidth parameter,

$$K_h(x - x_i) = \frac{1}{h}K\left(\frac{x - x_i}{h}\right) \tag{4}$$

and

$$K(z) = \frac{1}{\sqrt{2\pi}}\exp\left(-\frac{z^2}{2}\right), z \in (-\infty, +\infty) \tag{5}$$

is Gaussian kernel. Substituting Eqs. (2) and (3) into Eq. (1) gives

$$y(x) = \sum_{i=1}^{\mathcal{N}} \frac{K_h(x - x_i)}{\sum_{j=1}^{\mathcal{N}} K_h(x - x_j)} y_i, \tag{6}$$

which is Nadaraya-Watson estimator. WNW estimator is the slight variant of Eq. (6) by introducing the weight w_i for $K_h(x - x_i)$, $i = 1, 2, \cdots, \mathcal{N}$ as

$$y_{\mathrm{WNW}}(x) = \sum_{i=1}^{\mathcal{N}} \frac{w_i K_h(x - x_i)}{\sum_{j=1}^{\mathcal{N}} w_j K_h(x - x_j)} y_i, w_i \geq 0, i = 1, 2, \cdots, \mathcal{N}, \tag{7}$$

where the weights w_i satisfy the following constraints

$$\begin{cases} \sum_{i=1}^{\mathcal{N}} w_i = 1 \\ \sum_{i=1}^{\mathcal{N}} (x - x_i) w_i K_h(x - x_i) = 0 \end{cases}. \tag{8}$$

The Newton-Raphson iteration method [4] was used to optimize the weight $w_i, i = 1, 2, \cdots, \mathcal{N}$ in Eq. (7).

3   PSO-Based Weighted-Nadaraya-Watson Estimator

In fact, we find that the constraints $\sum\limits_{i=1}^{N} w_i = 1$ and $\sum\limits_{i=1}^{N} (x - x_i) w_i K_h (x - x_i) = 0$ as shown in Eq. (8) are unnecessary. Let

$$\lambda_i = \frac{w_i K_h (x - x_i)}{\sum\limits_{j=1}^{N} w_j K_h (x - x_j)}, \tag{9}$$

we can get

$$\sum_{i=1}^{N} \lambda_i = 1, \lambda_i \geq 0, i = 1, 2, \cdots, N \tag{10}$$

holds for any weight $w_i \geq 0$. The constraint $\sum\limits_{i=1}^{N} w_i = 1$ doesn't have any impact on Eq. (10). Moreover, the direct consequence of $\sum\limits_{i=1}^{N} (x - x_i) w_i K_h (x - x_i) = 0$ is

$$x = \sum_{i=1}^{N} \frac{w_i K_h (x - x_i)}{\sum\limits_{j=1}^{N} w_j K_h (x - x_j)} x_i$$
$$= \sum_{i=1}^{N} \lambda_i x_i \tag{11}$$

Based on Eqs. (7) and (9), we know WNW estimator predicts the output y of x as

$$y = \sum_{i=1}^{N} \lambda_i y_i. \tag{12}$$

Assume the non-linear relation between $\mathcal{X}$ and $\mathcal{Y}$ is $\mathcal{Y} = \mathcal{X}^2$. Giving two training instances $(2, 4)$ and $(3, 9)$ predicts the output of $x = 2.5$ as

$$y = 4\lambda + 9 (1 - \lambda). \tag{13}$$

Then, we can calculate $\lambda = 0.55$. According to Eq. (11), we can derive $0.55 \times 2 + 0.45 \times 3 = 2.45 \neq x$. This indicates that Eq. (11) is unreasonable. Thus, we give up the weight constraints as shown in Eq. (8) in our proposed PSO-WNW estimator. Algorithm 1 presents the procedure of PSO-WNW estimator.

In Algorithm 1, $U(0, 1)$ is the uniform distribution with minimum value 0 and maximum value 1. w_L and w_U are the lower and upper boundaries of particle position. v_L and v_U are the lower and upper boundaries of particle velocity. α is the inertia weight. c_1 and c_2 are the acceleration constants. $f(\boldsymbol{w}_j)$ is the fitness

---

Algorithm 1. PSO-WNW estimator

1: **Input:** The training data $\{(x_i, y_i) \,|\, x_i, y_i \in R, i = 1, 2, \cdots, \mathcal{N}\}$ and bandwidth parameter h;

2: **Output:** WNW estimator as shown in Eq. (7) with the optimal weights $w'_1, w'_2, \cdots, w'_\mathcal{N}$;

3: **for** each particle $\boldsymbol{w}_j = (w_{j1}, w_{j2}, \cdots, w_{j\mathcal{N}}), j = 1, 2, \cdots, \mathcal{M}$ in swarm $\mathcal{S}$ **do**

4:    **for** $i = 1; i <= \mathcal{N}; i{+}{+}$ **do**

5:       $w_{ji} = w_L + (w_U - w_L) \times U(0,1)$; % Initializing the position of particle $\boldsymbol{w}_j$

6:       $v_{ji} = v_L + (v_U - v_L) \times U(0,1)$; % Initializing the velocity of particle $\boldsymbol{w}_j$

7:       $pbest_{ji} = w_{ji}$; % Initializing the locally optimal solution of particle $\boldsymbol{w}_j$

8:    **end for**

9: **end for**

10: $gbest = (gbest_1, gbest_2, \cdots, gbest_\mathcal{N}) = \underset{\boldsymbol{w}_j \in \mathcal{S}, j=1,2,\cdots,\mathcal{M}}{\arg\min} [f(\boldsymbol{w}_j)]$; % Initializing the globally optimal solution of swarm $\mathcal{S}$

11: **while** the terminate condition isn't met **do**

12:    **for** each particle $\boldsymbol{w}_j = (w_{j1}, w_{j2}, \cdots, w_{j\mathcal{N}}), j = 1, 2, \cdots, \mathcal{M}$ in swarm $\mathcal{S}$ **do**

13:       **for** $i = 1; i <= \mathcal{N}; i{+}{+}$ **do**

14:          $v_{ji} = \alpha \times v_{ji} + c_1 \times U(0,1) \times (pbest_{ji} - w_{ji}) + c_2 \times U(0,1) \times (gbest_i - w_{ji})$;

15:          $w_{ji} = w_{ji} + v_{ji}$;

16:       **end for**

17:       **if** $f(\boldsymbol{w}_j) < f(pbest_j), pbest_j = (pbest_{j1}, pbest_{j2}, \cdots, pbest_{j\mathcal{N}})$ **then**

18:          $pbest_j = \boldsymbol{w}_j$; % Updating the locally optimal solution of particle $\boldsymbol{w}_j$

19:          **if** $f(pbest_j) < f(gbest)$ **then**

20:             $gbest = pbest_j$; % Updating the globally optimal solution of swarm $\mathcal{S}$

21:          **end if**

22:       **end if**

23:    **end for**

24: **end while**

25: $(w'_1, w'_2, \cdots, w'_\mathcal{N}) = (gbest_1, gbest_2, \cdots, gbest_\mathcal{N})$;

---

of particle $\boldsymbol{w}_j$, $i = 1, 2, \cdots, \mathcal{N}$, which is measured by the training **Root Mean Square Error (RMSE)** corresponding to the weight vector $\boldsymbol{w}_j$:

$$
f(\boldsymbol{w}_j) = \text{RMSE}(\boldsymbol{w}_j)
$$

$$
= \sqrt{\frac{\sum\limits_{k=1}^{\mathcal{N}} [y_{\text{WNW}}(x_k) - y_k]^2}{\mathcal{N}}}
$$

$$
= \sqrt{\frac{\sum\limits_{k=1}^{\mathcal{N}} \left[\left[\sum\limits_{i=1}^{\mathcal{N}} \frac{w_{ji} K_h(x_k - x_i)}{\sum\limits_{l=1}^{\mathcal{N}} w_{jl} K_h(x_k - x_i)} y_i \right] - y_k \right]^2}{\mathcal{N}}}. \tag{14}
$$

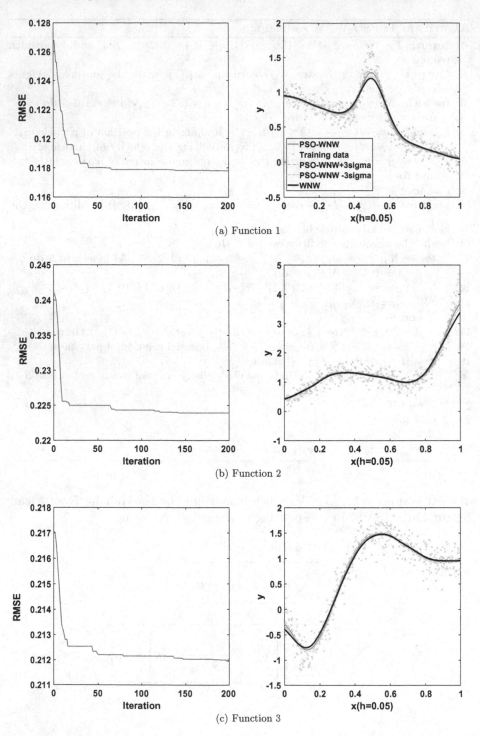

(a) Function 1

(b) Function 2

(c) Function 3

Fig. 1. The convergence and fitting capability of PSO-WNW (Color figure online)

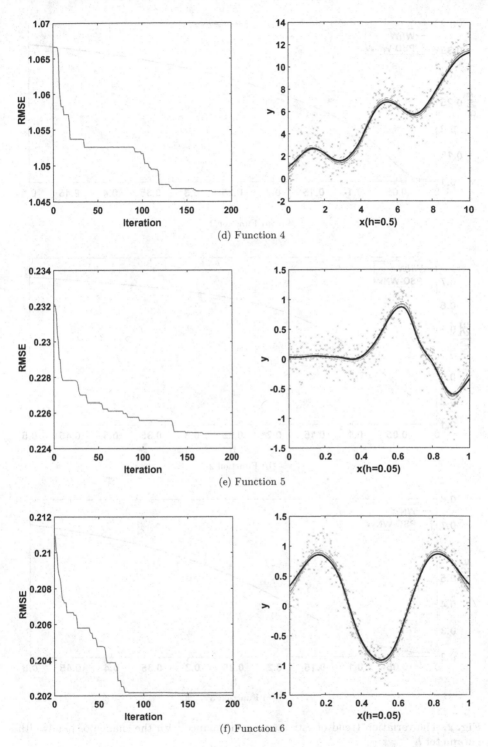

(d) Function 4

(e) Function 5

(f) Function 6

Fig. 1. (*continued*)

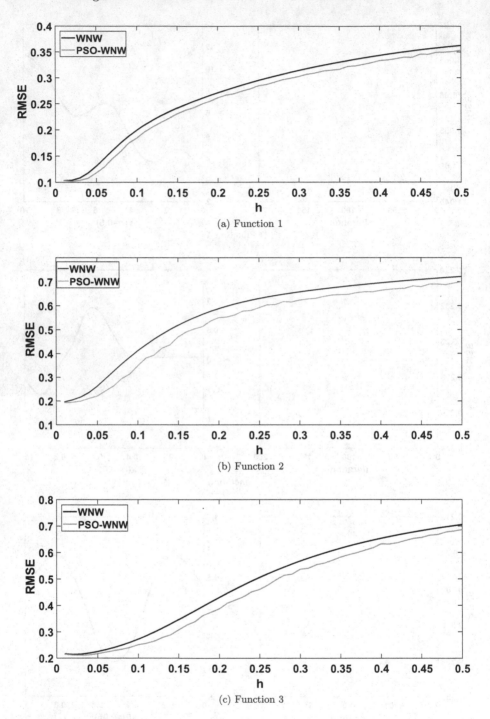

(a) Function 1

(b) Function 2

(c) Function 3

Fig. 2. The variation trend of estimation performance with the change of bandwidth parameter h

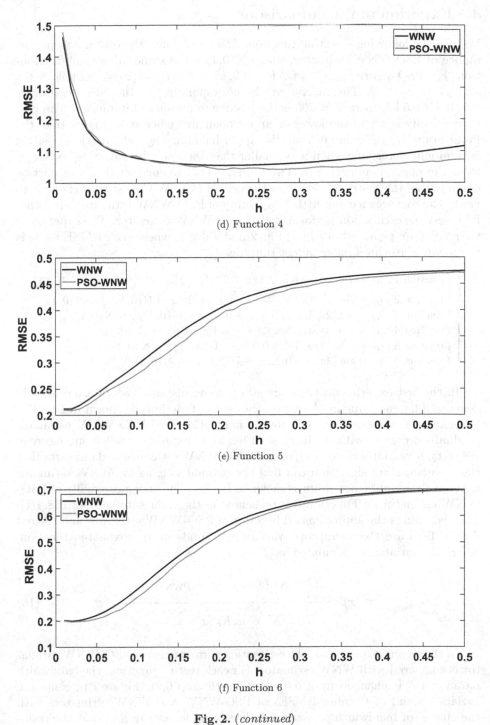

(d) Function 4

(e) Function 5

(f) Function 6

Fig. 2. (*continued*)

4　Experimental Comparisons

We use the following 6 testing functions [25] to evaluate the estimation performance of PSO-WNW estimator, where $N(0,1)$ is the standard normal distribution. For PSO algorithm, we set $\mathcal{M} = 100, w_L = 0, w_U = 1, v_L = -0.1, v_U = 0.1$ and $c_1 = c_2 = 2$. The inertia weight corresponding to the t-th iteration is $\alpha = 0.4 + 0.5\frac{t}{\mathcal{T}}$, where $\mathcal{T} = 200$ is the maximum number of iterations. The particle velocity is set to the lower or upper boundary when it is smaller than the lower boundary v_L or larger than the upper boundary v_U. The particle position is randomly determine when it is smaller than the lower boundary w_L or larger than the upper boundary w_U. Two experiments are conducted in this part to demonstrate the feasibility and effectiveness of PSO-WNW estimator: one is to verify the convergence and fitting capability of PSO-WNW estimator, the other is to test the estimation performance of PSO-WNW estimator. The experimental results are provided in Fig. 1, Fig. 2 and Table 1, where each RMSE value is the mean of 10-times independent running.

$$
\begin{cases}
\text{Function } 1 : y_1 = 1 - x + \exp\left[-200(x - 0.5)^2\right] + \varepsilon, x \sim U(0,1), \varepsilon \sim N(0,0.1) \\
\text{Function } 2 : y_2 = \sin(1.5\pi x) + \frac{\exp\left[(x+0.5)^3\right]}{5} + 0.2\varepsilon, x \sim U(0,1), \varepsilon \sim N(0,1) \\
\text{Function } 3 : y_3 = \sin\left[2\pi(1-x)^2\right] + x + 0.2\varepsilon, x \sim U(0,1), \varepsilon \sim N(0,1) \\
\text{Function } 4 : y_4 = x + 2\sin(1.5\pi x) + \varepsilon, x \sim U(0,10), \varepsilon \sim N(0,1) \\
\text{Function } 5 : y_5 = \left[\sin\left(2\pi x^3\right)\right]^3 + 0.2\varepsilon, x \sim U(0,1), \varepsilon \sim N(0,1) \\
\text{Function } 6 : y_6 = \sin(3\pi x) + 0.2\varepsilon, x \sim U(0,1), \varepsilon \sim N(0,1)
\end{cases}
\tag{15}
$$

In the first experiment, there are 400 training instances which are randomly generated for each running based on the testing functions as shown in Eq. (15). For example, we can find that the training RMSE of PSO-WNW estimator gradually decreases with the increase of iteration number in the left sub-figure of Fig. 1(f). It validates the convergence of PSO-WNW estimator and indicates that the standard PSO algorithm can find the optimal weights for WNW estimator without the weight constraints as shown in Eq. (8). The fitting capability of PSO-WNW estimator on Function 6 is presented in the right sub-figure of Fig. 1(f). The red solid is the approximated function of PSO-WNW estimator and the red dotted lines are the asymptotic variability bounds of approximated function, where the variance is calculated as

$$
\sigma^2(x) = \frac{\sum\limits_{i=1}^{\mathcal{N}} w_i K_h(x - x_i)\left[y_i - y_{\text{WNW}}(x)\right]^2}{\mathcal{N} \sum\limits_{i=1}^{\mathcal{N}} w_i K_h(x - x_i)}.
\tag{16}
$$

In the second experiment, the estimation performance of PSO-WNW estimator is compared with WNW estimator. For each testing function, the bandwidth parameter h is changed from 0.01 to 0.5 with step 0.01. Figure 2 presents the variation trend of training RMSEs of PSO-WNW and WNW estimators with the change of bandwidth parameter h. For example, we can find that the training RMSE of PSO-WNW estimator is smaller than WNW estimator for any

Table 1. Training and testing RMSEs of PSO-WNW and WNW estimators ($h = 0.05, 0.1, 0.2, 0.5$)

		h=0.05		h=0.1		h=0.2		h=0.5	
		Training RMSE	Testing RMSE	Training RMSE	Testing RMSE	Training RMSE	Testing RMSE	Training RMSE	Testing RMSE
Function 1	WNW	0.129±0.0025	0.1379±0.0121	0.1932±0.0061	0.1901±0.0224	0.2601±0.0052	0.2575±0.0206	0.3531±0.0114	0.3674±0.0238
	PSO-WNW	0.1161±0.0028	0.1239±0.0115	0.1802±0.0056	0.1766±0.0188	0.2483±0.0058	0.2447±0.019	0.3437±0.0114	0.3566±0.0244
Function 2	WNW	0.2594±0.0101	0.2625±0.0448	0.3882±0.0167	0.3825±0.0778	0.5798±0.0227	0.5184±0.0488	0.7191±0.0381	0.7437±0.1202
	PSO-WNW	0.2221±0.0097	0.228±0.0303	0.3285±0.0137	0.3275±0.0517	0.5325±0.0224	0.4874±0.039	0.6897±0.0374	0.7122±0.1069
Function 3	WNW	0.2098±0.0064	0.2147±0.0165	0.2664±0.0105	0.2712±0.0221	0.4271±0.014	0.4179±0.0337	0.6971±0.011	0.6918±0.071
	PSO-WNW	0.2014±0.0064	0.2081±0.0154	0.2421±0.0106	0.2422±0.0211	0.3871±0.0113	0.3795±0.029	0.6763±0.0121	0.6692±0.0571
Function 4	WNW	0.888±0.0265	1.0956±0.0577	0.9447±0.0446	1.0367±0.08	0.9709±0.0199	1.0085±0.0803	1.048±0.0252	1.0568±0.0464
	PSO-WNW	0.8754±0.0268	1.1202±0.0667	0.9352±0.0448	1.0582±0.0799	0.9612±0.0194	1.0101±0.0805	1.0048±0.0246	1.0243±0.0338
Function 5	WNW	0.2364±0.007	0.2345±0.0172	0.2911±0.0065	0.2906±0.0199	0.3982±0.0143	0.3961±0.0355	0.4817±0.0233	0.4676±0.0451
	PSO-WNW	0.2265±0.0068	0.2233±0.0141	0.2728±0.0068	0.2723±0.018	0.3815±0.0134	0.3798±0.0313	0.4783±0.023	0.4656±0.0441
Function 6	WNW	0.2198±0.0064	0.2263±0.018	0.3229±0.0071	0.329±0.0148	0.5383±0.0115	0.5478±0.0232	0.691±0.0086	0.7003±0.0411
	PSO-WNW	0.2062±0.0063	0.2139±0.0159	0.2919±0.0072	0.2971±0.015	0.5173±0.0113	0.5258±0.0219	0.6885±0.0084	0.6982±0.0423

given bandwidth parameter in Fig. 2(b). The comparative results indicates that the weight constraints as shown in Eq. (8) limit the selection of optimal weights of WNW estimator to some extent. Table 1 lists the representative comparisons corresponding to $h = 0.05, 0.1, 0.2, 0.5$ and shows that the testing RMSE of PSO-WNW estimator is basically smaller than WNW estimator. PSO-WNW estimator obtains the better prediction performance than WNW estimator.

5   Conclusions

A Particle Swarm Optimization-based Weighted-Nadaraya-Watson (PSO-WNW) estimator is proposed in this paper, which gives up the weight constraints of classical WNW estimator and uses the standard PSO algorithm to choose the optimal weights for WNW estimator. The experimental results on 6 well-known testing functions show that PSO-WNW estimator can further reduce the training and testing Root Mean Square Errors (RMSEs) of Nadaraya-Watson estimator for any given bandwidth parameter and demonstrate the feasibility and effectiveness of PSO-WNW estimator.

Acknowledgments. This paper was supported by National Natural Science Foundations of China (61503252 and 61473194), China Postdoctoral Science Foundation (2016T90799), Scientific Research Foundation of Shenzhen University for Newly-introduced Teachers (2018060) and Shenzhen-Hong Kong Technology Cooperation Foundation (SGLH20161209101100926).

References

1. Altman, N., Macgibbon, B.: Consistent bandwidth selection for kernel binary regression. J. Stat. Plan. Inference **70**(1), 121–137 (1998)
2. Bierens, H.J.: The Nadaraya-Watson kernel regression function estimator. In: Topics in Advanced Econometrics. Cambridge University Press, New York (1994)
3. Brockmann, M., Gasser, T., Herrmann, E.: Locally adaptive bandwidth choice for kernel regression estimators. Publ. Am. Stat. Assoc. **88**(424), 1302–1309 (1993)
4. Cai, Z.W.: Weighted Nadaraya-Watson regression estimation. Stat. Probab. Lett. **51**(3), 307–318 (2001)
5. Hardle, W., Vieu, P.: Kernel regression smoothing of time series. J. Time **13**(3), 209–232 (2010)
6. Hart, J.D.: Kernel regression estimation with time series errors. J. R. Stat. Soc. **53**(1), 173–187 (1991)
7. Hastie, T., Tibshirani, R., Friedman, J.: The Elements of Statistical Learning. SSS. Springer, New York (2009). https://doi.org/10.1007/978-0-387-84858-7
8. Herrmann, E.: Local bandwidth choice in kernel regression estimation. J. Comput. Graph. Stat. **6**(1), 35–54 (1997)
9. Kennedy, J., Eberhart, R: Particle swarm optimization. In: Proceedings of 1995 IEEE International Conference on Neural Networks, vol. 4, pp. 1942–1948 (1995)
10. Nadaraya, E.A.: On estimating regression. Theory Probab. Appl. **9**(1), 141–142 (1964)

11. Parzen, E.: On estimation of a probability density function and mode. Ann. Math. Stat. **33**(3), 1065 (1962)
12. Pawlak, M., Stadtmuller, U.: Kernel regression estimators for signal recovery. Stat. Probab. Lett. **31**(3), 185–198 (1997)
13. Poli, R., Kennedy, J., Blackwell, T.: Particle swarm optimization. Swarm Intell. **1**(1), 33–57 (2007)
14. Salha, R.B., Ahmed, H.: Reweighted Nadaraya-Watson estimator of the regression mean. Int. J. Stat. Probab. **4**(1), 138–147 (2015)
15. Schucany, W.R.: Adaptive bandwidth choice for kernel regression. Publ. Am. Stat. Assoc. **90**(430), 535–540 (1995)
16. Simonoff, J.S.: Smoothing Methods in Statistics. Springer, New York (1996). https://doi.org/10.1007/978-1-4612-4026-6
17. Slaoui, Y.: Plug-in bandwidth selector for recursive kernel regression estimators defined by stochastic approximation method. Stat. Neerl. **69**(4), 483–509 (2015)
18. Song, Y., Lin, Z., Wang, H.: Re-weighted Nadaraya-Watson estimation of second-order jump-diffusion model. J. Stat. Plan. Inference **143**(4), 730–744 (2013)
19. Takeda, H., Farsiu, S., Milanfar, P.: Kernel regression for image processing and reconstruction. IEEE Trans. Image Process. **16**(2), 349–66 (2007)
20. Wand, M.P., Jones, M.C.: Kernel Smoothing. Chapman & Hall/CRC, London (1995)
21. Wang, Y., Xiong, X.Z.: Re-weighted Nadaraya-Watson estimation of conditional density. Ann. Appl. Math. **1**, 63–76 (2017)
22. Watson, G.S.: Smooth regression analysis. Sankhyā: Indian J. Stat. Ser. A **26**(4), 35–372 (1964)
23. Zhang, K., Gao, X., Tao, D., Li, X.: Single image super-resolution with non-local means and steering kernel regression. IEEE Trans. Image Process. **21**(11), 4544–4556 (2012)
24. Zhang, Y.: Signal evaluation method based on generalized kernel regression model. J. Comput. Inf. Syst. **8**(18), 7511–7517 (2012)
25. Zhang, Y. M.: Bandwidth selection for Nadaraya-Watson kernel estimator using cross-validation based on different penalty functions. In: ICMLC 2014. CCIS, vol. 481, pp. 88–96 (2014)

Pacific Asia Workshop on Intelligence and Security Informatics (PAISI)

A Parallel Implementation of Information Gain Using Hive in Conjunction with MapReduce for Continuous Features

Sikha Bagui$^{(\boxtimes)}$, Sharon K. John, John P. Baggs, and Subhash Bagui

University of West Florida, Pensacola, FL 32514, USA
bagui@uwf.edu

Abstract. Finding efficient ways to perform the Information Gain algorithm is becoming even more important as we enter the Big Data era where data and dimensionality are increasing at alarming rates. When machine learning algorithms get over-burdened with large dimensional data with redundant features, information gain becomes very crucial for feature selection. Information gain is also often used as a pre-cursory step in creating decision trees, text classifiers, support vector machines, etc. Due to the very large volume of today's data, there is a need to efficiently parallelize classic algorithms like Information Gain. In this paper, we present a parallel implementation of Information Gain in the MapReduce environment, using MapReduce in conjunction with Hive, for continuous features. In our approach, Hive was used to calculate the counts and parent entropy and a Map only job was used to complete the Information Gain calculations. Our approach demonstrated gains in run times as we carefully designed MapReduce jobs efficiently leveraging the Hadoop cluster.

Keywords: Hadoop · Hive · Information Gain · Parallel implementation
Feature selection

1 Introduction

Big Data is increasingly becoming high dimensional. Data is being collected from text documents, images, videos, automated sensors and sensor networks, etc., all leading to high dimensionality. Dimensionality has gone into the hundreds [14], and even tens or hundreds of thousands of features [4, 11, 21]. It is becoming increasingly difficult for Data Mining algorithms to process this data due to the curse of dimensionality. To overcome this impediment, it is becoming increasingly important to perform feature selection [12]. Feature selection, also referred to as attribute selection or variable selection, is a process of selecting more relevant features, and removing irrelevant or less relevant features or noisy data. Feeding in just relevant features to machine learning algorithms allows for faster processing and more accurate predictability. Many techniques have been used for feature selection, including Information Gain, the Gini Index, uncertainty, and correlation coefficients [6], to name just a few. In this paper, we present a parallel implementation of Information Gain as a measure of feature selection or feature relevance analysis, in the Hadoop environment using MapReduce and Hive,

© Springer Nature Switzerland AG 2018
M. Ganji et al. (Eds.): PAKDD 2018, LNAI 11154, pp. 283–294, 2018.
https://doi.org/10.1007/978-3-030-04503-6_28

for continuous features. By performing feature relevance analysis via Information Gain, we attempt to quantify the relevance of a feature with respect to a given class or category. Less informative features (or dimensions as well as levels) are removed, keeping the more informative dimensions and levels for use in machine learning algorithms like decision trees [20], Naive Bayes algorithm [5, 16], Neural Networks [15], and many other algorithms.

1.1 The MapReduce Paradigm and Hive

The MapReduce paradigm is a parallel programming paradigm that facilitates processing of big data by exploiting parallelisms among data nodes of a cluster [6, 10]. It was originally built for commodity computing infrastructure leveraging the merits of distributed file processing with low cost computing components. MapReduce also spares the programmer from having to worry about failure since the architecture detects the failed map or reduce tasks and reschedules them with suitable replacements. Once a job is submitted for execution, the framework determines the number of input splits for a given dataset and creates a set of mappers to work on each of the input splits. The Job tracker, which is the daemon running on name node (otherwise known as master node), schedules the work to the available task tracker nodes (otherwise known as slave nodes) on the cluster. Once the mappers are done with the task assigned, the output of the mapper is passed on to the reducer, sorted by the key. This stage of the MapReduce job, where the output is sorted by key and transferred to the reducers as input is denoted as the shuffle and sort phase. Large amounts of intermediate data are shuffled and sorted before the output reaches the reducer. The size of the intermediate data generated depends upon the number of input splits generated for a given dataset.

Hadoop is an open source implementation of the MapReduce programming model [24]. Since the introduction of the model by Google [17], parallelizing data mining algorithms using the MapReduce model has received significant attention from the research community. In this study, we show how Information Gain implemented in Hadoop's parallel MapReduce environment is an ideal computing framework for massive and distributed datasets, especially for continuous attributes. Since the map-reduce programming model is at a very low level, certain high-level tasks are performed using the Hadoop component, Hive. Hive is a high-level open-source data warehousing solution built on top of Hadoop. It supports queries expressed in a SQL-like declarative language - HiveQL, which are compiled into MapReduce jobs that are executed using Hadoop. Hive is known for its fast execution of analytical queries. It does a good job on merging smaller files into larger files to avoid overflowing the HDFS metadata. Thus, the overhead of streaming the data in and out of HDFS is very negligible.

The rest of this paper is organized as follows. In Sect. 2 we present the related works. Section 3 presents the parallel implementation of our MapReduce algorithm. Section 4 presents the experimental results and Sect. 5 presents the conclusion and future works.

2 Related Works

There has been a lot of work done on parallel implementations of various data mining/machine learning algorithms [1] on Big Data, but there are very few works on the parallel implementation of Information Gain on Big Data. One of the main works is [25], where the authors perform parallel implementation of Information Gain using Hadoop and MapReduce. This work used Pig Latin scripts that are compiled into MapReduce jobs, which can then be executed on Hadoop clusters. Although our work might have similar goals, it is important to notice the difference in approach that we have taken to implement this classic feature selection algorithm. Their focus of research was more on how to get the maximum performance gain on the Hadoop clusters using Pig scripts, thereby specifying recommendations to use HBASE over HDFS and the overhead of pre-splitting the tables into regions for an optimized solution. We, on the contrary, suggest that Hive integrates well with both HDFS and HBASE, thereby giving users their choice. We also have sufficient control over the data flow even when the data set is stored on HDFS using the recommendations we have made on the data block size and map memory size to get an improved and optimized performance gain.

[20] presents a MR-tree that builds a decision tree in parallel and runs on MapReduce. This paper employs Information Gain to create the decision tree and mentions that it was used in a MapReduce job, but the details of the algorithms have not been presented. [3] presents the C4.5 algorithms for inducing classification rules in the form of decision trees. In this work, information gain is calculated in a map function, but again, details of the calculations of the parallel implementation of information gain have not been discussed. [2] presents an implementation of a decision tree through applets based on the entropies.

[7, 23] presents a forward feature selection heuristic that ranks features by their estimated effect on the resulting model's (logistic regression) performance. [22] proposes a data intensive parallel feature selection method where in each map node, a method is used to calculate the mutual information and combinatorial contribution degree to determine the number of selected features. [23] proposes a model that iteratively adds features to determine if it significantly improved the classification task. It is to be noted that Apache Mahout [18, 19] has only two algorithms related to feature selection and dimensionality reduction: Singular Value Decomposition (SVD) and Stochastic SVD.

3 Parallel Implementation of Information Gain

The Information Gain algorithm presented in [8] is used as a basis for this work. In this section we present our proposed parallel implementation of Information Gain using Hive and MapReduce. The first step is to divide the steps needed to calculate the gain of a feature into independent tasks that are performed in parallel on all features. To obtain initial count statistics from the data under consideration we used Hive, because Hive contains predefined functions to perform such tasks. If we did not use Hive, we would have to write a MapReduce job that would cost increased development and execution time. This will be evident from the conclusions derived in the experiments section. Figure 1 presents the process flow that was used.

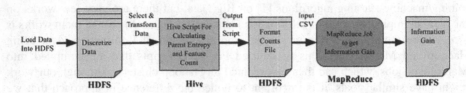

Fig. 1. Process flow of parallel implementation of Information Gain Using Hive and MapReduce

3.1   Hadoop's Components Employed for Parallelization

The frameworks that we employed to implement Information Gain are Hive and MapReduce. Hive comes in handy by providing a SQL-like query language that allows for data analysis and summarization on massive datasets. Hive was selected over Pig because HiveQL directly relates to database experts who are familiar with SQL. Also, Hive is more suitable than Pig for performing analytics on raw structured datasets stored on HDFS. Though there is no real-world battle of choosing between Pig and Hive, in this paper we chose the latter for the above-mentioned reasons [9, 13].

3.2   Why Hive Scripts Were Used to Calculate Counts and Parent Entropy

We used Hive to calculate the counts of instances and the parent entropy of the class feature. Hive was selected for this portion of the Information Gain algorithm due to the following benefits of Hive: (i) Like any query-based language (SQL, MySQL, etc.), HiveQL is extremely fast at retrieving data and performing aggregate tasks like counting the number of elements of a specific type; (ii) Also, using Hive, only one table has to be created in Hive's Metastore and the data is loaded into this table, this process has to be done only once – the first time the data is loaded into the table. If this section were done in MapReduce, the data file would have to be loaded each time to run the algorithm, hence there would be File IO overhead each time; (iii) In the case of dynamic datasets that grow every day or every hour, say from business transactions or in other kinds of datasets like weather data sets, Hive has to update only a small section of the table (by appending the data to the Hive table), whereas MapReduce would have to use File IO to re-load the entire data set. Hence Hive once again outperforms MapReduce. The sections below describe how the algorithm was implemented seamlessly chaining the output of Hive onto HDFS, which was then processed by a map only job to compute the gain.

Manipulating the Data in Hive

Hive was employed to derive the counts of instances and the parent entropy. This was achieved by a combination of subqueries, Group By and Case When statements. The Case When statements allow one to select rows based on the When clause, much like the

WHERE clause in regular SQL. Using a Case When class is equal to 0 and a Case When class is equal to 1 allowed for counts to be computed independent of each other. Once this was done, the counts were given an alias and formed together in a subquery. The aliases were then used in the outer query to calculate the entropy. To obtain results based on the class, we used the Group By clause. This retrieves the data in sections depending on the feature (class), hence we got the rows for class 0 followed by the rows for class 1. To calculate the counts of each instance of the classes, we used COUNT with the Case When statement. This gave us a count for each category of the instance for each variable. Each count instance for a feature is inserted into an array for that feature. After the counts were all calculated and collected, the arrays were split into two tables depending on the classes that is class 0 or class 1, thus transposing the columns to rows. This was done to output the counts in a format that would allow MapReduce to easily handle its task.

```
1.  CREATE tables
2.  LOAD DATA INTO tables
3.  SELECT sub.class, sub.a_1, sub.a_2,...,sub.a_N FROM
4.       (SELECT class,
5.           ARRAY(COUNT(F_1),
6.                 COUNT(CASE WHEN F_1 = C_1 END),
7.                 COUNT(CASE WHEN F_1 = C_2 END),...,
8.                 COUNT(CASE WHEN F_1 = C_N END)) as a_1
9.           ,...,


    F_i = Feature of data where i goes from 1 to n

    C_i = Categories of data where i goes from 1 to n
```

Algorithm 1: Counting Feature Categories

The final task of Hive is to output the data calculated and seamlessly feed it into MapReduce, so that MapReduce can finish the Information Gain algorithm.

```
1.  SELECT (Entropy Algorithm with total, z_count, and
2.            o_count) as Entropy
3.  FROM (SELECT COUNT(class) as total,
4.        COUNT(CASE WHEN class = 0.0 END) as z_count,
5.        COUNT(CASE WHEN class = 1.0 END) as o_count
6.        FROM tables) sub;
```

Algorithm 2: Calculating Parent Entropy

Having all the information of a feature on a single row makes it conducive for a Map-only job to execute the Information Gain algorithm in full parallelism without having to wait on receiving all the data. To perform this transformation, we used UNION ALL. As in regular SQL, the only drawback of using UNION ALL is that each query mandates the same number of column outputs. This was overcome with the use of hive's Array and Struct data types. Since Arrays act as singular data for output, we could use UNION ALL. Hence, we merged the two arrays for each feature. This output of the Hive script was then directed back into MapReduce to finish the Information Gain calculations.

```
1.  SELECT a_1, a_2,...,a_n From tables where class = 0.0;
2.  SELECT a_1, a_2,...,a_n From tables where class = 1.0;


1.  SELECT p.entropy, z.a_1, o.a_1 FROM Parent_table p, Zero_table
    z, One_table o
2.  UNION ALL
3.  SELECT p.entropy, z.a_2, o.a_2 FROM Parent_table p, Zero_table
    z, One_table o
4.  UNION ALL ... UNION ALL
5.  Select p.entropy, z.a_n, o.a_n FROM Parent_table p, ZERO_table
    z, One_table o
```

Algorithm 3: Separating arrays into two tables depending on class and outputting one feature per row

3.3 Calculating Information Gain Using MapReduce

Once Hive completed the task of counting instances, which is considered the most computationally expensive step, the table generated as output was stored on the HDFS. The table was exported in a .csv format.

Pseudocode for MapReduce Job

```
Input: File containing parent entropy and feature counts
Output: < Key', Value'> pair where key is the feature index and value is
the gain
  1.  FOR each value in features []
  2.    Parse the category and store corresponding counts and parent
         entropy
  3.  Total _count = class0_count + class1_count
  4.  TWCE =CalculateWeightedChildEntropy(category1_counts, Total_count
       )+ CalculateWeightedChildEntropy(category2_counts, Total_count
       )+.....
  5.  Gain = parent_entropy - TWC
  6.  Emit (FeatureIndex, Gain)


      4.1 CalculateWeightedChildEntropy(category counts, total)
        Category_total = category_count_class0 + category_count_class1
        P0 = category_count_class0 / category_total
        P1 = category_count_class1 / category_total
        P0LogP0   = (-P0*logP0)
        P1LogP1   = (-P1*logP1)
        H_Child   = P0LogP0 + P1LogP1
        Weight    = category_total/ total
        WCE       = weight * H_Child
        Return WCE
```

Category counts denote: category count of class0, category count of class1, class0 count and class1 count.
TWC   : Total Weighted Child Entropy
H_Child : Child Entropy
WCE   : Average Weighted Child Entropy

Algorithm 4. Map only job calculating Information Gain

This aided in easy processing for the MapReduce job. Having computed the parent entropy and the counts for each feature, this job only calculates the weighted child entropy for each feature and emits the corresponding gain. It is to be noted that the data, which spawned across 41 million rows, is now scrunched into 21 single rows containing meaningful information such that each row now maps onto a feature representing counts for each value and class. The data from this file was read by the MapReduce job to compute the gain for each feature. This job was a map only job, thus eliminating the shuffle and sort phase, which is considered the most expensive phase of a MapReduce job. As the mapper scanned through lines present in the data, it computed the average weighted child entropy for the categories present in each feature, emitting the feature index as the key and the gain as the value for the result.

Taking into consideration the art of designing MapReduce jobs, it is always best to minimize on the number of reduce tasks, thereby eliminating the shuffle and sort phase whenever possible to speed up the run time of the job. Segregating the calculation of gain into a Map only job saved on the time taken to setup, schedule and run reduce tasks.

4   Experimental Results

In this section we discuss the experiments that were performed on AWS clusters, varying the data size and recording the execution time at every phase of the algorithm to highlight how our hive script in combination with a map-only job performed in terms of efficiency and scalability. The following sections provide critical discussions of how our algorithm scaled well on the computationally expensive step over a MapReduce job and the need for a combinational model over a traditional MapReduce implementation. The dataset used in this work is [26].

4.1   Experimental Setup

The experiments were performed on an AWS EMR (Elastic MapReduce) cluster with one master node and varying slave nodes. All cluster nodes were of identical instance type m4.xlarge, which is a general purpose EC2 instance type that is used as default for all EMR clusters. The slave nodes were varied from five to a maximum of 25 nodes. It should be noted that a speedup was witnessed when we scaled from 1 to 5 nodes, so for brevity sake, the execution time is presented starting from 5 nodes. Each node had 8 virtual CPUs, 16 GB of memory, and an EBS (Elastic Block Storage) of 32 GB. Each cluster had 1 master node and then n − 1 core nodes. The experiments were run on three different data sizes: 6 GB, 12 GB, and 24 GB. The 6 GB file size had about 41 million rows of data and the 24 GB file had about 196 million rows of data. Each data file was used on all node sizes for the cluster, and the time for each run was recorded and is presented graphically.

4.2   Calculating Counts on Hive Vs. MapReduce

Calculating the counts for the feature categories per class was the expensive part of the experiments. The test was to see if Hive truly had an advantage when compared to MapReduce. Each of the advantages mentioned in section five are only expected to gain fruition during execution. Upon running the experiments, it became clear that when it comes to this kind of computation, Hive clearly outperforms MapReduce, as shown in Figs. 2 and 3. This was very apparent for smaller node counts. With a node size of five and a file size of 6 GB, Hive was more than twice as fast as MapReduce. In Fig. 3, when the file size was increased to 24 GB, the performance gain achieved using the Hive scripts was also tremendous. This discloses that as the file size increases, the performance of Hive over MapReduce still increases. For 6 GB data, the performance increase rate for Hive does not slow down until the final two runs, where the number of nodes was 20 and 25. It must be noted that the time Hive took to load the 6 GB file from HDFS to populate the table was only 66 s. Hence, the time taken to load the data is negligible and can be ignored, making Hive excel in the query execution time.

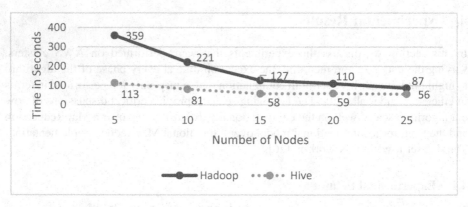

Fig. 2. Execution time for Hive vs MapReduce for calculating counts using 6 GB data

The convergence of the performance rate in Hive can also be attributed to the data block size that was used. The data block size limits the number of input splits generated for the data set. We should manipulate the data block size (for example from 32 to 512 MB) to arrive at an optimum data block size that suits the size of the data. In this manner, the number of map tasks that are to be scheduled can be easily controlled on Hive. This step is required to optimize the number of input splits generated such that we do not end up in a situation where many map tasks are created, each receiving small amounts of data. When this happens, the overhead of managing the map tasks will be more than performing the actual task at hand. Although it might take a few runs to figure out the optimum data block size for a given data set, we highly recommend it in order to witness a speedup in the execution time.

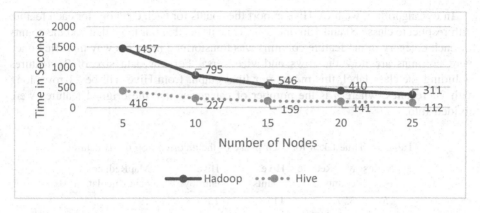

Fig. 3. Execution time for Hive vs MapReduce - calculating counts using 24 GB data

One can also change the map memory size as per the data set size to avoid this convergence in performance. Experiments were also conducted to realize how hive in union with a map-only job performed on the varied data sizes. Figure 4 proves that the suggested parallel implementation of the Information Gain algorithm has scaled well proving efficiency.

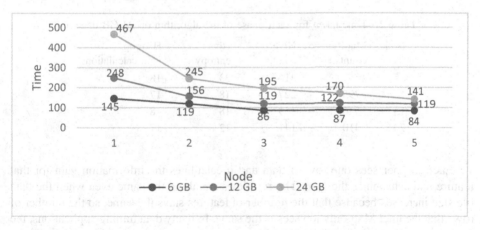

Fig. 4. Graph comparing total run time for our suggested model

4.3   Execution Time for Growing Data Size

The Information Gain algorithm is parallelizable by nature, thereby calculating the final Information Gain for each feature in the data is the last step in the overall algorithm. From Tables 1 and 2 we claim that this final step has a constant execution time only when the number of features can be accommodated in one input split. This is because our algorithm uses the best of both worlds with Hive and MapReduce.

In our algorithm, we have Hive export the counts for each category for each feature with respect to classes 0 and 1 in one row. This means that our table that had the counts for each category in the feature columns was transformed in such a way that what once were columns are now the rows and vice versa. For our data size of 21 features including the class label, this means our final output from Hive will be 21 rows long with $2n$ features, where n is the number of categories that the original features were divided into.

Table 1. Time taken for each phase of the algorithm on 6 GB dataset

Nodes	MapReduce counts	Hive counts	Hive entropy	MapReduce gain calculation
5	359	113	13	19
10	221	81	20	18
15	127	58	11	17
20	110	59	10	18

The algorithm was designed in this manner so that each mapper in the final Map-only job would have to see only one row to calculate the final information gain for that feature. If we kept the original view of the table, the mappers would have to wait for all the rows in the input file to be read before calculations could be done.

Table 2. Time taken for each phase of the algorithm on 24 GB dataset

Nodes	MapReduce counts	Hive counts	Hive entropy	Hadoop gain calculation
5	1457	416	33	18
10	795	227	18	17
15	546	159	18	18
20	410	141	12	17

Each mapper sees one row of data and it calculates the information gain for that feature and determines the gain. The run times are about the same even when the data file size increases because that the number of features stays the same, so the number of rows that the final Map-only job sees is the same. In many data mining applications, the features may scale up to thousands and even to several thousands, and in such cases the Map-only job will be very useful. With this being a map only job, calling the reducer(s) and the shuffle and sort phase is eliminated, allowing for performance gains even when highly scaled up.

It should be noted that, the traditional implementation of Information Gain on MapReduce would have required three Map Reduce jobs, and would have involved large amounts I/O overhead due to large amounts of intermediate data being shuffled and sorted since more number of reducers would have to be employed to perform the same task.

5   Conclusion and Future Work

In this paper, we have put forth a new approach for parallelizing the Information Gain algorithm using hive scripts in combination with a map-only job. The advantage of adopting this method at the enterprise level will surely reflect in a higher performance gain with reduced development and maintenance time. For the Hive portion, there is no necessity to stay with one reducer to calculate the counts. We can scale up as needed using the recommendations we stated earlier about manipulating the data block size and map memory size as needed. We have prudentially used the appropriate Hadoop components (Hive and MapReduce) that exploit the inherent merits present in each one of them. The suggested implementation method also allows for performing counts on close to real-time data, since a table present in the Hive warehouse could be easily updated by writing simple scripts. We can also generalize this implementation for discrete features present in the any dataset. Also, if one encounters varying categories for each feature, we can resolve this in Hive by padding zeros for missing categories. In addition to the performance gain encumbered due to using Hive for the appropriate tasks of the Information Gain calculations, usage of arrays in Hive to store all the category counts will help scale this algorithm well even if our features and category counts increase. This is because arrays in Hive grow dynamically and do not have to be iterated through. Each array can be exported as a whole regardless of the number of categories present in the features, and the final MapReduce job will be of computational complexity of O (1) provided the data size fits in a single input split.

As future work, we plan to implement this algorithm solely on Hive with the help of User Defined Functions (UDF's) and optimizations such as partitioning and bucketing will also be analyzed to reap better performance.

References

1. Bagui, S., Spratlin, S.: A review of data mining algorithms on Hadoop's MapReduce. Int. J. Data Sci. (2017, to appear)
2. Bhardwaj, R., Vatta, S.: Implementation of the ID3 algorithm. Int. J. Adv. Res. Comput. Sci. Softw. Eng. **3**, 845–851 (2013)
3. Dhai, W., Ji, W.: A MapReduce implementation of C4.5 decision tree algorithm. Int. J. Database Theory Appl. **7**, 49–60 (2014)
4. Dean, J., Ghemawat, S.: MapReduce: simplified data processing on large clusters. Commun. ACM **5**, 107–113 (2008)
5. Duda, R.O.T.: Pattern classification, 2nd edn. Wiley, New York (2001)
6. Forman, G.: An extensive empirical study of feature selection metrics for text classification. J. Mach. Learn. Res. **3**, 1289–1305 (2003)
7. Guillén, A., Sorjamaa, A., Miche, Y., Lendasse, A., Rojas, I.: Efficient parallel feature selection for steganography problems. In: Cabestany, J., Sandoval, F., Prieto, A., Corchado, Juan M. (eds.) IWANN 2009. LNCS, vol. 5517, pp. 1224–1231. Springer, Heidelberg (2009). https://doi.org/10.1007/978-3-642-02478-8_153. ISBN 978-3-642-02477-1
8. Han, J., Kamber, M., Pei, J.: Data Mining: Concepts and Techniques, 3rd edn. Morgan Kaufmann Publishers, Waltham (2011)

9. Huang, T.C., Chu, K.-C., Lee, W.-T., Ho, Y.-S.: Adaptive combiner for MapReduce. Cluster Comput. **17**, 1231–1253 (2014)
10. Kumar, V., Minz, S.: Poem classification using machine learning approach. In: Babu, B.V., et al. (eds.) Proceedings of the Second International Conference on Soft Computing for Problem Solving (SocProS 2012), December 28–30, 2012. AISC, vol. 236, pp. 675–682. Springer, New Delhi (2014). https://doi.org/10.1007/978-81-322-1602-5_72
11. Kumar, V., Minz, S.: Mood classification of lyrics using SentiWordNet. In: International Conference on Computer Communications and Informatics (2013)
12. Kumar, V., Minz, S.: Feature selection: a literature review. Smart Comput. Rev. **4**, 211–229 (2014)
13. Lam, C.: Hadoop in Action. Manning Publications Co., New York (2010)
14. Liu, H., Yu, L.: Toward integrating feature selection algorithms for classification and clustering. IEEE Trans. Knowl. Data Eng. **17**, 491–592 (2005)
15. Mitchell, T.M.: Machine Learning, 1st edn. McGraw-Hill Science/Engineering/Math, New York (1997)
16. Mladenic, D., Grobelnik, M.: Feature selection for unbalanced class distribution and Naive Bayes. In: Proceedings of the Sixteenth International Conference on Machine Learning (ICML). Morgan Kaufmann Publishers Inc., San Francisco, pp. 258–267 (1999)
17. NandaKumar, A.N., Yambem, N.: Survey on data mining algorithms on Apache Hadoop platform. Int. J. Emerg. Technol. Adv. Eng. **4**, 563–565 (2014)
18. Owen, S., Anil, R., Dunning, T., Friedman, E.: Mahout in Action. Manning Publications Co., Greenwich (2011)
19. Purdila, V., Pentiuc, S.-G.: MR-tree – a scalable MapReduce algorithm for building decision trees. J. Appl. Comput. Sci. Math. **16**, 16–19 (2014)
20. Quinlan, J.R.: C4.5: Programs for Machine Learning. Morgan Kaufmann Publishers Inc., San Francisco (1993). ISBN 1-55860-238-0
21. Sakr, S., Liu, A., Fayoumi, A.G.: The family of MapReduce and large-scale data processing systems. ACM Comput. Surv. (CSUR), **46** (2013)
22. https://2xbbhjxc6wk3v21p62t8n4d4-wpengine.netdna-ssl.com/wp-content/uploads/2012/06/Using_Tableau_with_Hortonworks_Data_Platform.v1.0.pdf
23. Singh, S., Kubica, J., Larsen, S., Sorokina, D.: Parallel large-scale feature selection for logistic regression. In: Proceedings of SIAM International Conference on Data Mining, pp. 1172–1183 (2009)
24. Sun, Z., Li, Z.: Data intensive parallel feature selection method study. In: International Joint Conference on Neural Networks (IJCNN), pp. 2256–2262 (2014). https://doi.org/10.1109/ijcnn.2014.6889409
25. Zdravevski, E., Lameski, P., Kulakov, A., Filiposka, S., Trajanov, D., Jakimovskik, B.: Parallel computation of information gain using Hadoop and MapReduce. In: Proceedings of the Federated Conference on Computer Science and Information Systems. ACSIS, vol. 5, pp. 181–192 (2015). https://doi.org/10.15439/2015f89
26. http://archive.ics.uci.edu/ml/datasets/HEPMASS?ref=datanews.io

Measuring Interpretability for Different Types of Machine Learning Models

Qing Zhou[1], Fenglu Liao[1], Chao Mou[1(✉)], and Ping Wang[2]

[1] The College of Computer Science, Chongqing University, Chongqing, China
chao_m@cqu.edu.cn
[2] School of Foreign Languages and Cultures, Chongqing University,
Chongqing, China

Abstract. The interpretability of a machine learning model plays a significant role in practical applications, thus it is necessary to develop a method to compare the interpretability for different models so as to select the most appropriate one. However, model interpretability, a highly subjective concept, is difficult to be accurately measured, not to mention the interpretability comparison of different models. To this end, we develop an interpretability evaluation model to compute model interpretability and compare interpretability for different models. Specifically, first we we present a general form of model interpretability. Second, a questionnaire survey system is developed to collect information about users' understanding of a machine learning model. Next, three structure features are selected to investigate the relationship between interpretability and structural complexity. After this, an interpretability label is build based on the questionnaire survey result and a linear regression model is developed to evaluate the relationship between the structural features and model interpretability. The experiment results demonstrate that our interpretability evaluation model is valid and reliable to evaluate the interpretability of different models.

Keywords: Structural complexity · Model interpretability
Interpretability evaluation model · Machine learning models

1 Introduction

Machine learning model performance evaluation is crucial in data mining process and commonly used metrics include accuracy, recall, precision, F_1-score and so on [1]. While in many applications, a good model should not only perform well in these prediction performance metrics, but also be interpretable to users [2, 19–21]. For instance, in medical diagnosis [2, 21], the model should be understandable so that the doctors can explain it and patients can accept it. Therefore, a trade-off between the prediction performance and model interpretability is of necessity. As for prediction performance, it can be accurately measured while model interpretability is a highly subjective concept and is difficult to be accurately evaluated [3, 16]. In addition, unlike prediction performance measurement by a general and well-recognized method, there is no such a method to evaluate model interpretability. Moreover, it is difficult to compare the interpretability of different models so as to select the best one(s). Therefore, it is necessary to propose a general method to evaluate the interpretability for different models.

© Springer Nature Switzerland AG 2018
M. Ganji et al. (Eds.): PAKDD 2018, LNAI 11154, pp. 295–308, 2018.
https://doi.org/10.1007/978-3-030-04503-6_29

The researches about model interpretability fall into two categories: white-box interpretability and black-box interpretability. White-box model is transparent and it's internals can be observed but usually not modified [22], such as decision tree and regression models; black-box model is an opaque model and only the input and output can be viewed while it's internal working is not known [23] ,such as Support Vector Machine (SVM) and artificial neural network (ANN). Researches on black-box model usually used some simple and interpretable rules (white-box model) to explain the process from input to output [8–10], thus researches on the black-box model interpretability be transformed to study the white-box model interpretability. Based on the above, the study on the white-box model interpretability is the foundation for the researches of the machine learning model interpretability. Therefore, this present study is focused on the interpretability of white-box model and two representative classifiers, decision tree and logistic regression model are considered. Although the white box model is simple and easy to understand, it is still unexplained when it is with high structural complexity [3, 4, 13].

Most studies on the interpretability of white-box model mainly focus on different factors and how they affect model interpretability through questionnaire survey, such as intelligence, prior knowledge, structural complexity and visual style [4, 5, 7, 13, 16]. As the subjective factors (e.g., user's prior knowledge) is strongly affected by individuals and hard to measure exactly, we only consider some objective factors regarding model structural complexity (e.g. the number of leaves, the depth in a decision tree model [3, 11] and the number of features in regression model [7, 12]). A lower structural complexity usually leads to a higher interpretability [5, 20]. In addition, in most researches, the usability of a model is approximate to its interpretability. In other words, a model is interpretable if the model can be used correctly to solve some problems in a certain period of time by humans [5, 16, 18]. However, it should be noted that prior works have some limitations. First, they mainly focus on exploring the factors and how these factors affect model interpretability without a deeper analysis of the components of model interpretability. Second, very few researchers have tried to propose a method to quantify the model interpretability, not to mention comparing the interpretability for different models.

In this paper, we propose a method to evaluate interpretability among different types of classification models in a unified framework. First, a questionnaire survey system consisting of five different tasks is developed to collect information about interpretability. Next, based on prior studies, three structural features are selected (i.e., number of leaves, average of depth and maximum of depth [16]), and their impact on model interpretability is analyzed respectively. Then an interpretability label is constructed and an interpretability evaluation model (i.e., linear regression model) is developed to evaluate the relationship between structural features and model interpretability. Finally, a case study is conducted to validate the proposed interpretability evaluation model. The main contributions of this paper can be summarized as follows:

- A framework to evaluate the interpretability for different models is developed. This assessment framework enables a generic measure of interpretability across different models.

- A practical method to compute an exact value of model interpretability is put forward. This method achieves the quantification of model interpretability and computes specific values.
- The components of model interpretability are identified and a general form of model interpretability is proposed. This form deep analyzes the formation of model interpretability and gives a unified expression of interpretability.

2 Related Work

The interpretability of a machine learning model plays a significant role in practical applications interpretability, thus attract many researchers' attention. Prior related works fall into several categories, such as the definition of model interpretability, methods employed to explore model interpretability, factors affecting model interpretability, applications of model interpretability, etc. Below is a brief review of prior works.

Classifier interpretability, referred to as comprehensibility or understandability [15]. is defined as "the ability to understand the logic behind a prediction model" [13]. Bibal and Frénay [15] hold the position that a model is interpretable only when it is interesting, usable, justifiable and acceptable for users. And Lipton [6] identified two properties of interpretable models: transparency (i.e., how does the model work?) and post-hoc interpretations (i.e., what can the build model tell us?).

Questionnaire survey is widely used to evaluate the interpretability of white-box model. Allahyari and Lavesson [5] conducted a questionnaire survey to examine the understandability of models with different complexity from users' point of view. Huysmans et al. [3] carried out a questionnaire survey consisting of three tasks: application classification, logical yes-no judgment of statements and model-figure matching judgment to investigate the relationship between model interpretability and structural complexity with three metrics considered (accuracy, response time and confidence of answer). Based on it, Piltaver et al. [18] measured the impact of tree properties on their comprehensibility via a questionnaire survey including 6 tasks, that are classification of instances collected though questionnaire; explanation of whether attributes' value should be changed when a label changes; validation in the form of logical yes-no questions; discovering the unusual attribute-value pair; rating of model interpretability in 5 levels; comparison of the interpretability of two classifiers. Based on aforementioned works, we design our questionnaire survey system.

Many researches indicate that model complexity corresponds to the interpretability [3, 11, 16]. Ustun et al. [7] put forward a Supersparse Linear Integer Model(SLIM) to sparse the features of linear regression model so as to produce an accurate and interpretable scoring system. Piltaver et al. [16] investigated how tree structural parameters (the number of leaves, the branching factor and the tree depth) impact model comprehensibility with 3 objective parameters considered (time-to-answer, answer-correctness and question-difficulty). Those work enlightened us how to select model structural features and construct the interpretability label.

Model interpretability is important in more and more applications, such as education, medicine, financial investment and so on [19–21]. Xing et al. [19] constructed the student activity theory containing 6 variables, and then employed Genetic algorithm (GP) to discover the interpretable rules. In such a way, an interpretable student activity model was obtained so that teachers know how each feature impacts students' study performance. Zhou et al. [20] put forward a filter feature selection method and use this selected features to predict the listing statuses of Chinese-listed companies. The generated predictive model was interpretable that can help financial analysts judge the model's practicability so as to increase the model's applicability.

3 Method

3.1 A General Form of Model Interpretability

Model interpretability is affected by many factors, such as structural complexity, intelligence, physical environment and so on [4, 5, 7, 13, 16]. With these factors considered, a structural of model interpretability is proposed, as in Eq. (1).

$$I = I_0 + \varepsilon_0 + \varepsilon_1 \tag{1}$$

where I_0 is the inherent interpretability primarily determined by the structural complexity of model, not affected by subjective factors. ε_0 refers to artificial interpretability associated with human factors, such as human intelligence, knowledge. ε_1 is environment interpretability, environment here refers to all the factors except inherent characteristics of model and human factors as mentioned above. There are wide personal variations in the two components ε_0 and ε_1, thus the inherent interpretability I_0 is the main concern in this paper.

3.2 Survey Design and Implementation

As the interpretability is inherently subjective [3, 13, 14] and is difficult to be measured directly [13], the usability of the model is usually used as the interpretability [16, 18]. Therefore, a questionnaire system is designed to investigate the usability of the model through collect respondents' answer about different classification models. With reference to the previous research [18], a questionnaire survey system consisting of 5 tasks (classification, explanation, validation, rating and comparison) is designed. The first four tasks in our method measure the model's usability [15], with the assumption that a model is interpretable only when it can be used successfully and quickly to solve problems. Model interpretability is evaluated by such parameters: time-to-answer, answer-correctness, question-difficulty and subjective-rating [3, 16]. Generally speaking, shorter time-to-answer, higher answer-correctness and lower question-difficulty contribute to a more interpretable model [16]. The comparison task is designed to compare the interpretability of decision tree models and logistic regression models from respondents' point of view. A description of the tasks in the survey is briefly recapitulated in the following:

- Classification task. Respondents are asked to classify an attribute-value instance referring to a given model then give the subjective judgment of question difficulty on the scale with 5 levels from "very easy" to "very difficult".
- Explanation task. Respondents are asked to assign a value to the attributes with reference to the given model. The values assigned by respondents should lead to a given label. Then they need to give the subjective judgment of question difficulty on the scale with 5 levels from "very easy" to "very difficult".
- Validation task. Respondents are asked to judge whether the input instance (only part of attribute-value are given) can lead to a predefined label (e.g. "A" = 16, "Aa" = 30, whether or not the final label is "1"?). Then they need to give the subjective judgment of question difficulty on the scale with 5 levels from "very easy" to "very difficult".
- Rating task. Respondents are asked to judge the acceptability of decision tree model and logistic regression model on the scale with 5 levels from "very easy" to "very difficult".
- Comparison task. Respondents are asked to choose a more interpretable model from decision tree model and logistic regression models.

As shown in Eq. (1), the model interpretability is consisting of inherent interpretability, artificial interpretability and environment interpretability. The artificial interpretability and environment interpretability vary from person to person, which is difficult to measure. So, the inherent interpretability is mainly concerned in this paper. To reduce the influence of user's knowledge and environment on interpretability, respondents have the same professional background and all other environment factors are set to the same.

WDBC (Wisconsin Diagnostic Breast Cancer) dataset from UCI Machine Learning Repository is used to generate different decision trees and linear regression models. The dataset consists of 569 instances, each representing a cell nucleus that can be classified into one of two classes (malignant or benign). Each instance is described with 30 features (e.g. mean, standard error, "worst" or largest of the cell nucleus' radius, texture, perimeter, area, smoothness, compactness, concavity, concave points, symmetry, fractal dimension, etc.). Since respondents may perform better when answering questions in familiar domains than that in unfamiliar domains [13, 17], the WDBC dataset is modified according to the purpose of our research. "1" and "0" are used to represent malignant label and benign label respectively. In addition, the 30 features are represented by letter from "A" to "Z", "Aa", "Ab", "Ac", "Ad" respectively.

Since prior studies suggest that model interpretability is negatively correlated with model structural complexity [5, 16], 40 models with different structures are generated to investigate the effect of structural complexity on model interpretability. There are 20 decision tree models and 20 logistic regression models. Decision tree models are generated from WDBC dataset. Logistic regression models are developed based on the nodes of the decision trees, specifically the nodes of a decision tree serve as the variables of a regression model[1]. Then 200 questions concerning 40 models involved in

---

[1] Data is available at http://pan.baidu.com/s/1eRSNWtW.

5 tasks are designed. Finally, an online survey system consisting of 200 questions is developed. A web link to our questionnaire survey is send to respondents and they can answer questions via this link.

3.3   Interpretability Evaluation Model

Interpretability Evaluation Framework. Figure 1 shows the overview of the interpretability evaluation framework we proposed. It consists of three main sections: feature selection, interpretability label constructing and interpretability evaluation model constructing.

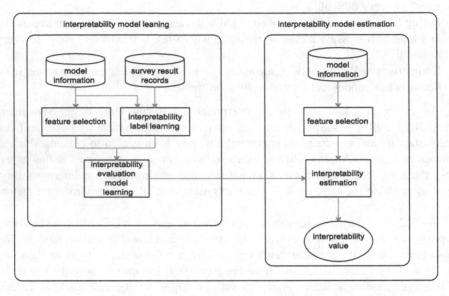

Fig. 1. The overview of interpretability evaluation framework

Feature Selection. Model structural complexity as indicated by the number of leaves and the depth in a decision tree [11], conditions in a rule set [3, 5] and the number of input variables in regression model [7, 12], is often used as reference to evaluate classifier interpretability [3, 5, 7, 16]. According to prior studies [3, 11, 16, 18], three structural features: number of leaves, average of depth and maximum of depth are selected. Table 1 shows the selected features and their formulation. Note that *variable* refers to the feature of regression model.

Interpretability Label Constructing. The interpretability of a model is measured by classification, explanation, validation and rating task in our study. So the model interpretability label was constructed as follows:

$$I = \frac{1}{4}\sum_{i=1}^{4} I_i \tag{2}$$

I_1, I_2, I_3 represent the interpretability of model measured by classification, explanation and validation tasks respectively which are formulated by three parameters: t (time-to-answer), c (answer-correctness) and d (question-difficulty). As shorter time-to-answer, higher answer-correctness, lower question-difficulty correspond to higher model interpretability [16], they are calculated as follows:

$$I_i = minmaxscaler\left(\frac{1}{3}\left(t_i^{-1} + c_i + d_i^{-1}\right)\right) \quad i \in \{1, 2, 3\} \tag{3}$$

And I_4 represents the interpretability of model measured by rating task is formulated by the respondent's subjective-rating r,

$$I_4 = minmaxscaler\left(r^{-1}\right) \tag{4}$$

The function $minmaxscaler()$ is used to normalize the value x to [0,1],

$$x = \frac{x - X.min}{X.max - X.min} \tag{5}$$

where $x \in X$.

Table 1. Structural features and their formulation

Features name	Decision tree	Logistic regression
Number of leaves (x_1)	$num.leaves = count(leave)$	$num.leaves = count(variable)$
Average of depth (x_2)	$ave.depth = \frac{\sum_{i=1}^{n} d_i}{num.leaves}$ (d_i : the depth of leave i; n : the number of leaves)	$ave.depth = 1 + count(variable)$
Maximum of depth (x_3)	$max.depth = max(D),$ $D = \{d_i\}$	$max.depth = 1 + count(variable)$

Interpretability Evaluation Model. After the feature selection and interpretability label constructing, the interpretability of a model is evaluated by the linear regression algorithm,

$$I = \omega^T X + b \tag{6}$$

X is the feature vector.

4 Result and Discussion

4.1 Survey Result

In total, 152 instances are collected and 93 instances remained after excluding samples with many missing or invalid value. The statistical data about the size of the survey are presented in Table 2. The classification, explanation and validation tasks include 2 types of question per model; one is about the objective task (e.g. classify instances, attribute-value assignment) and the other is about the subjective evaluation of question difficulty. Rating task includes one subjective question per model and the comparison task consists of one comparison question about the interpretability of two models. Concerning respondents' possible fatigue, relatively less questions were designed in explanation and validation tasks since these two tasks are more difficult for respondents.

Table 2. The statistical data about the size of the survey

Item	Classification	Explanation	Validation	Rating	Comparison
Number of model	40	40	40	40	40
Number of questions	80	80	80	40	20
Number of respondents	93	39	52	93	93
Number of answers	744	312	416	744	372

The descriptive statistics are presented in Table 3. As show in Table 3, classification task has the lowest question-difficulty and the highest answer-correctness among three tasks. By contrast, it is similar in explanation task except that respondents spend the lengthiest time-to-answer in this task. It is the opposite in validation task, respondents spend the shortest time, rate the highest question-difficulty and achieve the least correctness. Aforementioned statistics suggests that the task difficulty is in ascending order (classification < explanation < validation). As shown in Table 3, the mean subjective rating interpretability is 2.42, the model interpretability of decision tree is 2.31 while that of logistic regression model is 2.54. Since the lower the value, the more interpretable the model, thus it seems that the decision tree model is more interpretable (2.31 < 2.54). In comparison task, 232 out of 372 answers suggest that compared with logistic regression model with the same features, the decision tree is more comprehensible.

Table 3. The descriptive statistics about the 5 tasks of the survey

Item	Classification	Explanation	Validation	Rating	Comparison
Median time-to-answer(s)	28	29.5	23	/	/
Mean question-difficulty	1.96	1.97	2.24	/	/
Mean answer-correctness	0.96	0.95	0.56	/	/
Mean subjective-rating	/	/	/	2.42	/
Number of more interpretable (DT)	/	/	/	/	232
Number of more interpretable (LR)	/	/	/	/	140

4.2   The Impact of Structural Complexity on Model Interpretability

Figure 2 shows the impact of model structural features (number of leaves, average of depth and maximum of depth) on interpretability measured by classification, explanation and validation task. The red, green and blue dots represent the value of the classification, explanation and validation tasks. The three horizontal parameters are number of leaves, average of depth and maximum of depth. The vertical parameters are time-to-answer, question-difficulty and answer-correctness. From Fig. 2, we can see that time-to-answer grows with the increase in model structural complexity and the question-difficulty parameter shows the same tendency. The answer-correctness of classification and explanation tasks always stay near 1, while that of validation task decreases slightly while structural complexity increases.

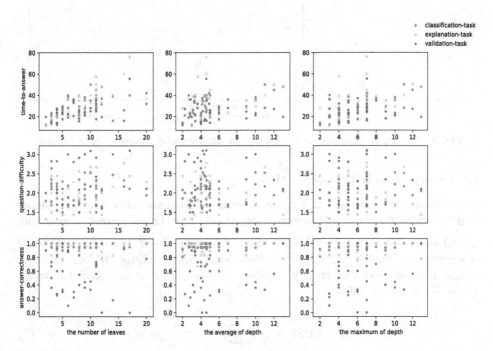

Fig. 2. The impact of model structural features on interpretability measured by classification, explanation and validation task

Figure 3 shows the impact of structural complexity on interpretability measured by rating task. The dot represents the subjective-rating value per model, and the lower the value, the better the model interpretability. As can been seen, when structure complexity increases, the model interpretability of rating also decreases.

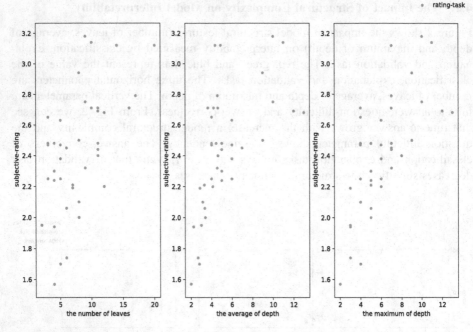

Fig. 3. The impact of structural complexity on interpretability measured by rating task

4.3 Interpretability Evaluation Model

Using the method proposed at 3.2, we generate the interpretability $I\epsilon[0,1]$, a higher value indicates that the model is more interpretable. Then Linear regression algorithm is used to develop the interpretability evaluation model f, $X \xrightarrow{f} I$. The metrics we adopt to measure the model performance is Root Mean Squared Error (RMSE), a lower RMSE indicates less discrepancy between the labeled value and predicted value. RMSE is defined as follows:

$$RMSE = \sqrt{\frac{1}{N}\sum_1^N (y - y')^2} \tag{7}$$

where y is the labeled value, y' is the corresponding predicted value, and N is the number of classifiers. We use 5-fold cross validation to measure the performance of interpretability evaluation model.

The RMSE achieves 0.093, (the interpretability $I \epsilon [0, 1]$) suggesting that our model can evaluate model interpretability well. This error may be induced by the artificial interpretability and environment interpretability.

The equation of the interpretability evaluation model we proposed is generated as follows:

$$I = -0.33 * x_1 - 0.25 * x_2 + 0.13 * x_3 + 0.59 \qquad (8)$$

Equation (8) suggests a strong correlation between model interpretability and the three structural features respectively (number of leaves x_1, average of depth x_2 and maximum of depth x_3). Number of leaves x_1 and average of depth x_2 are negatively correlated with model interpretability, that is to say, higher values of number of leaves and average of depth are associated with lower model interpretability. Maximum of depth x_3 is positively correlated with model interpretability; a higher value in Maximum of depth is associated with higher model interpretability. An increasing maximum of depth may lead to a more unbalanced model which is more likely to be understood.

4.4   Case Study

The Liver Disorders dataset from UCI Machine Learning Repository is used to construct four types of decision tree models with different structural complexity, as shown in Fig. 4. This dataset consists of 345 instances and each instance belongs to one of two classes: disorder or not disorder represented by 1 and 2 respectively. Instances are described with 6 features: mean corpuscular volume, alkaline phosphatase, alamine aminotransferase, aspartate aminotransferase, gamma-glutamyl transpeptidase, number of half-pint equivalents of alcoholic beverages drunk per day, represented by letter "a" to "f".

The interpretability evaluation model we build are used to measure the interpretability of 4 decision tree models. The structural features and predicted results are displayed in Table 4. As can be seen, the structural complexity of four models follows the following order: a < b < c < d; their interpretability is like this: a > c > b > d. Model a has the least structural complexity and the highest interpretability; while model d has the most structural complexity and the lowest interpretability, suggesting that structural complexity negatively correspond to model interpretability. While model c is more interpretable than model b and the reason may be that model b and model c have the same number of leaves, but they have different tree structures, that is, model c is unbalanced while model b is balanced. Therefore, the interpretability evaluation model we proposed can be used to measure a model's interpretability correctly.

Table 4. The structural features and predicted results

No.	Number of leaves	Average of depth	Maximum of depth	Interpretability
a	6	3.17	4	0.562
b	11	3.64	4	0.366
c	11	4.27	6	0.413
d	15	4.47	6	0.268

306   Q. Zhou et al.

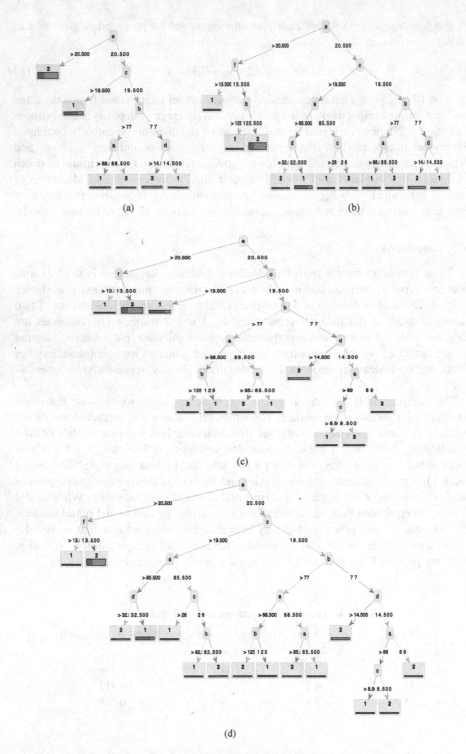

(a)

(b)

(c)

(d)

Fig. 4. Four types decision trees with different structure

5 Conclusion

In this paper, we developed an interpretability evaluation model to compute an exact value for model interpretability and to measure the interpretability for different models. We first designed a questionnaire survey system to collect information about respondents' understanding of different models. Secondly, we analyzed the questionnaire data and explored the relationship between model interpretability and structural complexity, and developed a linear regression model to quantify this relationship. Our model achieved higher accuracy (RMSE = 0.093). Finally, we used the interpretability evaluation model we build to measure the interpretability of models generated from a standard dataset. As shown in the experiment results, our interpretability evaluation model is valid and reliable for interpretability evaluation of linear regression model and decision three, two most popular white-box models. The evaluation method is obtained by training a specific set of data, different test samples may contribute different results. However, as the "no free lunch" theorem states any two optimization algorithms are equivalent when averaging all possible problems [24]. Since the interpretability of model is an open issue, it is reasonable to solve from user point of view. In future work, we hope to apply the method for computing model interpretability considering other classification models, especially black box models that usually involve the components of linear regression or decision tree.

References

1. Zhou, Z.H.: Machine Learning, 1st edn. Tsinghua University Press, Beijing (2016)
2. Jovanovic, M., Radovanovic, S., Vukicevic, M., et al.: Building interpretable predictive models for pediatric hospital readmission using Tree-Lasso logistic regression. Artif. Intell. Med. **72**, 12 (2016)
3. Huysmans, J., Dejaeger, K., Mues, C., et al.: An empirical evaluation of the comprehensibility of decision table, tree and rule based predictive models. Decis. Support Syst. **51**(1), 141–154 (2011)
4. Sim, K.C.: On constructing and analysing an interpretable brain model for the DNN based on hidden activity patterns. In: Automatic Speech Recognition and Understanding, pp. 22–29. IEEE (2016)
5. Allahyari, H., Lavesson, N.: User-oriented assessment of classification model understandability. In: Scandinavian Conference on Artificial Intelligence (2011)
6. Lipton, Z.C.: The Mythos of Model Interpretability (2016)
7. Ustun, B., Traca, S., Rudin, C.: Supersparse linear integer models for predictive scoring systems. In: AAAI Conference on Artificial Intelligence (2013)
8. Lu, T.K.P., Chau, V.T.N., Phung, N.H.: Extracting rule RF in educational data classification: from a random forest to interpretable refined rules. In: International Conference on Advanced Computing and Applications, pp. 20–27. IEEE (2015)
9. Turner, R.: A Model Explanation System: Latest Updates and Extensions (2016)
10. Ribeiro, M.T., Singh, S., Guestrin, C.: "Why should i trust you?": explaining the predictions of any classifier. In: ACM SIGKDD International Conference on Knowledge Discovery and Data Mining, pp. 1135–1144. ACM (2016)

11. Maimon, O., Rokach, L.: Data Mining and Knowledge Discovery Handbook. Springer, Berlin (2005)
12. Schielzeth, H.: Simple means to improve the interpretability of regression coefficients. Methods Ecol. Evol. **1**(2), 103–113 (2010)
13. Martens, D., Vanthienen, J., Verbeke, W., et al.: Performance of classification models from a user perspective. Decis. Support Syst. **51**(4), 782–793 (2011)
14. Freitas, A.A.: Comprehensible classification models:a position paper. ACM SIGKDD Explor. Newsl. **15**(1), 1–10 (2014)
15. Bibal, A., Frénay, B.: Interpretability of machine learning models and representations: an introduction. In: Proceedings ESANN, pp. 77–82 (2016)
16. Piltaver, R., Luštrek, M., Gams, M., et al.: What makes classification trees comprehensible? Expert Syst. Appl. **62**(10), 333–346 (2016)
17. Sweller, J.: Cognitive load during problem solving: effects on learning. Cognit. Sci. **12**(2), 257–285 (1988)
18. Piltaver, R., Luštrek, M., Gams, M., et al.: Comprehensibility of classification trees—survey design. In: International Multiconference Information Society—Is (2014)
19. Xing, W., Guo, R., Petakovic, E., et al.: Participation-based student final performance prediction model through interpretable Genetic Programming. Comput. Hum. Behav. **47**(C), 168–181 (2015)
20. Zhou, L., Si, Y.W., Fujita, H.: Predicting the listing statuses of Chinese-listed companies using decision trees combined with an improved filter feature selection method. Knowl.-Based Syst. **128** (2017)
21. Gorzałczany, M.B., Rudziński, F.: Interpretable and accurate medical data classification—a multi-objective genetic-fuzzy optimization approach. Expert Syst. Appl. **71**, 26–39 (2017)
22. White box. https://en.wikipedia.org/wiki/White_box_(software_engineering). Accessed 20 Nov 2017
23. Black box. https://en.wikipedia.org/wiki/Black_box. Accessed 20 Nov 2017
24. "no free lunch" theorem. https://en.wikipedia.org/wiki/No_free_lunch_theorem. Accessed 19 March 2018

In Search of Public Agenda with Text Mining: An Exploratory Study of Agenda Setting Dynamics Between the Traditional Media and Wikipedia

Philip T. Y. Lee[✉]

The University of Hong Kong, Lung Fu Shan, Hong Kong SAR
phil0127@hku.hk

Abstract. When Downs [1] proposed his famous *issue-attention cycle* in 1972, he thought that the mass media would report news and information that arouses people's interests. This thought, however, is prone to challenges. With the prevalence of the Internet and, perhaps more importantly, the concept of Web 2.0, Wikipedia becomes another major source of information for the public. Given that Wikipedia allows anyone to edit the content, the details about a particular event or issue on pages of Wikipedia can be considered as a quasi-public agenda. Understanding this quasi-public agenda may help us evaluate different models of policy cycles, including the Downs' famous issue-attention cycle. My study aims to assess the agenda setting dynamics among 5 major news outlets in the UK, as the traditional mass media, and Wikipedia, as a form of participatory journalism. By agenda, it refers to the choices of frames and sentiment. Using text mining techniques, my study assesses the choices of frames and sentiment adopted by the articles of the news outlets and the Wikipedia pages concerning with the issue Brexit. The timeline of the study is between the date when the Wikipedia page "Brexit" emerged and the date of the Brexit referendum. The study also explores the possible relationship between these agendas. Frame analysis of the news articles will be conducted through automatic text classification, whereas the frames on the Wikipedia pages will be analyzed through both text classification and clustering. Lexicon-based approach will be used for sentiment analysis. The relationship between the agendas will be explored through Granger-causality tests.

Keywords: Agenda setting · Media agenda · Participatory journalism Wikipedia · Text mining

1 Introduction

When Downs [1] proposed his famous *issue-attention cycle* in 1972, he thought that the mass media would report news and information that arouses people's interests. He implicitly suggested that the extent of media coverage on a particular issue reflects people's level of interest in that issue. Many subsequent papers concerning with policy cycles have taken in his thought and quantitatively measured media coverage as public views and interests [2, 3]. Maybe the mass media have their own interests to direct

© Springer Nature Switzerland AG 2018
M. Ganji et al. (Eds.): PAKDD 2018, LNAI 11154, pp. 309–317, 2018.
https://doi.org/10.1007/978-3-030-04503-6_30

people's thinking. In the recent Brexit referendum, the problem of the mass media disseminating loop-sided or misleading information has deteriorated [4, 5]. Some claimed that the problem may ultimately lead to "democratic deficit" [6] or even "failed democracy" [7].

With the prevalence of the Internet and, perhaps more importantly, the concept of Web 2.0, Wikipedia becomes another major source of information for the public. It is a form of participatory journalism in which citizens play an active role in the process of collecting, reporting, analyzing and disseminating news and information [8, 9]. Thanks to Google, the Wikipedia pages often appear to be among the top results for many keywords search, including "Brexit". These Wikipedia pages are free and convenient sources of information. This study is concerned with the agenda setting dynamics between the traditional mass media and Wikipedia, as a form of participatory journalism. Frames and sentiment are selected in my study as core elements of agendas. For an issue, what are the similarity and difference between the frames and sentiment shown by articles of the traditional mass media and user-generated content on Wikipedia pages across time? Is there any relationship (e.g. causation) between these agendas? These are the questions this study would like to address.

Media coverage is not a perfect reflection of public views and interests. To some researchers, the media are institutions with their own interests [10]. Readership may be one of their interests but not the only one. They have the ability to promote the salience of particular issues, and actually emphasize particular roots of a problem to direct readers' thinking [11]. In addition, they may adjust their news coverage for their advertising incomes [12, 13]. For example, they avoid following some of their major advertisers' irresponsible business behaviors for their income earning purposes. To some other researchers, a professional journalist should be ignorant to audience wants [14]. Audience desire should not be a factor for deciding what to be reported. What to be included in the news content should be evaluated on the basis of news values [15]. Downs' thought that media coverage represents public interest is prone to challenges.

Compared to the previous decades, the general public now benefit from the prevalence of the Internet. The Internet provides people with a free and convenient channel to disseminate and receive news and information. Some scholars have initiated the concept of participatory journalism [8, 16, 17]. Bowman and Willis [8], in their book *We Media: How audiences are shaping the future of news and information* defined the concept participatory journalism as:

> *"The act of a citizen, or group of citizens, playing an active role in the process of collecting, reporting, analyzing and disseminating news and information. The intent of this participation is to provide independent, reliable, accurate, wide-ranging and relevant information that a democracy requires"* (p. 9).

They argued that the participatory journalism involves bottom-up transmission of information. Citizens, as participants of the transmission, can change their roles among audience, publishers, editors, reporters and even advertisers. The information therefore is not filtered by any institution's control. Among other forms like Weblog and bulletin boards, Wikipedia is considered as the largest form of participatory journalism [9].

Given that Wikipedia allows anyone to edit the content, the details about a particular event or issue on pages of Wikipedia can be considered as a quasi-public agenda. Some users act as a reporter, adding content to the pages. Some users act as editor, revising existing content on the page. Every change of content indicates a user's interest. The content reveals users' choices of frames and sentiment about the event or the issue. Wikipedia keeps the *edit history* information for every page. The ample supply of user-generated data offers researchers a new way to further understand the public views and interests. Understanding this quasi-public agenda may help us evaluate models of policy cycles in this new era when the influential power of the traditional media is declining.

This study plans to assess the agenda setting dynamics among 5 major news outlets in the UK and Wikipedia, as a form of participatory journalism. By agenda, it refers to the choices of frames and sentiment. Using text mining techniques, the study assesses the choices of frames and sentiment adopted by the articles of the news outlets and the Wikipedia pages concerning with the issue Brexit. The timeline of the study is between the date when the Wikipedia page "Brexit" emerged and the date of the Brexit referendum. The study also explores the possible relationship (e.g. causation) between these agendas. Brexit is selected because it has drawn much attention from many people and the media in the first place. The result of the referendum will affect many people's daily lives. The "Brexit" page and other related pages on Wikipedia have been established and have involved many users' editing work. Moreover, Brexit involves a large number of aspects for evaluation including economics, immigration as well as the relations with the EU. The diversity of how to frame Brexit enriches the comparison among these agendas. Two software programs are developed for automatic data collection. One is to collect articles about Brexit from the websites of these 5 major news outlets. Another one is to collect the *edit history* of the "Brexit" Wikipedia page and the *edit history* of other pages of which hyperlinks are on the "Brexit" page. Frame analysis of the news articles will be conducted through automatic text classification, whereas the frames of the Wikipedia pages will be analyzed through both text classification and clustering. Lexicon-based approach will be used for sentiment analysis. The relationship between the agendas will be explored through Granger-causality tests.

2 Literature Review

In Downs' famous paper *Up and Down with Ecology—the "issue-attention cycle"* [1], he described news coverage as a form of entertainment that "competes with other types of entertainment for a share of each person's time". He thought that the media will shift their focus to other issues as soon as the media find their emphasis on the old problem bores a majority of the public. He implicitly suggested that the extent of media coverage on a particular issue reflects people's level of interest in that issue. Many subsequent papers concerning with policy cycles took in his thought and considered media coverage as public views. For example, Lodge and Hood [18] qualitatively investigated media coverage on dog attacks to demonstrate a typical pattern of attention cycle. Some other papers analyzing policy-making process quantitatively measure media coverage. Baumagartner and Jones [2] collected data of media coverage on different policy areas to evaluate the relationship between policy decisions and media attention. They reaffirmed their model punctuated equilibrium with this set of data. Howlett [3] collected newspaper coverage on nuclear energy and acid rain issues, and applied time series analysis to evaluate Downs' model and Jones and Baumagartner's model.

Some other researchers took one step further in quantifying public interests. They used text mining techniques with media coverage to analyze the media agendas and to explore the relationships with agendas of other stakeholders. One approach the scholars adopted in search of agendas is supervised classification, i.e. pre-defining key terms and categories before automatic classification. Altaweel and Bone [19] developed a text-mining software programme to explore how water and water-related issues were reported by online newspapers from Nebraska in the U.S. In essence, they pre-defined a set of categories (e.g. "water", "agriculture", and "temperature") and related key words. The software programme created semantic linkages between terms that co-occurred in the newspaper articles. They discovered that water issues were frequently reported in the context of agriculture but were less discussed in terms of water quality and habitat. Talamini et al. [20] analyzed how the government, scientists and journalists framed the problem of ethanol production between 1997 and 2006. They also explored possible causality between the production and these stakeholders' agendas with Granger-causality tests. They collected mass-media news articles, scientific papers and government documents related to ethanol. They predefined key macro-environmental frames, used sets of key words related to these frames to classify the texts, and aggregated the occurrence of every frame for analysis. They discovered that the agendas of these stakeholders differ. The journalists were gaining more influence on the policymaker's agenda across the period. The authors conducted similar studies related to ethanol production in Brazil and Germany [21, 22].

Texts can also be clustered through an unsupervised approach. One commonly used technique is K-means clustering [23]. It divides data into K number of clusters. The similarity of data within each cluster and the difference between each cluster center are maximized. Quinn et al. [24] applied K-means clustering to legislative speeches in the US Senate from 1997 to 2004 to explore the agendas of senators. Some researchers used probabilistic models and statistical models for text classification. Boussalis and Coan [25] employed a simple probabilistic model to determine text topics automatically. They also used natural language processing techniques to process the content of conservative think tanks on climate change from the mid-1990s. They analyzed how climate skepticism of these think tanks influenced people's perception of climate change. Grimmer [26] used a statistical model with the Congress members' press releases to estimate the Congress members' attention to different issues.

High-dimensionality is a typical problem that undermines performance of text representation. Term-based ontology mining is one possible method to address the problem. Silva et al. [27] developed an ontology of political entities to improve the performance of entity recognition for political opinion mining in Portuguese social media. Several researches also showed that incorporating domain knowledge can improve performance of political text classification [28–30].

Researchers have recently used sentiment analysis with news articles. Godbole et al. [31] suggested that assigning different lexicons to different news entities can improve sentiment analysis. They proposed a sentiment lexicon generation algorithm for seven different news dimensions (e.g. politics, business, and health) and showed improvement in identifying sentiment of newspaper articles within the same dimension.

Balahur et al. [32] showed that the separating positive or negative sentiment from good or bad news improved the sentiment analysis of news articles. Some recent studies have used social media content [28, 33–35] in search of politicians' popularity.

Prior studies examined the dynamics of agenda setting among various traditional media channels. Lim [36] explored the dynamics of agenda setting among leading online news media, secondary online news media and online wire services in South Korea. The agendas of the secondary online news media and the online wire services were influenced by the leading online newspapers. Dotson et al. [37] compared the media coverage of climate change by a conservative newspaper and a liberal newspaper in Chile. They showed that the liberal newspaper published articles related to climate change more frequently and had more illustrations about climate change than the conservative paper did.

Few studies examined the dynamics between the mass-media and different forms of participatory journalism. Meraz [38] examined how the traditional media setting news agenda in blogosphere. He used hyperlink analysis to demonstrate the traditional media's influence to agenda setting in political blogs of the leading newspaper and other top independent political blogs. He suggested that the traditional media no longer had the universal or singular agenda setting power in those independent political blogs. Wu et al. [39] examined the relationship of agenda setting between the newspapers and Sina Weibo in China. They shared similar conclusion with [38]. Ali and Fahmy [40] found that the traditional media still had a strong role as a gatekeeper. During uprisings in Iran, Egypt and Libya, citizen journalists managed to have their stories being consistent with the sources of the traditional media. Messner and Garrison [41] found the traditional media strongly influenced the agenda of blogs in the U.S. The traditional media were the dominant sources used by the blogs.

3 Method

A software programme is developed to collect the articles related to Brexit from the websites of five major UK news outlets including *BBC, Daily Mail, Guardian, Mirror* and *Telegraph*. The programme will collect the relevant articles from each website through the Google site filter. The time when each article published will be identified from the HTML codes of the article's webpage. Another software programme is developed to collect the *edit history* of the "Brexit" page of Wikipedia and other Wikipedia pages of which hyperlinks are on the "Brexit" page. Each revision of the page content represents a choice of frame and sentiment.

Both supervised and unsupervised approach will be adopted for frame analysis. Given that a substantial number of frames of Brexit are known, the supervised approach is suitable for classifying the frame of each article from the mass-media websites. This approach, however, inevitably limit some uncommon frames. Some rare frames may emerge on Wikipedia before they are removed from the pages. Thus, attempts of the unsupervised approach with the *edit history* in search of unseen frames are required. Some of the methods, including K-means clustering, terms-based ontology mining method, multi-label categorization and natural language processing, may be useful.

There is currently no optimal unsupervised method for all kinds of textual data [42]. Some trials and errors for finding good methods among them are necessary.

The text data collected will be pre-processed for sentiment analysis. According to Porter [43], a set of stop words and punctuations should be eliminated from the data. The data will subsequently be converted to lowercase and stripping suffixes before sentiment classification. The paragraphs of the news articles and the *edit history* of Wikipedia are usually well-written. Expressions of humor, sarcasm, irony and provocation are not frequently used. Machine learning approach may not be necessary [44]. This study will adopt lexicon-based approach for sentiment analysis.

Following Talamini et al. [20], the relationships between the agendas will be explored through Granger-causality tests. Granger-causality tests aim to explore the causation effect between the frequencies of frames and sentiment appeared on the news articles and on the Wikipedia pages. The Granger-causality test was proposed by Granger [45]. According to Granger's [45] definition, an independent variable is considered to be the cause of a dependent variable if lagged values of factor A are able to explain the regression of the dependent variable on lagged values of the dependent variable and the independent variable. The Granger-causality test has been commonly used in many studies exploring casual relationships. Some examples in the field of information systems include Lee et al. [46], Luo et al. [47] and Dutta [48].

4   Expected Result and Discussion

Readership of these 5 online news outlets differs, and therefore their agendas should not be the same. *BBC, Daily Telegraph* and *Guardian* are quality sources of information. They may share a common set of frames. Among them, reporting of *BBC* should be relatively politically neutral. If Wikipedia is politically neutral as it claims[1], its content should be most comparable to *BBC*'s. The coverage of these news outlets likely influences the agenda of the Wikipedia pages. On the other hand, the agenda of the Wikipedia pages may render the traditional media to report news originated from the social media.

Events happened across the study period may bring us some insights of the dynamics of agenda setting. After a Brexit-related event happens, all these news outlets likely share the same frames. One example of the events can be some political heavyweight disclosed their stances on Brexit for the first time. They revealed their worries about homeland security. The frame homeland security would subsequently be shared among all news outlets. The frequency of the frame homeland security adopted by the news outlets soared. The news outlets' choices of sentiment, however, may be different. The increasing occurrence of the frames may influence the content of the Wikipedia pages.

The pages of Wikipedia are expected to have more traffic after the event happens. These events attract public attention. More people may visit the pages and contribute to the content. Newcomers may join the reporting and editorial work on the pages of

---

[1] See https://en.wikipedia.org/wiki/Wikipedia:Neutral_point_of_view.

Wikipedia. Some newcomers may add some conspiracy news on the pages with negative sentiment. Did these newcomers generate the content independently? Are they influenced by tabloids, namely *Mirror* and *Daily Mail*? Are they influenced by quality newspapers instead? These questions remain to be answered via the relationship analysis.

5   Limitation

Similar to the media coverage, the content of Wikipedia is not a flawless reflection of people's views and interests. Active users of Wikipedia are not a perfect randomized sample of the entire population or online-user population. My study aims not to show that scholars should replace media coverage with the content of relevant Wikipedia pages or other forms of participatory when they examine public views and interests in policy cycles or related topics. Instead, my study aims to help the scholars recognize the agenda setting dynamics between the traditional news outlets and the forms of participatory journalism. Policy makers nowadays may not only focus on media coverage but also on social media attention. Understanding this agenda setting dynamics is important in the new era when the traditional media can no longer dictate information creation and dissemination.

Wikipedia is only one form of participatory journalism. My study does not examine other forms of participatory journalism (e.g. political blog). Agendas of political blogs may be largely different from that of Wikipedia. Also, the news outlets in the study are mainly leading newspapers, given that many previous studies on policy cycles examines newspaper articles. Nevertheless, Wikipedia is the largest form of participatory journalism. It should have higher similarity to the real public agenda, compared to other forms of participatory journalism. Also, my study does not examine the agenda setting of other forms of the mass-media such as online TV news broadcasts.

References

1. Downs, A.: Up and down with ecology: the issue-attention cycle. Polit. Am. Econ. Policy Mak. **48** (1996)
2. Baumgartner, F.R., Jones, B.D.: Agendas and instability in American politics. University of Chicago Press, Chicago (2010)
3. Howlett, M.: Issue-attention and punctuated equilibria models reconsidered: an empirical examination of the dynamics of agenda-setting in Canada. Can. J. Polit Sci. **30**(01), 3–29 (1997)
4. Beckett, C.: Deliberation, distortion and dystopia: the news media and the referendum (2016). http://www.referendumanalysis.eu/eu-referendum-analysis-2016/section-4/deliberation-distortion-and-dystopia-the-news-media-and-the-referendum/
5. Rowinski, P.: Mind the gap: the language of prejudice and the press omissions that led a people to the precipice (2016). http://www.referendumanalysis.eu/eu-referendum-analysis-2016/section-4/mind-the-gap-the-language-of-prejudice-and-the-press-omissions-that-led-a-people-to-the-precipice/

6. Fenton, N.: Brexit: inequality, the media and the democratic deficit (2016). http://www.referendumanalysis.eu/eu-referendum-analysis-2016/section-4/brexit-inequality-the-media-and-the-democratic-deficit/

7. Barnett, S.: How our mainstream media failed democracy (2016). http://www.referendumanalysis.eu/eu-referendum-analysis-2016/section-4/how-our-mainstream-media-failed-democracy/

8. Bowman, S., Willis, C.: We Media: How Audiences are Shaping the Future of News and Information. The Media Center at The American Press Institute, Arlington (2003)

9. Lih, A. (2004). Wikipedia as participatory journalism: reliable sources? Metrics for evaluating collaborative media as a news resource. Nature

10. Entman, R.M.: Punctuating the homogeneity of institutionalized news: abusing prisoners at Abu Ghraib versus killing civilians at Fallujah. Polit. Commun. 23, 215–224 (2006)

11. McCombs, M.E., Shaw, D.L.: The agenda-setting function of mass media. Public Opin. Q. 36, 176–187 (1972)

12. Baker, C.E.: Advertising and a Democratic Press. Oxford University Press, New York (1994)

13. Bennett, W.L.: News: The Politics of Illusion, 7th edn. Pearson, New York (2007)

14. Anderson, C.W.: Deliberative, agonistic, and algorithmic audiences: journalism's vision of its public in an age of audience transparency. Int. J. Commun. 5, 529–547 (2011)

15. Harcup, T., O'neill, D.: What is news? Galtung and Ruge revisited. Journal. Stud. 2(2), 261–280 (2001)

16. Lasica, J.D.: Blogs and journalism need each other. Nieman Rep. 57(3), 70–74 (2003)

17. Gillmor, D.: Moving toward participatory journalism. Nieman Rep. 57(3), 79–80 (2003)

18. Lodge, M., Hood, C.: Pavlovian policy responses to media feeding frenzies? Dangerous dogs regulation in comparative perspective. J. Conting. Crisis Manag. 10(1), 1–13 (2002)

19. Altaweel, M., Bone, C.: Applying content analysis for investigating the reporting of water issues. Comput. Environ. Urban Syst. 36(6), 599–613 (2012)

20. Talamini, E., Caldarelli, C.E., Wubben, E.F., Dewes, H.: The composition and impact of stakeholders' agendas on US ethanol production. Energy Policy 50, 647–658 (2012)

21. Talamini, E., Dewes, H.: The macro-environment for liquid biofuels in Brazilian science and public policies. Sci. Public Policy 39, 13–29 (2012)

22. Talamini, E., Wubben, E.F., Dewes, H.: The macro-environment for liquid biofuels in german science, mass media and government. Rev. Eur. Stud. 5(2), 33 (2013)

23. MacQueen, J.: Some methods for classification and analysis of multivariate observations. In Proceedings of the Fifth Berkeley Symposium on Mathematical Statistics and Probability, vol. 1, no. 14, pp. 281–297, June 1967

24. Quinn, K.M., Monroe, B.L., Colaresi, M., Crespin, M.H., Radev, D.R.: How to analyze political attention with minimal assumptions and costs. Am. J. Polit. Sci. 54(1), 209–228 (2010)

25. Boussalis, C., Coan, T.G.: Text-mining the signals of climate change doubt. Glob. Environ. Change 36, 89–100 (2016)

26. Grimmer, J.: A Bayesian hierarchical topic model for political texts: measuring expressed agendas in senate press releases. Polit. Anal. 18(1), 1–35 (2010)

27. Silva, M.J., Carvalho, P., Sarmento, L., de Oliveira, E., Magalhaes, P.: The design of OPTIMISM, an opinion mining system for Portuguese politics. In: New Trends in Artificial Intelligence: Proceedings of EPIA, pp. 12–15 (2009)

28. Thomas, M., Pang, B., Lee, L.: Get out the vote: determining support or opposition from congressional floor-debate transcripts. In Proceedings of the 2006 Conference on Empirical Methods in Natural Language Processing, pp. 327–335. Association for Computational Linguistics, July 2006

29. Taddy, M.: Measuring political sentiment on twitter: factor optimal design for multinomial inverse regression. Technometrics **55**(4), 415–425 (2013)
30. Malouf, R., Mullen, T.: Taking sides: user classification for informal online political discourse. Internet Res. **18**(2), 177–190 (2008)
31. Godbole, N., Srinivasaiah, M., Skiena, S.: Large-scale sentiment analysis for news and blogs. ICWSM **7**(21), 219–222 (2007)
32. Balahur, A., et al.: Sentiment analysis in the news. arXiv preprint arXiv:1309.6202 (2013)
33. Dang-Xuan, L., Stieglitz, S., Wladarsch, J., Neuberger, C.: An investigation of influentials and the role of sentiment in political communication on twitter during election periods. Inf. Commun. Soc. **16**(5), 795–825 (2013)
34. Park, S.J., Lim, Y.S., Sams, S., Nam, S.M., Park, H.W.: Networked politics on Cyworld: the text and sentiment of Korean political profiles. Soc. Sci. Comput. Rev. **29**(3), 288–299 (2011)
35. Tumasjan, A., Sprenger, T.O., Sandner, P.G., Welpe, I.M.: Predicting elections with twitter: what 140 characters reveal about political sentiment. ICWSM **10**, 178–185 (2010)
36. Lim, J.: A cross-lagged analysis of agenda setting among online news media. Journal. Mass Commun. Q. **83**(2), 298–312 (2006)
37. Dotson, D.M., Jacobson, S.K., Kaid, L.L., Carlton, J.S.: Media coverage of climate change in Chile: a content analysis of conservative and liberal newspapers. Environ. Commun. J. Nat. Cult. **6**(1), 64–81 (2012)
38. Meraz, S.: Is there an elite hold? Traditional media to social media agenda setting influence in blog networks. J. Comput. Mediat. Commun. **14**(3), 682–707 (2009)
39. Wu, Y., Atkin, D., Lau, T.Y., Lin, C., Mou, Y.: Agenda setting and micro-blog use: an analysis of the relationship between Sina Weibo and newspaper agendas in China. J. Soc. Media Soc. **2**(2), 8–25 (2013)
40. Ali, S.R., Fahmy, S.: Gatekeeping and citizen journalism: the use of social media during the recent uprisings in Iran, Egypt, and Libya. Media War Confl. **6**(1), 55–69 (2013)
41. Messner, M., Garrison, B.: Study shows some blogs affect traditional news media agendas. Newsp. Res. J. **32**(3), 112–126 (2011)
42. Grimmer, J., Stewart, B.M.: Text as data: the promise and pitfalls of automatic content analysis methods for political texts. Polit. Anal. **21**(3), 267–297 (2013)
43. Porter, M.F.: An algorithm for suffix stripping. Program **14**(3), 130–137 (1980)
44. Pang, B., Lee, L.: Opinion mining and sentiment analysis. Found. Trends Inf. Retr. **2**(1–2), 1–135 (2008)
45. Granger, C.W.: Investigating causal relations by econometric models and cross-spectral methods. Econom. J. Econom. Soc. 424–438 (1969)
46. Lee, S.Y.L., Gholami, R., Tong, T.Y.: Time series analysis in the assessment of ICT impact at the aggregate level–lessons and implications for the new economy. Inf. Manag. **42**(7), 1009–1022 (2005)
47. Luo, X., Zhang, J., Duan, W.: Social media and firm equity value. Inf. Syst. Res. **24**(1), 146–163 (2013)
48. Dutta, A.: Telecommunications and economic activity: an analysis of Granger causality. J. Manag. Inf. Syst. **17**(4), 71–95 (2001)

Modelling the Efficacy of Auto-Internet Warnings to Reduce Demand for Child Exploitation Materials

Paul A. Watters[✉]

La Trobe University, Melbourne, VIC 3068, Australia
p.watters@latrobe.edu.au

Abstract. A number of proposals have been made over the years to implement notification systems to modify user behaviour, by sensitising users to the fact that their activities are not anonymous, and that further consequences may follow from future detections of illicit activity. While these systems can be automated to a large extent, there is a degree of manual processing required, so the cost-effectiveness and potential user coverage of such controls is critical. In this paper, we consider the problem of sensitising entrenched paedophiles who search for and download large amounts of Child Exploitation Material (CEM). Some countries, like New Zealand, operate a centralised censorship system which could be used to issue notifications when entrenched paedophiles search for CEM. We develop a statistical model to determine how many notices would need to be sent to entrenched paedophiles to ensure that they receive at least one notification over a 12 month period. The estimate of CEM viewers is based on actual data from the New Zealand internet filter. The modelling results indicate that sending 9,880 notices would result in entrenched paedophiles receiving at least one notice; for average CEM users, 53.27% of users would receive at least one notice within 12 months.

Keywords: Child exploitation · Cost benefit analysis · Habituation

1 Introduction

Viewing Child Exploitation Material (CEM) is a serious crime in many countries. Not only does the offence create victims through the forced sexual servitude of the young people involved (Taylor and Quayle 2008), but it also results in secondary victimisation, as the same images and videos are downloaded by a small (but dedicated) cohort of viewers, often years after the original filming took place (Cale et al. 2017). Recent data (Quayle 2017) indicates that aggravated and unwanted sexual solicitations online are increasing, as are arrests for investigation of child exploitation offences (Wolak et al. 2012). Campbell (2016) suggests that we have now reached epidemic levels, with digital vigilantism supplementing the work of law enforcement in identifying offenders. There appears to be a CEM subculture arising, which presents challenges for law enforcement and society at large (Prichard et al. 2013). A study of CEM websites found no attempt to mask intentions (Westlake et al. 2016).

© Springer Nature Switzerland AG 2018
M. Ganji et al. (Eds.): PAKDD 2018, LNAI 11154, pp. 318–329, 2018.
https://doi.org/10.1007/978-3-030-04503-6_31

A significant amount of research has been conducted to investigate mechanisms for reducing the supply of CEM, including home and ISP-based content filters (Ho and Watters 2004). An alternative approach, based on situational crime prevention theory (Clarke 2008), asks whether reducing demand for the material may be a more effective approach, or whether both supply and demand reduction may be the best overall approach. For example, governments have increased taxes on tobacco products in recent years, which has led to a large increase in cigarette prices, leading to people smoking fewer cigarettes (Choi and Boyle 2017). In tandem, to reduce demand, plain packaging or explicit messages and photos illustrating the harms of smoking have also been very effective (Alamar and Glantz 2006). A combined strategy may be effective in reducing CEM, yet demand reduction has been under-studied.

One idea to reduce demand is to issue auto-internet warnings of some kind. This could range from letters being sent in the post (possibly with a Police letterhead), through to e-mails, or pop-up messages of various kinds. These messages could be generated by a range of bodies, including governments, ISPs, corporate networks, public libraries etc. There are two issues that need to be addressed in relation to the effectiveness of messaging – (1) is there a sound psychological basis for believing that messaging will reduce demand, and (2) is such a scheme cost-effective?

There are well-researched psychological models believing that messaging will result in behavioural change to reduce CEM demand. Downloading and sharing CEM has become a *habitual* behaviour for many users, since there are few adverse consequences resulting from the activity, and social attitudes to the behaviour are varied (Prichard et al. 2015). In psychology, this phenomenon is known as habituation (Bouton 2007). To modify the behaviour of users, it is necessary to *sensitise* them by notifying them that their *individual* behaviour is unacceptable. In general, multiple sensitisations are more effective than a single instance, since people's natural instinct is to form habits. Thus, ensuring that an intervention can sensitise users on multiple, sequential occasions is essential.

In other areas of electronic crime, sensitisation has been identified as a key strategy to modify the behaviour of users. Watters (2009), for example, identified how users could be sensitised to prevent them clicking on inappropriate links in phishing e-mails. In an ideal world, every time a user participates in an illegal activity, the account holder would receive a notification immediately or as soon as possible after the activity has begun. This could be in the form of a pop-up, but the widespread use of Secure Socket Layer (SSL) technology would render third-party injection impossible, unless an iFrame belonging to an interested party was available, such as an advertising banner. Another alternative would be to issue notices may be done manually, with the co-operation of ISPs, for example. However, manual processes – by their nature – are more resource and time intensive than an equivalent automated process.

If such schemes are psychologically sound, are they likely to be effective, and will they cost too much to implement? Intuitively, the cost of messaging has declined over the past two decades with the rise of "free" e-mail services. Yet these services are not free, they are paid for by advertising revenue generated from users viewing and clicking. In fact, the capital and operational costs are very significant. In a cognate area, reducing demand for music and film piracy, a scheme to issue notices via ISPs was recently disbanded, after it was found that the cost would be between $16 and $20 per notice (Whingham 2016). While civil losses are incurred in film piracy, the primary and secondary victimisation of children requires the careful consideration (and modelling) of the costs involved in reducing demand by issuing notices.

In this paper, the use of multiple, sequential notifications of CEM content downloading for users as a means to modify behaviour through sensitisation is considered. The cost of a notification system is modelled, such that a user is issued a notice from a centralised body, such as a censorship department, or an ISP, with a certain underlying rate of issuance. For example, if an ISP determined to send 200 notices per month, they could notify users from the first 200 detections, or randomly notify 200 from the pool of all detections during any month. An alternative would be to develop a pop-up warning system, subject to the caveats outlines above.

The key question is how many notices need to be sent every month so that CEM users will receive one or more messages within a reasonable timeframe, such as one year. Or, the goal can be phrased in terms of effectiveness: how likely is it for users to receive a message given a certainly monthly rate?

The model for the effective of auto-Internet warnings presented in this paper based on simple probabilities, and some assumptions about the underlying rates of downloading behaviour, especially those of the most dedicated (heavy) consumers. Where repeated attempts to modify behaviour fail, these consumers may then be termed *entrenched* users (Clark 1992). The goal of an intervention program should be to implement a treatment such that prospective heavy consumers are *prevented* from becoming entrenched users through sensitisation.

The proposed model has limitations: for example, it may be that entrenched users form a "hard core" element that simple messaging is unlikely to persuade. However, other categories of user, such as those who reach CEM sites through curiosity, are more likely to be receptive to internet warnings, and the model could forecast how many repeated warnings a user is likely to receive in a given timeframe.

Two research questions are posed on this paper:

1. How many notices need to be sent by an ISP on a monthly basis to ensure that an *average heavy consumer* would receive 1, 2, 3 and 4 notices over 12 months where an account holder would only ever receive one notice per month from their ISP, regardless of the amount of notices sent from law enforcement, and therefore regardless of the number of CEM downloads which may be identified to their account? ($\rightarrow$ RQ1)
2. What is the probability of an *average heavy consumer* receiving 1, 2, 3 and 4 notices over a 12 month period using the following volume of notices sent to an ISP per month: 75, 100, 200, 500, 1,000, or 2,000 where an account holder would only ever receive one notice per month from their ISP, regardless of the amount of notices sent from law enforcement and therefore regardless of the number of CEM downloads which may be identified to their account? ($\rightarrow$ RQ2)

To answer these questions about heavy CEM consumers, we firstly need to answer the same questions about average users, and then determine a method to map the average user results to heavy consumer results.

2 Methods

The number of downloads per user are assumed to be normally distributed with a mean X and standard deviation σ. A realistic scenario for downloading for an average user might be that they download roughly one new CEM movie per week – perhaps on a Saturday night – so we assume X = 5 (since the majority of months have 5 weeks). This is a gross simplification, since users are likely to engage in a wide range of behaviours involving CEM, such as viewing images, live cams etc., yet it offers the attractiveness of streamlining the modelling.

We can use the characteristics of the normal distribution to interpret the characteristics of an "average" user, as well as users at the extremities (including heavy consumers). According to the "68-95-99.7 Rule" (Fraenkel and Wallen 2005), we would anticipate the following characteristics of users:

- 68.27% would be within one standard deviation from the mean. Thus, for X = 5 and σ = 1.5, 68.27% of users would be downloading between 3.5 and 6.5 titles per month.
- 95.45% would be within two standard deviations from the mean. Thus, for X = 5 and σ = 1.5, 95.45% of users would be downloading between 2 and 8 titles per month.
- 99.73% would be within three standard deviations from the mean. Thus, for X = 5 and σ = 1.5, 99.73% of users would be downloading between 0.5 and 9.5 titles per month.

There are two good reasons to select the normal distribution as the basis for this model:

- Many distributions of human behaviour have an excellent "fit" to the normal (Gaussian) distribution, including human intelligence (Neisser 1997)
- The lower bound where x < 1.5 (being less than one download on average per month) is a hard limit (i.e., if user is downloading less than one title per month, they're not really downloading anything!). This, in turn, allows us to infer that the upper bound of x > +3σ is 9.5 titles per month, i.e., we define heavy consumers as those that download at the rate of three standard deviations above the mean, i.e., five or more titles per month on average.
- Since 0.135% of users will be downloading 5 or more titles per month, it seems reasonable to define an entrenched user as one whose activity is three standard deviations above the mean.

In the following analysis, we use the average downloads per month per user and the normal distribution to estimate the proportion of an average user's sharing activity, and the probability of an average user receiving a certain number of notices (on average), and the probability of a user receiving a certain number of notices where the number of notices varies between 75–10,000 per month. The number of users is derived from counts of incidents and estimated behaviour (assumed) and not from looking at IP addresses.

The analytical process is described in the following steps:

1. Estimate the detected suspected downloads for the ISP ($\sum N$) over the sample period.
2. Determine the number of months in the sample period ($M = 12$).
3. Estimate the average monthly downloads per user ($X = 5$).
4. Estimate the dispersion of average monthly downloads per user ($\sigma = 1.5$).
5. The total downloads for an average user are given by the product of the number of months and the average downloads per month ($M \times X = 90$).
6. An average user's downloads as a proportion of the total downloads can then be derived by dividing the total average user downloads (90) by the total number of downloads for the ISP (ρ).
7. Conversely, the total number of sharing users can be derived by the inverse, i.e., by dividing the total number of downloads by the average user downloads (ρ^{-1}).
8. The number of notices that needs to be sent to the *average* user to ensure that they would receive 1, 2, 3 and 4 notices (one per month over the sample period of one year) is a multiple of the number of notices and the total number of users, i.e., $1 \times \rho^{-1}, 2 \times \rho^{-1}, 3 \times \rho^{-1}, 4 \times \rho^{-1}$.
9. The probability of an average user receiving a notices with b notices sent per month is then given by dividing the number of notices received by the number of users, further divided by the number of notices sent ($a \div \rho^{-1} \div b$). As the number of notices sent increases, the likelihood of receiving a notice increases, while the likelihood of receiving a larger number of notices decreases in proportion to the number of notices to be received.
10. The probability of receiving a notice is limited to 100% (i.e. 1 notice as a maximum can be received by an account holder in any calendar month).

We then repeat the analysis for heavy consumers, by using the ratio of titles downloaded by an average user ($X = 5$) to the titles downloaded by a heavy consumer ($X = 5 + 3\sigma = 9.5$). The estimates determined for average users can then be multiplied by this ratio (9.5:5, or 1.9). Generally, as a heavy consumer downloads more titles, they are proportionally more likely to receive a notice under the same conditions as an average user. Conversely, since heavy consumers have a greater chance of receiving a notice, you need to send proportionally fewer to have the same impact as on an average user.

The different variables used in the analysis are summarised in Table 1.

Table 1. Model variables and their derivation.

Item	Variable	Value
Sample Period (months)	M	12
Number of downloads per month, per ISP	$N_1, N_2, \ldots N_M$	
Total Downloads per ISP	$\sum N$	
Assumed mean downloads per user per month	X	5
Downloads for any user	X	
Assumed standard deviation of downloads	Σ	1.5
Users within 1 SD from mean	$-\sigma < x < \sigma$	68.27%
Users within 2 SDs from mean	$-2\sigma < x < 2\sigma$	95.45%
Users within 3 SDs from mean	$-3\sigma < x < 3\sigma$	99.73%
Total average user downloads	$M \times X$	60
Proportion of average user's sharing activity	$\frac{M \times X}{\sum N} = \rho$	
Number of sharing users	$\frac{\sum N}{M \times X} = \rho^{-1}$	
Number of notices (1 notice per user on average, M)	$1 \times \rho^{-1}$	
Number of notices (2 notice per user on average, M)	$2 \times \rho^{-1}$	
Number of notices (3 notice per user on average, M)	$3 \times \rho^{-1}$	
Number of notices (4 notice per user on average, M)	$4 \times \rho^{-1}$	
Probability of an average user receiving a notices over the sample period with b notices sent per month	$\frac{a \div \rho^{-1}}{b}$	
Downloads per month for heavy consumers (3 SDs above mean)	$X + 3\sigma$	9.5
Ratio of heavy to average downloads	$\frac{X + 3\sigma}{X}$	1.9
Number of notices for heavy consumers (1 notice per user on average, M)	$\frac{1 \times \rho^{-1}}{1.9}$	
Number of notices for heavy consumers (2 notice per user on average, M)	$\frac{2 \times \rho^{-1}}{1.9}$	
Number of notices for heavy consumers (3 notice per user on average, M)	$\frac{3 \times \rho^{-1}}{1.9}$	
Number of notices for heavy consumers (4 notice per user on average, M)	$\frac{4 \times \rho^{-1}}{1.9}$	
Probability of an average heavy consumer receiving a notices over the sample period with b notices sent per month	$\frac{a \div \rho^{-1}}{b} \times \frac{1}{1.9}$	

3 Data

Understandably, getting reliable, hard data on downloads for CEM is very challenging. One industry estimate suggests that 116,000 searches per year are for CEM[1], but no detailed methodology is provided. The best estimate for New Zealand comes from a

---

[1] http://www.webroot.com/nz/en/home/resources/tips/digital-family-life/internet-pornography-by-the-numbers.

study that revealed 13,515,639 searches for child exploitation material were blocked by the New Zealand internet filter[2]. Given that during the release of Harry Potter movies, searches for CEM on isohunt.com were more popular (Prichard et al. 2011), this seems to be a great underestimate. But for the purposes of our simulation, this serves as a very conservative baseline estimate.

4   Results

Results here are presented for the assumed case of $\sum N$ = 13,515,639, ρ^{-1}=150,174.

RQ1 (Average Users). How many notices need to be sent by an ISP on a monthly basis to ensure that an average user would receive 1, 2, 3 and 4 notices over 12 months where an account holder would only ever receive one notice per month from their ISP, regardless of the amount of notices sent which may be related to their account?

The average user will download 5 files per month over 12 months, giving 60 downloads over the year. Further, the number of CEM consumers can then be inferred by dividing the number of suspected CEM downloads by the average number of titles per year:

$$\frac{\sum N}{M \times X} = \frac{13,515,639}{60} = 225,261 \text{ CEM consumers} = \rho^{-1}$$

Thus, the number of notices to be sent over 12 months is the product of the number of notices to be sent and the total number of CEM consumers. To determine the number of monthly notices to be sent, we divide by 12, as shown in Table 2.

Table 2. Notices to be sent to ISP each month to ensure that the average user receives 1–4 notices over 18 months where an average heavy consumer can receive a maximum of 1 notice per month.

Notices to be sent to an ISP to ensure that an 'average user' (as defined above) receives 1–4 notices over a 12 month time frame	Number of notices sent to ISP (monthly)
1st notice	18,772
2nd notice	37,543
3rd notice	56,315
4th notice	75,087

RQ2. What is the probability of an *average* user (defined above) receiving 1, 2, 3 and 4 notices over a 12 month period using the following volume of notices sent to an ISP per month: 75, 100, 200, 500, 1,000, or 2,000 where an account holder would only

---

[2] http://www.3news.co.nz/nznews/search-for-child-pornography-becoming-normalised-2011111317#axzz3XuPUBLLN.

ever receive one notice per month from their ISP, regardless of the amount of notices sent from law enforcement and therefore regardless of the number of suspected CEM downloads which may be identified to their account?

For one notice, this can be computed by dividing the number of users by the number of notices desired to be received over the sample period, e.g.,

$$p = a \div \rho^{-1}$$

where a is the number of notices sent over the sample period and ρ^{-1} is the number of users. Thus, 75 notices sent per month gives 900 over the sample period, thus the probability of an average user receiving one notice over 12 months with 75 notices sent per month is:

$$p = (75 \times 12 \div 225,261) = 0.0333$$

The results for all combinations of notices received and notices sent are shown in Table 3.

Table 3. Notices to be sent to ISP each month to ensure that the average account user receives 1–4 notices over 18 months where an average user can receive a maximum of 1 notice per month.

			Notices received		
	75	0.40%	0.20%	0.13%	0.10%
	100	0.53%	0.27%	0.18%	0.13%
Monthly	200	1.07%	0.53%	0.36%	0.27%
Notices	500	2.66%	1.33%	0.89%	0.67%
Sent	1000	5.33%	2.66%	1.78%	1.33%
	2000	10.65%	5.33%	3.55%	2.66%
	10000	53.27%	26.64%	17.76%	13.32%

RQ1 (Heavy Users). How many notices need to be sent by an ISP on a monthly basis to ensure that an *average heavy consumer* would receive 1, 2, 3 and 4 notices over 12 months where an account holder would only ever receive one notice per month from their ISP, regardless of the amount of notices sent from law enforcement, and therefore regardless of the number of suspected CEM downloads which may be traced to their account?

This can be estimated by dividing the number of notices per average user by the ratio of heavy to average downloads (9.5:5). Thus, the number of notices to be sent (once per month) is shown in Table 4.

Table 4. Notices to be sent to users each month to ensure that the average heavy consumer receives 1–4 notices over 12 months where an average heavy consumer can receive a maximum of 1 notice per month.

Notices to be sent to an ISP to ensure that an 'heavy consumer' (as defined above) receives 1–4 notices over a 12 month time frame	Number of notices sent to ISP (12 months)
1	9,880
2	19,760
3	29,640
4	39,519

RQ2. What is the probability of an *average heavy consumer* receiving 1, 2, 3 and 4 notices over a 12 month period using the following volume of notices sent to an ISP per month: 75, 100, 200, 500, 1,000, or 2,000 where an account holder would only ever receive one notice per month from their ISP, regardless of the amount of notices sent from law enforcement and therefore regardless of the number of suspected downloads which may be associated with their account?

As for RQ1, this can be estimated by using in the number of notices per average user by the ratio of heavy to average downloads (9.5:5). The numbers of notices to be sent are shown in Table 5.

Table 5. Notices to be sent to users each month to ensure that the average heavy consumer receives 1–4 notices over 12 months where an average heavy consumer can receive a maximum of 1 notice per month.

				Notices Received	
	75	0.76%	0.38%	0.25%	0.19%
	100	1.01%	0.51%	0.34%	0.25%
Total	200	2.02%	1.01%	0.67%	0.51%
Notices	500	5.06%	2.53%	1.69%	1.27%
Sent	1000	10.12%	5.06%	3.37%	2.53%
	2000	20.24%	10.12%	6.75%	7.59%
	10000	101.22%	50.61%	33.74%	37.96%

5   Discussion

In this paper, we have developed a simple model for determining the potential reach of auto-Internet warnings for CEM users, based on data from the New Zealand censorship filter. We used the theory of sensitisation as a means to modify behaviour, since repeated notices received by heavy users should alert them to the potential consequences following future detections. We have presented a means to determine the cost effectiveness in terms of potential coverage of users who download CEM, ensuring that heavy users receive up to four notices over a 12 month period, and to estimate the likelihood of heavy users receiving between 1–4 notices over a 12 month period, given a range of different notices being sent (75–10,000).

In summary, issuing approximately 10,000 notices would result in all entrenched users receiving at least one notification message over a 12 month period, and 53.27% of average users. This modelling clearly shows the effectiveness of such schemes in sensitising both average and entrenched users. However, we acknowledge that "hard core" entrenched users are unlikely to respond to internet warnings with the same response rate as average users. If an ISP administered the scheme manually, there would be an obvious cost. Even if the system could be automated, such as using advertising banners, there would also be a cost, depending on whether "cost per click" or "cost per view" placements were used. Using the estimate of $16–20 to generate a notice, the scheme would cost around $160,000–$200,000 to implement.

There are limitations to any methodology, and we note the following:

- Although the normal distribution is an ideal one for modelling, because its properties are well known and well established, it is possible that the behaviour of heavy consumers does not follow the patterns anticipated by this distribution. For example, heavy consumers might follow the pattern of the Pareto distribution (otherwise known as the 80-20 rule, where 80% of activity is due to 20% of users). This would mean that we may be under (or over) estimating the impact of notices being sent to this particular user group. Further empirical research is required to determine the actual distribution of real consumers.
- We have assumed that heavy consumers are 3σ above the mean; this is a subjective assessment, and even assuming that the data are normally distributed, it may be more realistic to consider other definitions, such as the Top 10% being classed as heavy.
- An alternative would be to use a simulation methodology, such as a Monte Carlo simulation, that provided deeper insight into specific cases, such as the difference between heavy and normal consumers. Such a model might take into account the actual variation of individual user behaviour. For example, we currently only assume that an average user consistently downloads the same number of titles every month. In reality, there is bound to be much more variation at the level of an individual user. Indeed, some heavy users may have a heavy downloading month and potentially download nothing during the following month. This variation is likely to have an impact on the results presented in this study.
- The data excludes non-monitored and unknown CEM titles; thus, it is only a subset of the overall CEM downloading activity taking place. Unsupervised learning techniques could be employed to increase the overall detection rate of previously unknown CEM titles.
- The methodologies used in this report are based on simple probability and some assumptions about the mean and dispersion of file downloading behaviour. In real life, these assumptions may not be met. For example, it is entirely possible that an assumed mean of 5 is either too high or too large. In this case, the results would change depending on whether this mean was increased or decreased respectively.

- The methodology assumes that notices are to be sent then selected essentially a random. This is not the most effective strategy for a cost effective intervention, given that we know that heavy consumers download more than normal users. Further research is required to further automate the notification process, even if some techniques still require user intervention. Numerous machine learning technologies could be applied to enhance the selection of heavy consumers, in much the same way as those technologies are used to detect spam e-mail, malware (Alazab et al. 2011), phishing e-mails (Layton et al. 2010) etc.

The direct impact of changes in user behaviour resulting from the receipt of notices has not been factored in; downloading behaviour is simply averaged across the entire sampling period. Again, Monte Carlo simulation could be used to estimate the impact of receiving notices, possibly using well-established mathematical models of habituation and sensitisation as a guide to parameterisation.

Acknowledgments. This project is supported by DP160100601 "Automated internet warnings to prevent viewing of minor-adult sex images" - with J.Prichard, C. Spiranovic, T. Krone and R. Wortley. I would like to acknowledge the valuable feedback and contributions of Drs Prichard, Spiranovic, Krone and Wortley in the preparation of this manuscript.

References

Alamar, B., Glantz, S.A.: Effect of increased social unacceptability of cigarette smoking on reduction in cigarette consumption. Am. J. Public Health **96**(8), 1359–1363 (2006)

Alazab, M., Venkatraman, S., Watters, P.A., Alazab, M.: Zero-day malware detection based on supervised learning algorithms of API call signatures. In: Proceedings of the 9th Australian Data Mining Conference (2011)

Bouton, M.E.: Learning and Behavior: A Contemporary Synthesis. Sunderland, Sinauer, MA (2007)

Cale, J., Burton, M., Leclerc, B.: Primary prevention of child sexual abuse: applications, effectiveness, and international innovations. In: Winterdyk, J.A. (ed.) Crime Prevention: International Perspectives, Issues, and Trends, pp. 91–114. CRC Press, Boca Raton (2017)

Campbell, E.: Policing paedophilia: assembling bodies, spaces and things. Crime Media Cult. **12**, 345–365 (2016). https://doi.org/10.1177/1741659015623598

Choi, K., Boyle, R.G.: Changes in cigarette expenditure minimising strategies before and after a cigarette tax increase. Tob. Control **27**(1), 99–104 (2017)

Clark, T.: Entrenched paedophiles: treating the untreatable. Aust. J. Forensic Sci. **24**, 31–34 (1992)

Clarke, R.: Situational crime prevention. In: Wortley, R., Mazerolle, L. (eds.) Environmental Criminology and Crime Analysis. Willan Publishing, Cullompton (2008)

Fraenkel, J., Wallen, N.: How to Design and Evaluate Research in Education. McGraw Hill, New York (2005)

Ho, W.H., Watters, P.A.: Statistical and structural approaches to filtering internet pornography. In: 2004 IEEE International Conference on Systems, Man and Cybernetics, vol. 5, pp. 4792–4798. IEEE (2004)

Layton, R., Watters, P.A., Dazeley, R.: Automatically determining phishing campaigns using the USCAP methodology. In: Proceedings of the 5th APWG E-crime Research Summit (2010)

Neisser, U.: Rising scores on intelligence tests. Am. Sci. **85**, 440–447 (1997)

Prichard, J., Spiranovic, C., Gelb, K., Watters, P.A., Krone, T.: Tertiary Education Students' Attitudes to the harmfulness of viewing and distributing child pornography. Psychiatry Psychol. Law (ahead-of-print), **23**(2), 224–239 (2015)

Prichard, J., Watters, P., Spiranovic, C.: Internet subcultures and pathways to the use of child pornography. Comput. Law Secur. Rev. **27**(6), 585–600 (2011)

Prichard, J., Spiranovic, C., Watters, P., Lueg, C.: Young people, child pornography, and subcultural norms on the internet. J. Am. Soc. Inf. Sci. Technol. **64**(5), 992–1000 (2013)

Quayle, E.: Over the internet, under the radar: online child sexual abuse and exploitation–a brief (2017)

Taylor, M., Quayle, E.: Criminogenic qualities of the internet in the collection and distribution of abuse images of children. Ir. J. Psychol. **29**(1–2), 119–130 (2008)

Watters, P.A.: Why do users trust the wrong messages? A behavioural model of phishing. In: Proceedings of the 5th eCrime Researchers Summit, pp. 1–7 (2009). https://doi.org/10.1109/ecrime.2009.5342611

Westlake, B.G., Bouchard, M., Girodat, A.: How obvious is it? The content of child sexual exploitation websites. Deviant Behav. **38**(3), 282–293 (2016)

Wolak, J., Finkelhor, D., Mitchell, K.J.: Trends in arrests for child pornography possession: the third national juvenile online victimization study (NJOV-3). Crimes against Children Research Center, University of New Hampshire, Durham, NH (2012)

Whingham, N.: Rights holders abandon 'three strikes' notice scheme as fresh piracy fight looms (2016). http://www.news.com.au/technology/online/piracy/rights-holders-abandon-three-strik es-notice-scheme-as-fresh-piracy-fight-looms/news-story/a0590bf35b9fc1c6d0e847b12b2cacf1

Data Mining for Energy Modeling and Optimization (DaMEMO)

Comparison and Sensitivity Analysis
of Methods for Solar PV Power Prediction

Mashud Rana[1]($\boxtimes$), Ashfaqur Rahman[2], Liwan Liyanage[3],
and Mohammed Nazim Uddin[4]

[1] Data61, CSIRO, Sydney, NSW, Australia
mdmashud.rana@data61.csiro.au
[2] Data61, CSIRO, Sandy Bay, TAS, Australia
ashfaqur.rahman@data61.csiro.au
[3] School of Computing, Engineering and Mathematics,
Western Sydney University, Sydney, Australia
l.liyanage@westernsydney.edu.au
[4] Department of Computer Science and Engineering, East Delta University,
Chittagong, Bangladesh
nazim@eastdelta.edu.bd

Abstract. The variable nature of solar power output from PhotoVoltaic
(PV) systems is the main obstacle for penetration of such power into the electricity grid. Thus, numerous methods have been proposed in the literature to
construct forecasting models. In this paper, we present a comprehensive comparison of a set of prominent methods that utilize weather prediction for future.
Firstly, we evaluate the prediction accuracy of widely used Neural Network
(NN), Support Vector Regression (SVR), k-Nearest Neighbours (kNN), Multiple Linear Regression (MLR), and two persistent methods using four data sets
for 2 years. We then analyze the sensitivities of their prediction accuracy to 10–
25% possible error in the future weather prediction obtained from the Bureau of
Meteorology (BoM). Results demonstrate that ensemble of NNs is the most
promising method and achieves substantial improvement in accuracy over other
prediction methods.

Keywords: Solar power prediction · Sensitivity analysis · Neural networks
Support Vector Regression · Nearest neighbours · Regression

1 Introduction

Solar power is one of the prominent sources of renewable energy and expected to
contribute to a major share of electricity generation in future. Many countries have been
installing large-scale solar PV plants and connecting them to grid to meet their electricity demand. For example, in 2015–2016 solar power generation in Australia
increased by 24% and accounted for 3% of its total energy generation [1]. By the year
2050, Australia also aims to generate 29% of the electricity from solar PV systems [2].

However, unlike the traditional sources of energy, power output from solar PV
plants fluctuates because of its dependency on meteorological conditions. This fluctuation imposes substantial challenges on achieving high level penetration of solar

© Springer Nature Switzerland AG 2018
M. Ganji et al. (Eds.): PAKDD 2018, LNAI 11154, pp. 333–344, 2018.
https://doi.org/10.1007/978-3-030-04503-6_32

power into the electricity grid [3]. A rapid unexpected change in the PV output power creates grid operational issues and negatively affects the security of supply. Reliable prediction of solar PV power output at different horizons is therefore critical to compensate the negative consequences related to variability in generation.

The prominent methods for solar power prediction are based on machine learning techniques such as NN (e.g. [4, 5]), SVR (e.g. [6]), and kNN (e.g. [7]); and statistical methods such as MLR and Autoregressive methods [8, 9]. Pedro and Coimbra [10] studied the performance of several methods for 1 and 2 h ahead prediction using a data set consisting of only previous solar power data. Although various weather information (such as solar exposure, temperature, and rainfall) have significant influence on the power output, they excluded the weather information and used only previous power data as inputs as their primary goal was to develop the baselines for further evaluation.

Long et al. [11] evaluated the performance of four methods using a data set that included weather parameters in addition to previous power as the inputs. However, they considered forecasting the cumulative power output for a day. Predicting daily total power output has very limited use in practical applications for real-time grid operation as the power output largely varies at different times of a day (see Fig. 1 showing the variability of solar power outputs at different times during the days) depending on the weather condition.

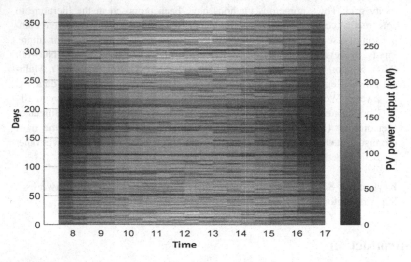

Fig. 1. Variability of solar PV power output at different times from 7:00 am to 5:00 pm for all days in 2015 (sample data set name UQC)

A review of the relevant literature suggests the predictive performance of the existing methods were evaluated on different data sets and for different forecasting scenarios (such as different prediction horizons). Hence, their computational accuracy reported in the literature is not readily comparable and does not convince to demonstrate the superiority of any single method over others. Therefore, it can be concluded that despite numerous approaches proposed and many notable achievements cited systematic comparison of different methods utilizing weather data accompanied by analysis of their sensitivities to error in future weather prediction is very limited. In this paper, we address this deficit in the literature and provide a comprehensive evaluation of a group of solar power forecasting methods that utilize weather prediction for future. In contrast to the previous work our contributions can be summarized as follows:

We compare a set of prominent methods for the task of predicting the output power profile for a day - i.e. predicting all the power outputs at half-hourly intervals for a day. Forecasting daily power profile is crucial for real time unit dispatching, gird reliability, and supporting energy trading in the deregulated markets. Firstly, we evaluate the 4 state-of-the-art and 2 baseline methods using 4 different data sets for two years 2015–2016 collected from the largest flat-panel grid connected PV plant in Australian. We then analyze the sensitivities of those methods to error in weather prediction for future by considering 10–25% noise in weather data obtained from BoM. This has been done to evaluate the robustness of the prediction methods in dealing the uncertainty associated with weather prediction. To the best of our knowledge this has not been investigated before in the literature.

2   Data Sets and Problem Statement

2.1   Solar Power Data

We use the data from the largest flat-panel PV system in Australia which has been located at the St Lucia campus of University of Queensland, Brisbane. It has a capacity of 1.22 MW and consists of about 5000 solar panels installed across four different sites: University of Queensland Centre (UQC), Sir Llew Edwards Building (LEB), Car Park 1 (CP1) and Car Park 2 (CP2). We use data from all four sites and consider the data from each site as a separate case study.

For each site, we collect the solar power output data in 1 min interval for 2 years – 2015 and 2016. For each day we use the data from 7:00 am to 5:00 pm since the solar power outputs have been recorded as either zero or not available outside this 10-h window, due to the absence (or very small amount) of solar irradiation. This leads to $2 \times 365 \times 600 = 438,000$ 1-min measurements for each site. The 1-min data is publicly available at [12]. The 1-min data was then aggregated into 30-min intervals by averaging every 30 consecutive measurements, resulting in 20 measurements per day and $2 \times 365 \times 20 = 14,600$ measurements in total for each site.

2.2   Weather Data

We also collect the meteorological data for the four variables – solar exposure, rainfall, sunshine hours, temperature and wind speed, for the same time period and from the nearest weather station of the PV sites, from BoM Australia [13]. These are among the widely cited weather variables affecting the solar power output from PV systems.

All data (PV power, solar exposure, rainfall, sunshine hours, temperature, and wind speed) has been normalized to the range of [0, 1].

2.3   Problem Statement

Given

1. a time series of previous solar power outputs up to the day d: $P = [P_1, P_2, P_3, \ldots, P_d]$, where $P_i = [p_i^1, p_i^2, p_i^3, \ldots, p_i^n]$ represents the power profile for day i, i.e. n observations of the power outputs measured at half-hourly intervals ($n = 20$ for our case);
2. a time series of previous weather data from BoM up to day d: $W = [W_1, W_2, W_3, \ldots, W_d,]$, where W_i is the weather data for day i. W_i is a 6-dimensional vector of the daily global solar exposure (SE), rainfall(R), sunshine hours (SH), maximum wind speed (WS), and maximum and minimum temperature (T), $W_i = [SE^i, R^i, SH^i, WS_{max}^i, T_{max}^i, T_{min}^i]$.
3. predicted weather data W_{d+1} for day $d + 1$.

   Goal: forecast P_{d+1} - half-hourly solar PV power profile for the next day $d+1$ using predicted weather data W_{d+1} for day $d + 1$ as input. It is important to note that unlike other methods, only kNN requires: (1) previous weather data up to day d to select the k neighbours (days) nearest to predicted weather data W_{d+1} for target day $d + 1$, and (2) previous power outputs of the k selected neighbours (days) to compute the prediction for P_{d+1}.

3   Methods

For comparison, we consider four most prominent methods in the literature: NN, SVR, kNN, and MLR, and two persistent methods as baselines. All the methods use forecasted weather profile as input to predict the daily solar power curve for next day.

3.1   NN

To build NN based prediction model, we use multilayer perceptron NN. Approaches based on such NN are the most popular for solar power prediction. NN can learn and estimate complex functional relationships between the input and output variables from examples. However, the performance of NN depends on the network architecture and the random initialization of weights. To reduce this sensitivity, we apply an ensemble of NNs. Ensembles are shown to be more accurate than a single NN in previous works (e.g. [14]) using the same data sets.

   To develop ensemble of NNs we follow [14, 15]. We first construct V different NN structures by varying the number of neurons in hidden layer from 1 to V. For each structure $S_{v \in V}$, we then build an ensemble E_v that consists of n NNs (we used $n = 10$). Each member of the ensemble E_v has the same structure S_v, i.e. the same number of hidden neurons, but is initialized to different random weights. Each of the n members of ensemble E_v has been trained separately on the training data using the Levenberg Marquardt (LM) algorithm. The NN training has been stopped when there is no improvement in the error for 20 consecutive epochs or a maximum number of 1000 epochs is reached.

To predict the half-hourly power outputs for a given day, the individual predictions of the ensemble members are combined by taking their median value, i.e. the prediction for time (half an hour) h for day $d + 1$ is: $\hat{P}_h^{d+1} = median\left(\hat{P}_{h,NN_1}^{d+1}, \ldots, \hat{P}_{h,NN_n}^{d+1}\right)$, where $\hat{P}_{h,NN_j}^{d+1}$ is the prediction for h generated by an ensemble member NN_j, $h = 1, \ldots, 20$ and $j = 1, \ldots, n$.

The performance of each ensemble E_v was evaluated on the validation set. The best performing ensemble, i.e. the one with the lowest prediction error, is then selected and used to predict testing data.

3.2   SVR

SVR is an advanced machine algorithm that has shown excellent performance for solar power forecasting [6, 16]. The key idea of SVR is to map the input data into a higher dimensional feature space using a non-linear transformation and then apply linear regression in the new space. The task is formulated as an optimization problem. The main goal is to minimize the error on the training data, but the flatness of the line and the trade-off between training error and model complexity are also considered to prevent overfitting.

Solving the optimization problem requires computing dot products of input vector in the new space which is computationally expensive in high dimensional spaces. To help with this, kernel functions satisfying the Mercer's theorem are used - they allow the dot products to be computed in the original lower dimensional space and then mapped to the new space.

Since SVR can have only one output, we divide the daily load curve prediction task into 20 subtasks (i.e. predicting power output for each half-hourly time separately) and build a separate SVR prediction model for each subtask.

The selection of kernel function is important for SVR, and is done by empirical evaluation. After experimenting with different kernel functions, we selected the Radial Basis Function (RBF) kernel as it achieved the best performance on the validation data.

3.3   kNN

kNN is an instance based method for forecasting. The main concept of kNN is selecting a subset of training examples whose inputs are similar to the inputs of test example, and use the outputs of that training subset to predict the outputs for the test example.

To forecast the power outputs for the next day $d + 1$, kNN firstly obtains the weather profile W_{d+1} for $d + 1$ from the weather forecast report. It then finds the k nearest neighbors of $d + 1$. This is done by matching the weather profile of all previous days ending with d and finding the k most similar days. This leads to a neighbour set $NS = \{q_1, \ldots, q_k\}$ where $q_1, \ldots, q_k$ are the k closest days to day $d + 1$, in order of closeness computed using a distance measure between weather profiles of the neighbors to that for $d + 1$. The prediction for the new day is the weighted linear combination of the power outputs for the k nearest neighbors:

$$\hat{P}_{d+1} = \frac{1}{\sum_{s \in NS} \alpha_s} \cdot \sum_{s \in NS} \alpha_s . P_s^h.$$

The weights α_s are computed by following [17]: $\alpha_s = \frac{dist\left(W_{q_k}, W_{d+1}\right) - dist(W_s, W_{d+1})}{dist\left(W_{q_k}, W_{d+1}\right) - dist\left(W_{q_1}, W_{d+1}\right)}$, where *dist* is the Euclidian distance measure, and k is the number of neighbors. The optimal value for k can be set by applying kNN method on the training data set, i.e. it is the one that minimizes the prediction error for training data set.

3.4   MLR

MLR is a classical statistical method for forecasting. It assumes linear relationship between the predictor variables and the variable that is predicted, and uses the least square method to find the regression coefficients. In this work we apply Weka's implementation of linear regression. It has an inbuilt mechanism for input variable selection based on the M5 method. This method firstly builds a regression model using all inputs and then removes the variables, one by one, in decreasing order of their standardized coefficient until no improvement is observed in the prediction error given by the Akaike Information Criterion (AIC). Similar to SVR, for MLR we train one model for each half-hourly time since MLR can have only one output.

3.5   Persistent Methods

We also implement two persistent methods as baselines for comparison.

The first persistent method (B_{msday}) firstly selects the most similar previous day (s) in the historical data based on the weather profile where the similarity is measured by Euclidean distance between the weather profiles of $d + 1$ and s. It then uses the power output for the day s as the predictions for the day $d + 1$. This means the prediction for $\hat{P}_{d+1} = \left(\hat{P}_{d+1}^1, \hat{P}_{d+1}^2, \ldots, \hat{P}_{d+1}^{20}\right)$ is given by $P_s = \left(p_s^1, p_s^2, \ldots, p_s^{20}\right)$. Obviously B_{msday} is same as kNN if only a single neighbour is considered in kNN.

The second persistent method (B_{pday}) considers the power outputs from the previous day d as the predictions for the next day $d + 1$, i.e. the predictions for $\hat{P}_{d+1} = \left(\hat{P}_{d+1}^1, \hat{P}_{d+1}^2, \ldots, \hat{P}_{d+1}^{20}\right)$ are given by $P_d = \left(p_d^1, p_d^2, \ldots, p_d^{20}\right)$.

4   Simulation Settings

4.1   Training and Testing Data

We divide the data for each case study into two non-overlapping subsets – *training* and *testing*. The training set consists of all the samples from the year 2015 and has been used to build the prediction models. This applies to all the models expect NN and SVR. For NN and SVR 90% of the samples from the training set has been used to train the models and remaining 10% (validation set) has been used for parameter selection (such as selecting the number of hidden neurons for NN and kernel for SVR). During the training phase we used actual weather data W_{d+1} for target day $d + 1$ as input since training was performed offline and we do not have access to the historical weather prediction.

On the other hand, testing set consists of all the samples from the year 2016 and has been used to evaluate the accuracy of the prediction models.

4.2 Evaluation Metrics

To evaluate the accuracy of forecasting models, we use Mean Relative Error (MRE). MRE is one of the widely cited measure for the accuracy of solar power prediction and defined as: $MRE = \frac{1}{D}\frac{1}{H}\sum_{d=1}^{D}\sum_{h=1}^{H}\left|\frac{p_d^h - \hat{P}_d^h}{R}\right| \times 100\%$.

where p_d^h and $\hat{P}_d^h$ are the actual and predicted power outputs for day d at time h, respectively; D is the number of instances (days) in the testing data; H is the total number of predicted power outputs for a day ($H = 20$ for our task), and R is the range of the power output.

For comparison of prediction models, we also compute the improvement in accuracy between two prediction models A and B as: $improvement(A, B) = \frac{|MRE_A - MRE_B|}{MRE_B} \times 100\%$.

5 Results and Discussion

Table 1 presents the accuracy (MRE) of the prediction models using actual weather data for target day as input as we do not have access to the historical weather prediction. The statistical significance for differences in accuracy for each pair of prediction models are also shown Table 2. Results show that ensemble of NNs is the most accurate model and outperforms all other prediction models including two baselines in all data sets, except SVR for EBD data set. The overall improvements of accuracy for NNs ensemble are 0.18–2.94% over SVR, 23.77–56.06% over kNN, 1.04–3.14% over MLR, 5.34–30.35% over B_{msday}, and 14.82–39.25% over B_{pday}. All the improvements of NNs ensemble are also statistically significant at $p \leq 0.05$ except the difference between NNs ensemble and SVR for CP2 data (see Table 2).

Table 1. Accuracy results (MRE) of all prediction models evaluated on four data sets.

	NN	SVR	kNN	MLR	B_{msday}	B_{pday}
CP1	9.43	9.50	19.19	9.53	11.72	11.95
CP2	7.75	7.77	17.65	7.98	11.13	13.46
EBD	12.01	11.89	15.76	12.24	12.69	14.10
UQC	8.45	8.70	16.38	8.72	11.26	13.90

SVR and MLR come next in the ranking with SVR being slightly better. Even though SVR and MLR shows similar performance, the difference in accuracy between them is statistically significant for all data sets except UQC. Besides, although NN shows the highest accuracy, the performance of SVR and MLR is not too far behind: MRE = 7.75–12.01% for NNs ensemble vs 7.77–11.89% for SVR and 7.98–12.24%

for MLR. In addition, both SVR and MLR outperform kNN and two baselines and their improvements over kNN and baselines are statistically significant too (see Table 2).

On the other hand, kNN provides the lowest accuracy among all the prediction models – they even unexpectedly outperformed by the baselines B_{msday} and B_{pday}. From the poor performance of kNN compared to B_{msday}, it can be said the using more than 1 similar day to compute the prediction next day is not quite beneficial.

Table 2. Statistical significance test (two-sample t-test) of pair-wise differences in accuracy for the prediction models: T = statistically significant at $p \leq 0.05$ and F = not statistically significant. Four letters in a cell indicates results on the four data sets – CP1, CP2, EBD and UQC respectively; for example T, T, F, T in row #1, column #2 indicate that difference in accuracy between NN and SVR are statistically significant for CP1, EBD, and UQC data sets, but not significant for CP2 data set.

	NN	SVR	kNN	MLR	B_{msday}	B_{pday}
NN		T, F, T, T	T, T, T, T	T, T, T, T	T, T, T, T	T, T, T, T
SVR			T, T, T, T	T, T, T, F	T, T, T, T	T, T, T, T
kNN				T, T, T, T	T, T, T, T	T, T, T, T
MLR					T, T, T, T	T, T, T, T
B_{msday}						T, T, T, T

Table 3. Accuracy of all prediction models after adding noise in weather data.

	NN	SVR	kNN	MLR	B_{msday}	B_{pday}
With 10% added noise in weather data						
CP1	9.32	9.72	18.98	9.85	12.60	11.95
CP2	8.17	8.39	18.31	8.72	12.14	13.46
EBD	12.50	12.50	17.05	12.70	14.37	14.10
UQC	8.90	9.49	17.55	9.57	12.35	13.90
With 15% added noise in weather data						
CP1	9.91	10.14	18.04	10.38	12.62	11.95
CP2	8.83	9.09	17.80	9.55	12.34	13.46
EBD	12.48	13.08	17.40	13.14	14.78	14.10
UQC	9.43	10.26	17.24	10.41	12.83	13.90
With 20% added noise in weather data						
CP1	10.14	10.72	17.52	11.12	12.62	11.95
CP2	9.42	9.90	17.55	10.54	12.20	13.46
EBD	13.16	13.82	17.52	13.74	14.37	14.10
UQC	10.16	11.12	17.04	11.42	12.69	13.90
With 25% added noise in weather data						
CP1	10.47	11.40	17.37	12.02	12.84	11.95
CP2	9.88	10.76	17.48	11.65	12.67	13.46
EBD	13.73	14.63	17.62	14.47	14.73	14.10
UQC	10.49	12.03	17.01	12.53	13.13	13.90

It is important to note that results in Table 1 have been computed using measured weather data for future as we do not have access to historical weather prediction. However, in practical applications, the models require predicted weather data W_{d+1} for day $d + 1$ from BoM. The performance of the solar power forecasting models substantially depends on the accuracy of such weather prediction. Therefore, to check the robustness of the prediction models and analyse their sensitivity to the error in weather prediction, we evaluate their accuracy by adding 10–25% random noise to measured weather data.

Table 3 presents accuracy of all prediction models after adding different level of noise in weather information in the test data sets. Figures 2, 3, 4 and 5 shows the comparison of how much the accuracy (MRE) for different models are affected by error in future weather prediction for CP1, CP2, EBD and UQC data sets respectively (kNN are excluded for better visualization since its MRE range is much higher). We can see that only the accuracy of B_{pday} is unaffected by the error in weather data since it considers the power outputs from previous day as the outputs for next day irrespective of the changes in weather.

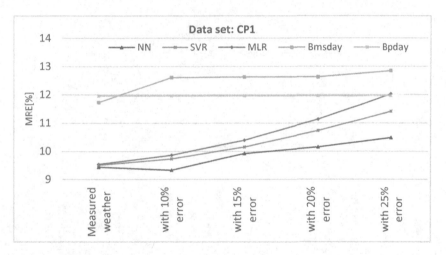

Fig. 2. MRE of prediction models for different level of error in weather data: CP1 data set

The prediction error (MRE) for all other models (except kNN) rises and shows an increasing trend as the error in the future weather information increases from 10–25%. The MRE after adding noise in the weather data reaches to the range of 8.17–13.73% for NNs ensemble, 8.39–14.63% for SVR, 8.72–14.47% for MLR, and 11.13–14.78% for B_{msday}.

On the other hand the MRE of kNN shows slight improvement or remains similar as error in weather goes up. Although it is unexpected and quite opposite to the case for remaining models, it does not make any difference in the ranking of models as the MRE of kNN is still far behind than two baselines.

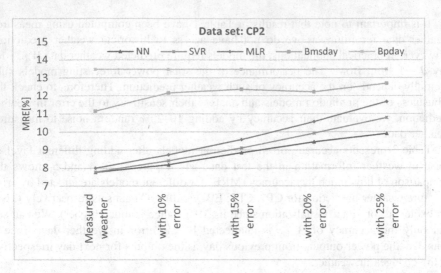

Fig. 3. MRE of prediction models for different level of error in weather data: CP2 data set

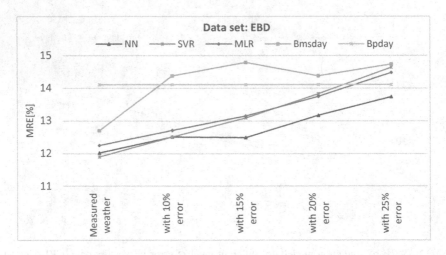

Fig. 4. MRE of prediction models for different level of error in weather data: EBD data set

Moreover, comparison of MRE results from Tables 1 and 3 shows that the ranking of all prediction models after adding noise to actual weather data follows the same order as it was before adding the noise: NNs ensemble being the most accurate followed by SVR, MLR, baselines, and kNN. Despite the performance of the all prediction models (except kNN) deteriorates after adding noise to weather data, the difference in accuracy between the NNs ensemble and other models becomes more prominent as the noise goes from 10–25% (see Figs. 2, 3, 4 and 5). This indicates that NN model is more robust in handling the error in weather prediction for future and able to forecast the solar power outputs for next day even the weather profile for next day does not exactly matches with the weather prediction obtained from BoM.

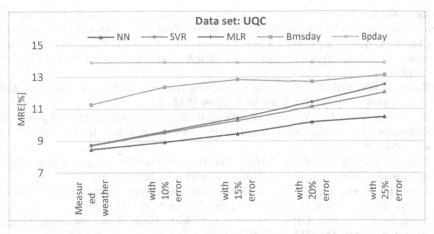

Fig. 5. MRE of prediction models for different level of error in weather data: UQC data set

In summary, considering the overall accuracy and ability to handle the uncertainty associated with the weather data for future, ensemble of NNs is the most effective among all the models used for comparison.

6   Conclusion

In this paper, we present a comprehensive assessment of a set of prominent methods for forecasting day ahead solar power output profile. We evaluate the performance of ensemble of NNs, kNN, SVR, MLR and two baselines using 4 different sets of data collected for 2 years. The presented results show that ensemble of NNs is the most accurate prediction method and achieves considerable improvement of accuracy over all other methods. Ensemble of NNs also has been found to be very successful in dealing the error in weather prediction - its performance is less sensitive to inaccuracies in weather prediction for future. Although the performance of SVR and MLR also found comparable to NNs, the difference in accuracy between any of these two models and NNs increases significantly as the error in weather forecast increases. Therefore, we conclude that ensemble of NNs is more viable for practical application for forecasting solar power outputs from PV systems.

References

1. Department of the Environment and Energy, Australian Government: Australian Energy Update (2017). https://www.energy.gov.au/sites/g/files/net3411/f/energy-update-report-2017.pdf. Accessed 05 May 2018
2. European Photovoltaic Industry Association: Connecting the Sun-Solar Photovoltaics on the Road to Large Scale Grid Integration. http://www.pvtrin.eu/assets/media/PDF/Publications/other_publications/263.pdf. Accessed 01 Aug 2017

3. Rana, M., Koprinska, I., Agelidis, V.G.: Univariate and multivariate methods for very short-term solar photovoltaic power forecasting. Energy Convers. Manag. **121**, 380–390 (2016)
4. Chu, Y., Urquhart, B., Gohari, S.M.I., Pedro, H.T.C., Kleissl, J., Coimbra, C.F.M.: Short-term reforecasting of power output from a 48 MWe solar PV plant. Sol. Energy **112**, 68–77 (2015)
5. Rana, M., Koprinska, I.: Neural network ensemble based approach for 2D-interval prediction of solar photovoltaic power. Energies **9**, 829–845 (2016)
6. Shi, J., Lee, W.-J., Liu, Y., Yang, Y., Wang, P.: Forecasting power output of photovoltaic systems based on weather classification and support vector machines. IEEE Trans. Ind. Appl. **48**, 1064–1069 (2012)
7. Wang, Z., Koprinska, I., Rana, M.: Clustering based methods for solar power forecasting. In: International Joint Conference on Neural Networks (IJCNN) (2016)
8. Yang, C., Thatte, A., Xie, L.: Multitime-scale data-driven spatio-temporal forecast of photovoltaic generation. IEEE Trans. Sustain. Energy **6**, 104–112 (2015)
9. Yang, D., Ye, Z., Lim, L.H.I., Dong, Z.: Very short term irradiance forecasting using the lasso. Sol. Energy **114**, 314–326 (2015)
10. Pedro, H.T., Coimbra, C.F.: Assessment of forecasting techniques for solar power production with no exogenous inputs. Sol. Energy **86**, 2017–2028 (2012)
11. Long, H., Zhang, Z., Su, Y.: Analysis of daily solar power prediction with data-driven approaches. Appl. Energy **126**, 29–37 (2014)
12. UQ Solar Photovoltaic Data. http://solar.uq.edu.au/user/reportPower.php. Accessed 01 Aug 2017
13. Climate Data Online. http://www.bom.gov.au/climate/data/. Accessed 01 Aug 2017
14. Rana, M., Koprinska, I., Agelidis, V.G.: Forecasting solar power generated by grid connected PV systems using ensembles of neural networks. In: International Joint Conference on Neural Networks (IJCNN), Killarney, Ireland (2015)
15. Rana, M., Koprinska, I., Agelidis, V.G.: Solar power forecasting using weather type clustering and ensembles of neural networks. In: International Joint Conference on Neural Networks (IJCNN), Canada (2016)
16. Rana, M., Koprinska, I., Agelidis, V.G.: 2D-interval forecasts for solar power production. Sol. Energy **122**, 191–203 (2015)
17. Lora, A.T., Santos, J.M.R., Exposito, A.G., Ramos, J.L.M., Santos, J.C.R.: Electricity market price forecasting based on weighted nearest neighbors techniques. IEEE Trans. Power Syst. **22**, 1294–1301 (2007)

An Improved PBIL Algorithm
for Optimal Coalition Structure
Generation of Smart Grids

Sean Hsin-Shyuan Lee[1], Jeremiah D. Deng[1(✉)], Martin K. Purvis[1],
Maryam Purvis[1], and Lizhi Peng[2]

[1] Department of Information Science, University of Otago, Dunedin, New Zealand
sean.hslee@postgrad.otago.ac.nz,
{jeremiah.deng,martin.purvis,maryam.purvis}@otago.ac.nz
[2] Shandong Provincial Key Laboratory of Network Based Intelligent Computing,
University of Jinan, Jinan, China
penglizhi.jn@gmail.com

Abstract. Coalition structure generation in multi-agent systems has long been a challenging problem because of its NP-hardness in computational complexity. In this paper, we propose a stochastic optimization approach that employs a modified population based incremental learning algorithm and a customized genotype encoding scheme to find the optimal coalition structure for smart grids with renewable energy sources. Empirical results show that the proposed approach gives competitive performance compared with existing solutions such as genetic algorithm and dynamic programming.

Keywords: Coalition structure generation
Smart grids · Optimization · Dynamic programming
Population-based incremental learning

1 Introduction

For a cooperative game with multiple agents, in particular when a grand coalition is impractical or less effective, a major interest is find out how to split agents into disjoint coalitions so that the system can enhance its operation, or obtain a maximized total profit from the consequent coalitions [4,11]. As a paradigm in cooperative game theory, coalition structure generation (CSG) [9], also known as "coalitional game in partition form", has received much attention in the literature. Specifically, CSG provides a mathematical framework for coalition formation on smart grids, where multiple types of renewable energies sources and various power consumption profiles make it a challenging task to find optimal coalition partitions.

However, the main contest for applying CSG is its NP-hardness in computational complexity. Probing an optimal solution by exiting approaches, e.g. Dynamic Programming (DP) [13], can only be practical when solving problem

© Springer Nature Switzerland AG 2018
M. Ganji et al. (Eds.): PAKDD 2018, LNAI 11154, pp. 345–356, 2018.
https://doi.org/10.1007/978-3-030-04503-6_33

of small scales. Many studies have been devoted to improve the efficiency of DP. For instance, Sandholm et al. [10] presented a partial search algorithm that guarantees the solution to be within a bound from the optimum; Changder et al. [5] used some heuristics to select optimal values from some subproblems, and chose the remained unassigned agents from other subproblems; Michalak et al. [7] combined DP with a tree-search algorithm to avoid the redundant processes in DP, which was reported to be the fastest exact algorithm for complete set partitioning. Even so, it is shown that DP variants such as Björklund et al. [3] and Michalak et al. [7] run in $O(2^n)$ and $O(3^n)$ time respectively, which restricts their application to problems with limited numbers of agents.

For large scale CSG, it remains impractical to search for a global optimum using the aforementioned approaches. Alternatively, stochastic optimization (SO) algorithms for CSG may provide a promising solution with great efficiency. In the field of energy system optimization, techniques such as particle swarm optimization have been employed to generate SO-based solutions for maximizing the energy production and meeting power demands with minimal costs and maximal reliability [8]. Although PSO has demonstrated the potentials for solving large-scale optimization problems, it is not suitable for CSG as it is designed for continual variable spaces.

So far there are few studies [9] in applying SO algorithms to CSG. Specifically, based on the population based incremental learning (PBIL) algorithm [1], we have proposed an improved algorithm, top-k merit weighting PBIL (PBIL-MW) [6], for solving CSG problems. The result has indicated the algorithm's promising ability both in accuracy and efficiency in comparison with other algorithms. Based on our previous approach, we have developed a new genotype encoding scheme for the improved PBIL algorithm. The new approach outperforms a few SO counterparts, such as GA and the original PBIL. Moreover, in comparison with DP, our approach has largely reduced the memory consumed in computation and shortened the running time significantly.

The rest of this paper is organized as follows. In Sect. 2 we first give the mathematical framework for forming a coalition structure of agents in smart grids. We then propose our SO-based solutions, especially the new algorithm, for exploring the optimal partition scheme. Some results of experiment are shown in Sect. 4, with the algorithm's performance compared with DP, PBIL and GA in terms of convergence speed and computational efficiency for larger scale optimization. In the end, we conclude the paper and point some possible further directions.

2   Coalition Model for Smart Grids

In comparison with other approaches for CSG in smart grids, we adopt our previous model [6] which assumes every agent to take part in the coalition should have sufficient renewable energy to meet its own demand in general. However, due to the irregularity of renewable energy, every agent could expose shortage frequently. Consequently, holding together with an agreement to share the surplus energy among others is a more economic solution in comparison with measures to expand the facility, install larger backup capacity or deal with power companies. Since the demand and supply are both dynamic, the model needs to engage a flexible mechanism to derive the optimal coalition formation endlessly. We assume all agents exchange their power consumption needs and power generation capabilities regularly, e.g., on an hourly basis.

2.1 Coalition Criterion

In every hourly period any agent a_i with extra power can share its surplus with shortage ones within the cooperative union. The target of that union is to maximize the total profit by forming coalitions. For the stability of power grids, the union enjoins every feasible coalition must have a power excess.

For instance, a_1 and a_3 each has 1.5 and 0.6 kWh excess accordingly, but a_2 has a shortage of 1.1 kWh. In terms of the requirement, a_1 and a_2 can team a feasible coalition $\{a_1, a_2\}$. To the contrary, a coalition such as $\{a_2, a_3\}$ is not rewarding. Therefore, a grand coalition cannot always be feasible and a model of CSG [11] should be constructed and needs to be resolved. Furthermore, our study is focused on local coalition, such as within a community, therefore transmission costs and power losses are ignored consequently.

2.2 Coalition Evaluation

Let S denote a set of n cooperative agents, a coalition C_k is a subset of S. A coalition structure CS is a collection of coalitions, where CS $= \{C_k\}$ such that $\cup_k C_k = S$, $C_k \neq \emptyset$, and $C_k \cap C'_k = \emptyset$ if $k \neq k'$. The coalition ID k goes from 1 to Bell(n), the Bell number. The Bell number has a generating function in the form [12]:

$$B(x) = e^{e^x} - 1. \tag{1}$$

To show its extraordinary growth rate, the first 16 Bell numbers are: 1, 2, 5, 15, 52, 203, 877, 4140, 21147, 115975, 678570, 4213597, 27644437, 190899322, 1382958545, 10480142147.

A coalition structure CS is said to be globally optimal for a characteristic function game G, if CS gives the maximal overall characteristic value [4]:

$$v(\text{CS}) = \sum_{C_k \in \text{CS}} v(C_k). \tag{2}$$

In our study, $v(C_k)$ is the fitness function of C_k given by

$$v(C_k) = \begin{cases} 0 & \text{if } |C_k| = 1, \\ Q_d \times P_r & \text{if } |C_k| > 1 \text{ and } Q(C_k) \geq 0, \\ -9999 & \text{otherwise}, \end{cases} \tag{3}$$

where $|C_k|$ denotes the size of coalition C_k, P_r represents the price difference between getting power from coalition and trading with power utilities, Q_d is the total power need for deficit agents in the C_k, and $Q(C_k)$ is the net surplus within the C_k. Furthermore, for giving penalty to an infeasible coalition, we let $v(C_k) = -9999$.

The goal is to arrive at the best CS and to maximize total profit. Certainly, an exhausted search (ES) for searching the optimal CS could be a solution, but will be impracticable for a large number of agents. Thus, as in [9], other options are essential for acquiring the optimum, or a near-optimal coalition structure practically.

3   Algorithmic Solutions

3.1   PBIL

First, we briefly introduce the original PBIL algorithm [1,2].

Let $\mathbf{p}$ denote the probability vector with random values within $[0, 1]$ for each component p_j. The length of $\mathbf{p}$, denoted by L, is equal to that of the genotype. In PBIL, every component p_j represents its threshold of obtaining a value 1 in the j-th component of the genotype.

At first, the probability vector, denoted by $\mathbf{p}^{(0)}$, is simply set with $p_j = 0.5$. During each iteration step, an $N \times L$ matrix $\mathbf{G}$ with entries of random numbers within $[0, 1]$ is chosen, where N is the size of population's samples. Let $\mathbf{G}_i$ denote the i-th genotype, corresponding to the i-th CS. Since $\mathbf{p}$ and $\mathbf{G}$ are given, binarization follows, where every entry G_{ij} in $\mathbf{G}_i$ is assigned as 1 if $G_{ij} < p_j$, otherwise $G_{ij} = 0$. Let f_i denote the fitness value of $\mathbf{G}_i$, which is evaluated by Eq. (3) with G_{ij} given, such that $f_i = \sum_{C_k \in CS_{(i)}} v(C_k)$.

Therefore, the top K genotypes in $\mathbf{G}$ with the best fitness values are found. The rule for updating the new $\mathbf{p}^{(t+1)}$ is be given by

$$p_j^{(t+1)} = (1 - \gamma)p_j^{(t)} + \frac{\gamma}{K} \sum_{i=1}^{K} G_{ij} , \tag{4}$$

where γ denotes the learning rate. Note $\mathbf{G}_i$'s are sorted according to their fitness values.

The pseudocode of PBIL is shown in Algorithm 1.

Algorithm 1. PBIL

1: Initialize probability vector $\mathbf{p}^{(0)}$
2: **repeat**
3:    Generate a population $\mathbf{G}_n$ from $\mathbf{p}^{(t)}$;
4:    Evaluate the fitness f_i of each member $\mathbf{G}_i$;
5:    Sort $\{\mathbf{G}_i\}$ according to their fitness;
6:    Update $\mathbf{p}^{(t+1)}$ according to Eq. (4);
7: **until** termination condition has been met

3.2   PBIL-MW

As we can see from the PBIL algorithm, the probability vector $\mathbf{p}^{(t+1)}$ is updated based on top K elements $\mathbf{G}_i$ chosen from $\mathbf{G}$. In fact, K bests are utilized for updating $p_j^{(t+1)}$ with equal weighting, i.e. $\frac{1}{K}$. However, it is reasonable that the fitness values $\{f_i\}$ may be used to tune the weighting when updating $\mathbf{p}^{(t+1)}$.

Our proposed adaptive algorithm, PBIL with merit weighting (PBIL-MW), incorporates this idea [6]. The first steps are the same as in PBIL until every fitness individual f_i has been evaluated in first iteration. Then f_i is ranked and the weights are given by

$$w_i = \frac{f_i'}{\sum_{i=1}^k f_i'} \ , \qquad 2 \le K \le N \tag{5}$$

where $f_i' = f_i - \min_{j \in i}(f_j), \forall f_j \ge 0$, and K is the number of chosen particles with the highest fitness values. Now that every w_i has been obtained, the probability vector $\mathbf{p}^{(t+1)}$ is given by

$$\mathbf{p}^{(t+1)} = (1 - \gamma) \, \mathbf{p}^{(t)} + \gamma \sum_{i=1}^K w_i \mathbf{G}_i \ . \tag{6}$$

Note that, every fitness f_i is considered and its weight w_i is given accordingly. The pseudocode of PBIL-MW is shown in Algorithm 2.

Algorithm 2. PBIL-MW

1: Initialize probability vector $\mathbf{p}^{(0)}$
2: **repeat**
3:    Generate a population $\mathbf{G}_n$ from $\mathbf{p}^{(t)}$;
4:    Evaluate and rank the fitness f_i of each member $\mathbf{G}_i$;
5:    Obtain w_i from Eq. (5);
6:    Update $\mathbf{p}^{(t+1)}$ according to Eq. (6);
7: **until** termination condition has been met

3.3   Set-ID Encoding Scheme for SO Algorithms

Encoding Scheme. The algorithms of SO used in our study are GA, PBIL and PBIL-MW. To incorporate SO algorithms for CSG, we propose a new encoding scheme for giving the initial probability of population which are described below.

In our previous study [6], we proposed a binary encoding scheme to represent the connection status among agents as the genotype, and the probability vector is used to generate the connection scheme, hence forming a CS. Furthermore, we have shown that for a conventional probability threshold set as 0.5, the possibility of having $n - 1$ connections in the network potentially forming a grand coalition is quite significant. To avert this we suggest $1/n$ as the threshold for initializing the probability vector.

The drawback of this connection-based encoding scheme is its vector length, which takes a quadratic form: $L = n(n - 1)/2$. Hence, when the number of agents becomes larger, the length of the probability vector and the genotypes will increase rapidly. Therefore, we propose a novelty encoding scheme by using the coalition IDs to allocate agents into different groups during the process of searching for an optimal coalition structure.

For example, in an 8 agents scenario, the bit-length is 3, the binary vector $[0, 0, 0]$ represents the no. 0 set, and $[0, 0, 1]$ represents no. 1 etc. Hence, if the ID array for agents 1 to 8 is $[3, 2, 3, 7, 2, 2, 4, 0]$ then the coalitions will be

Coalition 0: $\{a_8\}$;
Coalition 2: $\{a_2, a_5, a_6\}$;
Coalition 3: $\{a_1, a_3\}$;
Coalition 4: $\{a_7\}$;
Coalition 7: $\{a_4\}$.
Therefore, the CS for this ID set is
$\{\{a_1, a_3\}, \{a_2, a_5, a_6\}, \{a_4\}, \{a_7\}, \{a_8\}\}$.

Consequently, agents with the same ID suggest that they are in the same coalition. For a set of n agents, the maximum coalition is n (all singletons) and the minimum coalition is 1 (the grand coalition). Therefore, $L = n \times m$ bits ($m = \lceil \log_2 n \rceil$) will be a sufficient length for the probability vector to represent all possible coalition structures.

Initial Probability Threshold for Set-ID Encoding. Using the set-ID scheme, we have found that an initial probability threshold of 0.5 will make the algorithms tend to choose coalition structures with combination of smaller grouping sizes for coalitions, which leads to much more iterations to find a good solution. Therefore, we have examined a series of initial probabilities from 0.5 to 0.05 with 0.05 interval, and the results suggest that a 0.1 threshold is the best initial value for 20-agent cases, which leads to optimal, or competitive suboptimal solutions within fewer iterations. Consequently, it is reasonable to set a threshold away from 0.5 such that coalitions with a variety of sizes can be resulted in the coalition structure.

4   Experiment

4.1   Setup

To assess the potentials for our approach to be utilized in real-world smart grids, we follow the same approach as in our previous work to construct a realistic dataset, which are composed of two different sources. The first part is power consumption of smart-meter readings in New Zealand. The other part is power generated by commercialized facilities of wind turbines and solar panels which are coupled with meteorological data of New Zealand[1]. The power conditions of all agents are then given by subtracting demand from supply. The power(kWh) used are on an hourly basis, and the price $P_r = 20$ ($\text{¢}/kWh$). Moreover, we know that once the union has a power surplus at a given hour then the grand coalition will make a trivial solution. Thus, we consider only cases with overall power deficits. In our data, we have one year of hourly power demand and supply for 240 agents. Among those, four cases with power deficits are randomly chosen. The statistics of these four cases are shown in Table 1.

Table 1. Statistics of power status for 20 agents in 4 cases.

Case	S_{no}	Surplus	Mean-S_{no}	D_{no}	Deficit	Mean-D_{no}
I	7	5.394	0.771	13	−12.741	−0.980
II	5	2.659	0.532	15	−7.948	−0.530
III	10	4.646	0.465	10	−10.87	−1.087
IV	10	5.024	0.502	10	−7.663	−0.766

S_{no}: No. of agents with Surplus power, D_{no}: No. of agents with Deficit power.

4.2 Results

To demonstrate the advantages of PBIL-MW with the set-ID scheme, we compare its efficiency and accuracy with different approaches.

Effectiveness. For the comparison on effectiveness, as shown in Table 2, there are six approaches which are ES, DP, PBIL-MW with the connection scheme [6], and PBIL-MW with the set-ID scheme. The number of agents in the experiments are 12, 13, 16, 20 respectively. For SO-based algorithms the population size and number of iterations used are 200 and 500 accordingly.

From Table 2 it is obvious that a grid size of 12 agents actually takes over 200 s to complete the ES, while for 16 agents the time required increases up to more than 140 h, which becomes impractical for larger smart grids.

Even though DP runs faster than ES apparently, for 20-agents it takes over 2.8 h, also becoming impractical.

For PBIL-MW using the connection coding scheme, it is clearly faster than DP, but compared with its counterpart with the set-ID scheme it consumes more memory and takes longer while the grid size gets larger.

Accuracy. Although DP consumes more time to obtain the result, it remains an important approach, for it is an exact algorithm and guarantees to find the global optimization. To compare with other algorithms, we choose 4 cases of 20 agents run by DP to be the ground-truth (Tables 3 and 4).

---

[1] Meteorological data obtained from "CliFlo: NIWA's National Climate Database on the Web", https://cliflo.niwa.co.nz/.

Table 2. Average time needed by different approaches.

Number of agents	12	13	16	20
Exhaustive search	207.1	1,215.3	504,125.2	na
Dynamic programming	8.7	65.3	136.5	22,367.7
PBIL-MW(C)	6.2	12.4	30.2	44.6
GA(ID)	7.5	10.2	13.5	30.6
PBIL(ID)	5.3	6.5	**9.6**	**19.2**
PBIL-MW(ID)	**5.1**	**6.3**	9.7	**19.2**

-The best result are highlighted in bold. Unit: sec
*(C) indicates the connection scheme, (ID) indicate the set-ID
scheme.

Table 3. Average best results by different approaches compared to Optimum.

Case	**Optimum**	PBIL-MW(ID)	PBIL-(ID)	GA	PBIL-MW(C)
I	**53.16**	**53.16**	**53.16**	51.8	**53.16**
II	**92.86**	**92.86**	**92.86**	81.79	92.85
III	**100.34**	100.3	100.29	87.192	99.9
IV	**107.86**	**107.86**	**107.86**	100.79	107.85

-The best result are highlighted in bold. Unit: ¢

Table 4. Average iterations which hit the ground-truth.

Case	PBIL-MW(ID)	PBIL-(ID)	GA	PBIL-MW(C)
I	**98**	115	na	171
II	196	**184**	na	345(85%)*
III	**228(80%)***	276(80%)*	na	Na
IV	**159**	186	na	283(90%)*

-The best result are highlighted in bold.
*Average Iterations for the Percentage, denotes by (%), of
20 runs which hit the ground-truth.

For an extensive feasible study for GA, we use three crossover parameters
(single-point, uniform and segment), five mutation probabilities (0.05, 0.08, 0.1,
0.2), 200 samples, 500 iterations, 20 runs and three initial probabilities (0.1, $1/n$
(=1/20) and 0.03). Among these, we found the best one is with these parameters:
segment crossover, mutation = 0.03, and initial probability = 0.05.

Similar to GA experiments, PBIL and PBIL-MW are run by the same parameters of population, iterations and runs. Beside this, there are only three parameters (number of top-best, the learning rate and the initial probability) that need
to be assigned for both approaches. Summarizing the best results, we have fixed
the number of top-bests as $K = 2$, and the learning rate as 0.005. The initial

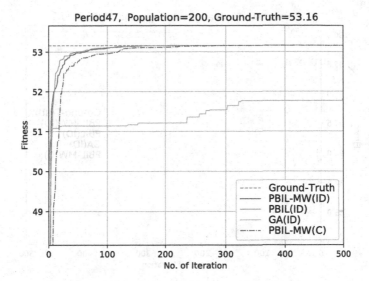

Fig. 1. Comparative results of PBIL-MW for Case-I

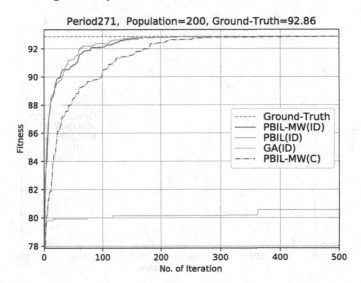

Fig. 2. Comparative results of PBIL-MW for Case-II

probability used in the connection scheme is $1/N$ (0.05) and 0.1 for the set-ID scheme.

Figures 1, 2, 3 and 4 present the results of the four cases, where (ID) and (C) in the legend denote the schemes based on set-ID and connections accordingly.

From the four figures we can see that PBIL-MW(ID) outperforms all others both in converging speed and accuracy. For case I, II and IV, the optima of these cases hit the ground-truth within fewer iterations. Except for case III, the

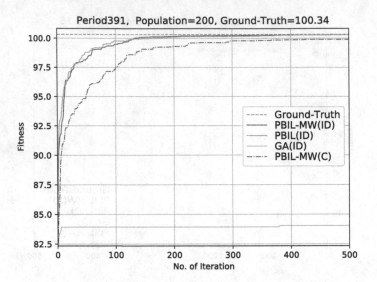

Fig. 3. Comparative results of PBIL-MW for Case-III

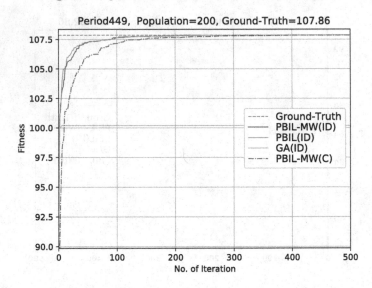

Fig. 4. Comparative results of PBIL-MW for Case-IV

mean of the best fitness for 20 runs is 100.32, which is slightly less than the ground-truth at 100.34. Actually, among those 20 runs, there are 17 runs whose best results hit the optimum, therefore we can say that the accuracy to find the exact optimal CS is 85% in this case. However, we have also tried to increase the number of iterations to 1000 for case III, and found that the PBIL-MW(ID) always reaches the optimum for all 20 runs. The average iterations for new experiments to hit the ground-truth is 238, the shortest and longest iterations are 82 and 512 respectively.

Furthermore, we also use a measurement, named "solution quality" [7], to examine the current best solution found at a given time while running the optimization process. The solution quality is given by $v(cs)/v(cs^*)$, where $v(cs)$ and $v(cs^*)$ denote the current and the optimal solution respectively. The quality solution for PBIL-MW(ID) after 500 iterations is over 99.98% for case III and 100% for other three cases.

In general, our experiment results show that PBIL-MW(ID) is a fast and reliable approach for solving the CSG problem in smart grids.

4.3   Conclusions

In this paper, we have demonstrated the ability of our PBIL-MW algorithm with a new scheme to encode PBIL genotype vectors for solving the CSG problem in smart grids. The number of agents used in this study are 20, giving 51.72 trillion possible CSs to be evaluated by DP algorithms. We have designed a stochastic optimization based solution to this problem. Our results have brought evidence to confirm that once the appropriate population size and number of iterations have been adequately set, the optima can be reachable by the new algorithm. Consequently, our approach has demonstrated a great potential for dealing with large-scale smart grids and finding optimal or near-optimal CSG, which remains almost infeasible for DP approaches.

In general, our result shows that PBIL-MW(ID) outperforms DP, PBIL and GA algorithms when searching for the optimal solution. Obviously, the faster response of our approach meets the requirement of periodic power exchange for smart grids operating in a cooperative mode, enabling a decentralized power system and improving the penetration of renewable energy.

Finally, this scope of our study is limited in the sense that it only considers local smart grids without power transmission via transformers, hence the power losses have been neglected. For future work, we will expand the current study to regional applications with scenarios of more agents and long-distance power sharing, taking into account the power losses during transmission.

References

1. Baluja, S.: Population-based incremental learning. a method for integrating genetic search based function optimization and competitive learning. Technical report No. CMU-CS-94-163, Carnegie-Mellon University, Department of Computer Science (1994)
2. Baluja, S., Caruana, R.: Removing the genetics from the standard genetic algorithm. In: Machine Learning: Proceedings of the Twelfth International Conference, pp. 38–46 (1995)
3. Björklund, A., Husfeldt, T., Koivisto, M.: Set partitioning via inclusion-exclusion. SIAM J. Comput. 39(2), 546–563 (2009)

4. Chalkiadakis, G., Elkind, E., Wooldridge, M.: Computational aspects of cooperative game theory. Synth. Lect. Artif. Intell. Mach. Learn. **5**(6), 1–168 (2011)
5. Changder, N., Dutta, A., Ghose, A.K.: Coalition structure formation using anytime dynamic programming. In: Baldoni, M., Chopra, A.K., Son, T.C., Hirayama, K., Torroni, P. (eds.) PRIMA 2016. LNCS (LNAI), vol. 9862, pp. 295–309. Springer, Cham (2016). https://doi.org/10.1007/978-3-319-44832-9_18
6. Lee, S.H.S., Deng, J.D., Peng, L., Purvis, M.K., Purvis, M.: Top-k merit weighting PBIL for optimal coalition structure generation of smart grids. In: Liu, D., Xie, S., Li, Y., Zhao, D., El-Alfy, E.S. (eds.) International Conference on Neural Information Processing, pp. 171–181. Springer, Cham (2017). https://doi.org/10.1007/978-3-319-70093-9_18
7. Michalak, T., Rahwan, T., Elkind, E., Wooldridge, M., Jennings, N.R.: A hybrid exact algorithm for complete set partitioning. Artif. Intell. **230**(C), 14–50 (2016). 10.1016/j.artint.2015.09.006
8. Mohamed, M.A., Eltamaly, A.M., Alolah, A.I.: PSO-based smart grid application for sizing and optimization of hybrid renewable energy systems. PLoS ONE **11**(8), e0159702 (2016)
9. Rahwan, T., Michalak, T.P., Wooldridge, M., Jennings, N.R.: Coalition structure generation: a survey. Artif. Intell. **229**, 139–174 (2015)
10. Sandholm, T., Larson, K., Andersson, M., Shehory, O., Tohmé, F.: Coalition structure generation with worst case guarantees. Artif. Intell. **111**(1–2), 209–238 (1999)
11. Shoham, Y., Leyton-Brown, K.: Multiagent sYstems: Algorithmic, Game-Theoretic, and Logical Foundations. Cambridge University Press, Cambridge (2008)
12. Wilf, H.S.: Generatingfunctionology, 2nd edn. Academic Press, Cambridge (1994)
13. Yeh, D.Y.: A dynamic programming approach to the complete set partitioning problem. BIT Numer. Math. **26**(4), 467–474 (1986)

Dam Inflow Time Series Regression Models Minimising Loss of Hydropower Opportunities

Yasuno Takato[✉]

Yachiyo Engineering Co., Ltd, Asakusabashi 5-20-8, Taito-Ku, Tokyo, Japan
tk-yasuno@yachiyo-eng.co.jp

Abstract. Recently, anomalies in dam inflows have occurred in Japan and around the world. Owing to the sudden and extreme characteristics of rainfall events, it is very difficult to predict dam inflows and to operate dam outflows. Hence, dam operators prefer faster and more accurate methods to support decision-making to optimise hydroelectric operation. This paper proposes a machine learning method to predict dam inflows. It uses data from rain gauges in the dam regions and upstream river level sensors from previous three hours and predicts the dam inflow in the next three hours. The method can predict the time of rise to the peak, the maximum level at the peak, and the hydrograph shapes to estimate the volume. The paper presents several experiments applied to inflow time series data for 10 years from the Kanto region in Japan, containing 55 floods events and 20,627 time stamped dam in-flow points measured at 15-min intervals. It compares the performance of four regression prediction models: generalised linear model, additive generalised linear model, regression tree and gradient boosting machine and discusses the results.

Keywords: Dam inflow time series · Upstream river sensor
Time series regression · Hydropower opportunities loss

1 Objective

1.1 Related Papers

In recent years, extreme flooding has caused massive damages worldwide [1, 2]. In Japan, sudden rainfalls and extreme river floods have occurred in regions such as Joso City at the Ibaraki prefecture [3]. Owing to such extreme and sudden rainfall events, it is difficult to predict the dam inflow and to operate the dam outflow. There are several approaches for water resources time series prediction modelling using statistical and machine learning methods. For example, artificial neural network (ANN) models are applied to monthly reservoir inflow time series [4, 5]. Furthermore, the data are 500 monthly streamflows covering a period of 40 years, including climate and land cover data where rainfall-runoff modelling simulates the streamflow of watersheds. Here, several models are compared with the Gaussian linear regression, Gaussian generalised additive models (GAMs), multivariate additive regression splines (MARSs), ANN models, random forest (RF), and regression tree models [6]. They suggest that GAMs and RF can effectively capture some non-linear relationships. Regarding daily lake

© Springer Nature Switzerland AG 2018
M. Ganji et al. (Eds.): PAKDD 2018, LNAI 11154, pp. 357–367, 2018.
https://doi.org/10.1007/978-3-030-04503-6_34

water level forecasting, the data were collected at five water gauges at the lake for 50 years, and several models were compared, such as the RF, support vector regression, ANNs, and the linear model. The results suggested that the RF can obtain a more reliable and accurate lake water level based on daily forecasting than others [7]. The variable selection approach can improve the daily reservoir discharge forecasting efficiency by machine learning methods. The data collected were 2,854 daily records over a period of eight years. After training and testing five methods, such as the ANN, the instance-based classifier, k-nearest-neighbour classifier, RF, and the random tree, key variables that influence the reservoir water level were selected and better models were built. These experimental results indicate that the RF forecasting model, when applied to variable selection with full variables, has better performance than other models [8]. Furthermore, regarding monthly reservoir inflow forecasting, hybrid processes are proposed where they are combined with a linear model and a non-linear model; for example, the former is the seasonal autoregressive integrated moving average (SARIMA) and the latter is the ANN and the genetic programming (GEP). The data were collected at the hydrometric stations and calculated as 360 monthly inflow series for a period of 30 years. After modelling and predicting, a comparison of the two hybrid models indicated the superiority of SARIMA-GEP over the SARIMA-ANN [9].

In these studies, the aim was decision support for water resource management and planning on monthly and daily terms. Monthly dam inflows prediction is important for adopting a reservoir water control plan to meet agricultural and hydropower demands. However, short term daily dam inflows prediction could be crucial for real-time reservoir operations to prevent extreme floods and ensure a consistent water supply by maintaining a safe reservoir level.

1.2   Dam Inflow Prediction for Hydropower Operation

When the dam region for hydropower is narrow, it takes a short time for the main river stream to flow from upstream to the dam inflow. Extreme rainfall events occur suddenly at real-time streams whose data range is hourly or every minute. Recent high-level dam inflow events are different from normal inflow patterns and are rarely observed. Owing to these extreme and sudden dam inflow events, it is difficult to predict dam inflows and operate dam outflows for hydropower generation. To support decisions concerning outflow operations in response to sudden and extreme high-level inflow events, there is a need for more accurate dam inflow time series modelling and for hourly interval predictions.

This paper proposes a method to predict dam inflows for a target 3-h forecast using time series data with a unit interval of 15 min. It will enable decision-makers to predict the time of rise to the peak, maximum level at the peak, and hydrograph shapes to estimate the volume. Furthermore, it will facilitate knowledge of clustering dam inflows and minimise the loss of hydropower opportunities. This paper discusses several experiments applied to the 20,000 time stamped recorded dam inflow time series data. The inputs are the hundreds of time series data features, such as rain gauges in the dam region and upstream river level sensors, and their time windows.

2 Modelling

2.1 Minimising the Loss of Hydropower Opportunities

Using MetaCost Minimise Hydropower and Flood Damage. Based on the inflow classification of the annotated boundary between high and low water levels, the task is to build a classification model. Using the classification model we can calculate a confusion matrix that describes the overall accuracy, precision, and recall for each inflow classification.

Table 1 lists two types of dam inflow prediction errors. Minimising the under-prediction error (shown on the upper right side of the confusion matrix) is important for minimising the risk of flood damage [10, 11]. Minimising the under-prediction error (shown on the bottom left side of the confusion matrix) is important for minimising the loss of hydropower opportunities influenced by the excess outflow operations [12]. Incorporating these costs, the MetaCost algorithm proposed a method for creating cost-sensitive classifiers [13, 14].

Table 1. Two types of dam inflow prediction errors

<table>
<tr><td colspan="2" rowspan="2"></td><th colspan="2">True, actual value</th></tr>
<tr><th>Low water</th><th>High water</th></tr>
<tr><td rowspan="2">Prediction</td><td>Low water</td><td>True Low water</td><td>**Under-prediction Error** (Flood Damage Risk)</td></tr>
<tr><td>High water</td><td>**Over-prediction Error** (Hydropower Opportunity Loss)</td><td>True High water</td></tr>
</table>

Loss Index to Compute Over-prediction Error for Hydropower. This paper provides the indexes to compute the loss of hydropower opportunities and the risk of flood damages. First, the over-prediction error at time t is indicated as:

$$ope_t = \hat{y}_t - y_t, \quad \text{if positive then } \hat{y}_t > y_t. \tag{1}$$

Here, $\hat{y}_t$ is the dam inflow prediction at time t, and y_t is the actual inflow value. The sum of over-prediction errors among a term $t \in \{1, \ldots, T\}$ is formulated as follows:

$$sope = \sum_{t=1}^{T} \max(\hat{y}_t - y_t, 0). \tag{2}$$

This positive value can be computed as the predicted case when there are only excess inflows over the actual level. If over-predictions frequently occur, i.e. the value of $sope_t$ is large, then the operator may carry out an excessive outflow more than an actual required level. These errors would correspond to the loss of hydropower opportunities.

Risk Index to Compute Under-Prediction Error for Flood Damage. Next, the under-prediction error at time t is indicated as:

$$upe_t = \hat{y}_t - y_t, \quad \text{if negative then } \hat{y}_t < y_t. \tag{3}$$

This situation appears to be an optimistic forecast at time t, where $\hat{y}_t$, the dam inflow prediction, is less than y_t, the actual inflow value. The sum of under-prediction errors among a term $t \in \{1, \ldots, T\}$ is formulated as follows:

$$supe = \sum_{t=1}^{T} \min(\hat{y}_t - y_t, 0) \tag{4}$$

This negative value can be computed as the predicted case when there are only lower inflows under the actual level. If under-predictions frequently occur, i.e. the absolute value of $supe_t$ is due largely to these negative values, then the mitigation policy may be insufficient for the risk of an over-flow scenario. These errors would correspond to the risk of flood damages.

2.2   Inflow Prediction Model

Windowing Features and Inflow Moving Average Filter. To predict 3-h forecasts, the models are applied to windows [15] of data from the previous three hours containing the rain gauge and river level sensor features. Incorporating the trends in dam inflow movements, the models are applied to invert the dam inflow moving average filer [16, 17] for vibration reduction.

Generalised Linear Model and Additive GLM Regression. This paper discusses several applications via the linear model regression approach, such as the generalised linear model (GLM) [18–20]. The GLM with Gaussian family is useful for predicting a dam inflow time series. Furthermore, they extend the additive regression model based on the Gaussian GLM [21, 22].

Regression Tree and Gradient Boosting Machine. This paper discusses some applications via non-linear models, such as the regression tree [23, 24] and the gradient boosting machine (GBM) [25–28]. They enable optimising parameters like the number of trees and the maximum depth of a tree using grid searches.

3 Applied Results

3.1 Dataset of Target Inflow and Features

Figure 1 shows the case study of a dam and a river region where three river height sensors are located. The purpose of the dam is hydropower generation and the dam type is PG which means concrete gravity. The height is 32 m and the reservoir capacity is 1.5 million m³. The dam river region has an area scale of 112 km². It has a yearly electrical capacity of 279 MWh. The total arrival time takes 90 min from sensor-1 to the dam. In addition, the middle arrival takes time 60 min from sensor-2 to the dam. Third, the short arrival time takes time 20 min from sensor-3 to the dam.

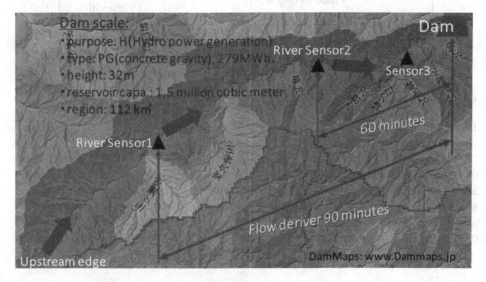

Fig. 1. Case of dam and river region with three sensors (http://www.dammaps.jp/?d=0699)

Table 2 lists the target dam inflow variable and several features, such as rain gauges and river level sensors. Here it is targeted to predict the forward time horizon to set at 12, i.e. the 3-h forward forecast. Regarding the features, each of these window sizes is set at 12, i.e. they are the historical data of the past three hours. The unit time intervals are always 15 min.

Table 2. Data name, role, and variable profile

Data name	Data role and feature profile
Dam-inflow time series	Target variable. The volume of dam-inflow time series from the upstream region. The interval is 15 min
Rain gauge	Rain features. The quantity of rainfall measured at nine points over the upstream region
River height sensor	Main stream river features. The level of the height sensed at the three points within the main river

3.2   Training Data and Test Set

The case study area is in the Kanto region in Japan. The data are from 10 years of data collected from 2006 to 2015, containing the 55 flood events with high water flows. The number of target inflow value data is 20,627 in unit intervals of 15 min. Figure 2 shows an image of the training data and the test set split with linear sampling. The training set contains the 52 flood events from 2006 to 2014. In contrast, the test set contains the three flood events that recently occurred in 2015.

Training dataset	Test set
Flood events # 52	Flood events # 3 (53th-55th)
2006 2014	2015

Fig. 2. Training data and test set split with linear sampling

3.3   Prediction Results

The case study included several numerical experiments using RapidMiner Studio version 8.1 in a 64-bit computing environment with 8 GB RAM.

Figure 3 shows the accuracy comparison among the four prediction models based on the root mean square error (RMSE) and the correlation between each prediction and actual inflow. Table 3 lists the 5-fold cross validation results, each split under linear sampling. Both results suggested that the GLM extended additive regression model is more preferable than the other models, such as original GLM, the regression tree, and the gradient boosting machine.

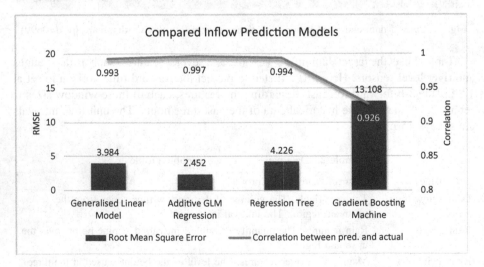

Fig. 3. Compared accuracy among prediction models based on root mean square error and correlation between prediction and actual inflow

Table 3. Cross validation results among prediction models

Prediction model	Root mean square error	Correlation between pred. and actual	TOC (time of computation)
Generalised linear model (Gaussian family, identity link)	3.984 ± 0.638	0.993 ± 0.004	2 s
Additive GLM regression (iteration = 10, shrinkage = 1.0)	**2.452 ± 0.594**	**0.997 ± 0.001**	**20 s**
Regression tree (number of depth = 15)	4.226 ± 1.894	0.994 ± 0.001	18 s
Gradient boosting machine (max depth = 10, ntree = 2,000)	13.108 ± 4.128	0.926 ± 0.044	8 m 54

The RMSE of the additive GLM regression is 2.452 ± 0.594 with both the lowest mean error and the smallest deviation. Furthermore, the correlation between the actual and predicted value is the highest value at 0.997 ± 0.001. The execution time of 20 s is faster and practical. Hence, additive GLM regression is both an accurate and practical fast inflow prediction method for the applications compared in the experiments in this study.

Figures 4 through 6 show the plots of actual and prediction values using classification models, such as the original GLM under the Gaussian distribution, additive GLM regression, regression tree, and gradient boosting machine.

The GLM selected the best family compared with others such as the Poisson, gamma family, and so forth. The Tweedie family has the same accuracy as the Gaussian; however, the author selected the Gaussian distribution as the most usable and acceptable. Although GLM approximates good output like the complete hydrograph shape, we should be concerned about the weakness where the several jumped up points do not match the peak level far from the actual inflow, exactly at Fig. 4.

The additive GLM regression model could almost approximate the same shape as the actual inflow series. We also focus on the strong points where the maximum inflow predictions are always below the actual peak inflow without over-prediction. This model only required 20 s for 5-fold cross validation. This result maintains that the additive GLM regression model is practical for dam inflow time series prediction tasks without over-prediction. In addition, such an advantage would correspond to minimising the loss of hydropower opportunities. Furthermore, there is little under-prediction owing to the smoothness of the additive function and the flexible formulation [21].

The gradient boosting machine (GBM) has parameters where the number of trees is 2000 and maximum depth is 10. This model only set two parameters to maximise the accuracy using grid search (e.g. the number of trees in an interval from five to 20 and maximum depth from 100 to 2000). This result means that the GBM could approximate almost the same shape of the actual inflow time series, although there are more biases than the additive GLM regression model because it has many fluctuations at the high

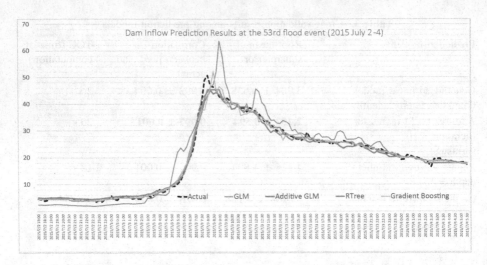

Fig. 4. The 53rd event of actual and prediction value using dam inflow time series regression

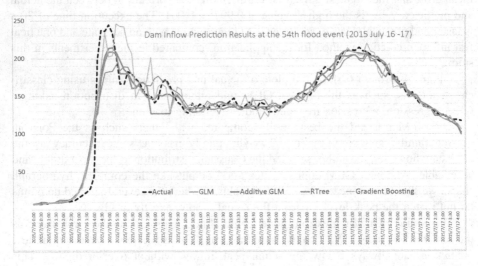

Fig. 5. The 54th event of actual and prediction value using dam inflow time series regression

inflow level close to the peak. The regression tree has little over-prediction; however, there are some under-predictions that are less than the actual inflows, as shown in Fig. 5. These under-predictions would correspond to the risk of flood damages.

Table 4 lists the calculated results of the prediction error indexes regarding over-prediction and under-prediction errors. In the left side three columns are the sum of over-prediction error, where the additive GLM regression model had very low scores. Therefore, this model had the advantage of the smallest loss of hydropower opportunities. On the right-side column is the sum of under-prediction error, where the additive GLM regression model is preferable at the lower scores based on the absolute values. Hence, this model had the advantage of a minimal risk of flood damages.

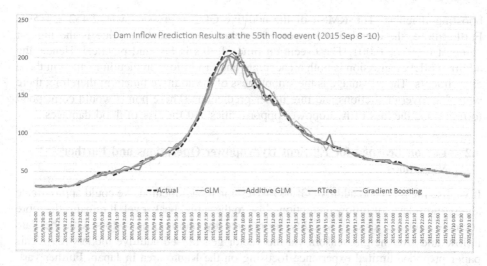

Fig. 6. The 55th event of actual and prediction value using dam inflow time series regression

Table 4. Computed results of loss indexes regarding over/under-prediction

Test data of flood event	Sum of over prediction error (sope)			Sum of under perdition error (supe)		
	53th	54th	55th	53th	54th	55th
Generalised linear model (Gaussian family, identity link)	250.8	490.7	172.3	−110.6	−609.8	−194.1
Additive GLM regression (iteration = 10, shrinkage = 1.0)	**42.3**	400.5	**114.9**	−92.3	**−428.7**	**−104.3**
Regression Tree (number of depth = 15)	45.3	**385.1**	169.7	**−53.0**	−758.8	−206.9
Gradient Boosting Machine (max depth = 10, ntree = 2,000)	60.0	704.2	177.5	−73.9	−607.6	−286.7

4   Concluding Remarks

4.1   Prototyping Dam Inflow Data Mining Process for Prediction

This paper proposed a method to predict dam inflows for the target of a 3-h forecast using a time series data with a unit interval of 15 min. This study implemented numerical experiments applied to the 20,000 time stamped recorded dam inflow time series data. They contained the input of hundreds of time series data features, such as nine rain gauges in the dam region and three upstream river level sensors. These experiment results suggested that the GLM extended additive regression model is more preferable than other models such as the original GLM, regression tree and gradient boosting machine. The RMSE of the additive GLM regression was 2.452 ± 0.594. Furthermore, the correlation between the actual and prediction value is the highest value at 0.997 ± 0.001. The execution time of 20 s is fast and practical. Hence, the additive GLM regression is both an accurate and fast inflow prediction method in these experiments. The advantage is the smoothness of the additive function; therefore, there were few over-predictions and rare under-predictions. These points would correspond to minimise the loss of hydropower opportunities and the risk of flood damages.

4.2 Lesson Learned for Efficient Hydropower Operations and Further Variations

If we use the proposed additive GLM Regression model method, we could approximate a maximum level of inflow prediction close to the actual peak level. It remains the more accurate and the faster possibility formulated by multiple predictors with the dam inflow and river height sensors via the vector generalised additive models [29]. This paper provided limited experience focusing on the Kanto area in Japan. Further variations will be obtained in cases where dam inflow time series mining and predictive machine learning are used to operate more efficient and safe hydropower generation. This case study contained only three main river sensors; however, there are opportunities to measure sensitive locations such as the edge of the upstream river which are broadly spanned at each sub rivers owing to reinforcement learning for hydropower generation performances. Further, we need research concerning the threshold for high and low inflow levels to predictively detect the boundary between normal inflows and anomaly inflows. For example, the one-class support vector machine [30] enhances the outlier inflows and computes the outlier scores to classify anomaly inflows and other flows. Furthermore, long-term 30-year data mining needs to be continued so that we can develop another accurate and fast dam inflow prediction application.

Acknowledgements. I wish to thank the DaMEMO committee and referees for their time and valuable comments. I also wish to thank Yachiyo Engineering Co., Ltd. for various support based on big data and AI project outcomes since 2012. I am also grateful to Mizuno Takashi for providing development opportunities and Amakata Masazumi for providing domain knowledge in river and dam engineering.

References

1. UN News Centre: EM-DAT International Disaster Database (2016). www.emdat.be & Reliefweb-reliefweb.int/disaster
2. Munich, R.: Geo risks research: NatCatSERVICE. https://www.iii.org/fact-statistic/facts-statistics-global-catastrophes. Accessed 31 December 2017
3. Japan floods: city of Joso hit by 'unprecedented' rain BBC, 10 September 2016 http://www.bbc.com/news/world-asia-34205879. Accessed 31 December 2017
4. Othman, F., et al.: Reservoir inflow forecasting using artificial neural network. Int. J. Phys. Sci. **6**(3), 434–440 (2011)
5. Attygalle, D., et al.: Ensemble forecast for monthly reservoir inflow: a dynamic neural network approach (2016). https://doi.org/10.5176/2251-1938_ors16.22

6. Shortridge, J.E., et al.: Machine learning methods for empirical streamflow simulation: a comparison of model accuracy, interpretability and uncertainty in seasonal watersheds. Hydrol. Earth Syst. Sci. **20**, 2611–2628 (2016)
7. Li, B., et al.: Comparison of random forests and other statistical methods for the prediction of lake water level: a case study of the Poyang Lake in China. Hydrol Res. **47**(S1), 69–83 (2016)
8. Yang, J.-H., et al.: A time-series water level forecasting model based on imputation and variable selection method. Comput. Intell. Neurosci. 1–11 (2017)
9. Moeeni, H., et al.: Monthly reservoir inflow forecasting using a new hybrid sARIMA genetic programming approach. J. Earth Syst. Sci. **126**, 18 (2017)
10. Hofmann, M., Klinkenberg, R.: RAPID MINER: Data Mining Use Cases and Business Analytics Applications. Data Mining and knowledge Discovery Series. CRC Press, Cambridge (2014)
11. Kotu, V., Deshpande, B.: Predictive Analytics and Data Mining: Concepts and Practice with Rapidminer. Morgan Kaufmann, New York (2015)
12. International Hydropower Association (IHA): cost-benefit and economic performance. Hydropower Sustainability Assessment Protocol. Accessed 16 February 2018
13. Mattmann, M., et al.: Hydropower externalities: a meta-analysis. Energy Econ. **57**, 66–77 (2016)
14. International Renewable Energy Agency (IRENA): renewable energy technologies: cost analysis series. Hydropower vol. 1: Power Sector (2012)
15. Domingos, P.: MetaCost: A General framework for making classifiers cost-sensitive. In: ACM Conference on Knowledge Discovery in Databases, pp. 165–174 (1999). 10.1.1.15.7095
16. Ling, C.X., et al.: Cost-sensitive learning and the class imbalance problem. In: Sammut, C. (ed.) Encyclopedia of machine learning. Springer, Berlin (2008)
17. Pavlyshenko, B.M.: Linear, machine learning and probabilistic approaches for time series analysis. In: IEEE International Conference on Data Stream Mining and Processing (2016)
18. Wedderburn, R.W.M.: Quasi-likelihood functions, generalized linear models and the Gauss–Newton method. Biometrika **61**(3), 439–447 (1974)
19. McCullagh, P., Nelder, J.: Generalized Linear Models. Monographs on Statistics and Applied Probability. Chapman & Hall, London (1983)
20. Hardin, J.W., Hilbe, J.M.: Generalized Linear Models and Extensions, 3rd edn. Stata Press, College Station (2012)
21. Hastie, T., Tibshirani, R., Friedman, J.: The Elements of Statistical Learning; Data Mining, Inference and Prediction, 2nd edn. Springer, Berlin (2009)
22. Hastie, T., Tibshirani, R.: Generalize additive models. Stat. Sci. **1**, 297–318 (1986)
23. Brieman, L., Friedman, J. et al.: Classification and Regression Trees. Wadsworth (1984)
24. Ripley, B.D.: Pattern Recognition and Neural Networks. Cambridge University Press, Cambridge (1996)
25. Friedman, J., Hastie, T., Tibshirani, R.: Additive logistic regression: a statistical view of boosting (with discussion). Ann. Stat. **28**, 307–337 (2000)
26. Friedman, J.: Greedy function approximation: a gradient boosting machine. Ann. Stat. **29**(5), 1189–1232 (2001)
27. Scholkopf, B., Freund, Y.: Boosting: Foundations and Algorithms. MIT Press, Cambridge (2012)
28. Efron, B., Hastie, T.: Computer Age Statistical Inference: Algorithms, Evidence and Data Science. Cambridge University Press, Cambridge (2016)
29. Yee, T.W., Wild, C.J.: Vector generalized additive models. J. R. Stat. Soc. Ser. B **58**(3), 481–493 (1996)
30. Amer, M., Goldstein, M., et al.: Enhancing one-class support vector machines for unsupervised anomaly detection. In: 11th ODD'2013, Chicago (2013)

Author Index

Printed in the United States
By Bookmasters